KU-686-517

## ABOUT ISLAND PRESS

Island Press is the only nonprofit organization in the United States whose principal purpose is the publication of books on environmental issues and natural resource management. We provide solutions-oriented information to professionals, public officials, business and community leaders, and concerned citizens who are shaping responses to environmental problems.

In 2004, Island Press celebrates its twentieth anniversary as the leading provider of timely and practical books that take a multidisciplinary approach to critical environmental concerns. Our growing list of titles reflects our commitment to bringing the best of an expanding body of literature to the environmental community throughout North America and the world.

Support for Island Press is provided by The Nathan Cummings Foundation, Geraldine R. Dodge Foundation, Doris Duke Charitable Foundation, Educational Foundation of America, The Charles Engelhard Foundation, The Ford Foundation, The George Gund Foundation, The Vira I. Heinz Endowment, The William and Flora Hewlett Foundation, Henry Luce Foundation, The John D. and Catherine T. MacArthur Foundation, The Andrew W. Mellon Foundation, The Moriah Fund, The Curtis and Edith Munson Foundation, The New-Land Foundation, Oak Foundation, The Overbrook Foundation, The David and Lucile Packard Foundation, The Pew Charitable Trusts, The Rockefeller Foundation, The Winslow Foundation, and other generous donors.

The opinions expressed in this book are those of the author(s) and do not necessarily reflect the views of these foundations.

## ABOUT THE OCEAN CONSERVANCY

The Ocean Conservancy is the world's largest and oldest nonprofit organization dedicated solely to protecting the world's oceans. We envision a world of healthy, protected oceans with wild and flourishing ecosystems, free of pollution, and filled with diverse and abundant marine wildlife. With more than half a million members and volunteers, The Ocean Conservancy works to final lasting solutions to issues affecting our oceans and marine life.

The Ocean Conservancy works to promote a healthy global system of reefs, estuaries, bays, and oceans and to prevent damage from wasteful and destructive practices that threaten the viability of ocean life and human life. Through science-based research, public education, and advocacy, The Ocean Conservancy informs, inspires, and empowers people to speak and act on behalf of the oceans, our shared responsibility. In all its work, The Ocean Conservancy strives to be the world's foremost advocate for the oceans.

The Ocean Conservancy Headquarters, 1725 DeSales St. NW, Suite 600, Washington, DC 20036. (202) 429-5609. www.oceanconservancy.org, info @oceanconservancy.org

Alaska Regional Office, 425 G Street, Suite 400, Anchorage, AK 99501, (907) 258-9922 New England Regional Office, 371 Fore Street, Suite #301, Portland, ME 04101, (207) 879-5444

Pacific Regional Office, 116 New Montgomery Street, San Francisco, CA 94105, (415) 979-0900

SE U.S. Reg'l Office, 449 Central Avenue, Suite 200, St. Petersburg, FL 33701, (727) 895-2188

U.S. Virgin Islands Office, PO Box 1287, Cruz Bay, St. John, USVI 00831, (340) 776-4701

Pollution Prevention and Monitoring Office, 1432 N. Great Neck Road, Virginia Beach, VA 23454, (757) 496-0920

# Marine Reserves

# Marine Reserves

A GUIDE TO SCIENCE, DESIGN, AND USE

Jack A. Sobel
Craig P. Dalgren, Ph.d.

ISLAND PRESS

WASHINGTON
COVELO
LONDON

*Library of Congress Cataloging-in-Publication data.*
Sobel, Jack (Jack A.)
Marine reserves : a guide to science, design, and use / Jack Sobel
p. cm.
Includes bibliographical references and index.
1. Marine parks and reserves. I. Title.
QH91.75.A1S63 2004
333.95′616—dc22 2003021205

*British Cataloguing-in-Publication data available.*

Printed on recycled, acid-free paper

Design by Kathleen Szawiola

Manufactured in the United States of America
10 9 8 7 6 5 4 3 2 1

# Contents

# Preface

My own first exposure to marine reserves came in the mid-1980s. As a neophyte graduate student, I was directing a research project in the town of San Pedro, on Ambergris Caye, Belize, and got entrained in the development of a local marine reserve there, which eventually became the country's first, the Hol Chan Marine Reserve (see chapter 10). Though the local community, the reserve itself, and my role in its development were relatively small; the exciting process leading to its creation, the community dynamics surrounding it, and its successful outcome provided me with a great appreciation for the conservation potential of marine reserves, the challenges of establishing them, and the approaches needed to overcome these challenges. The basic tenets of strong science and design, good public process, active community involvement, and careful development of public and governmental support that I first learned there I have repeatedly seen as essential to successful reserve development across a diversity of settings, both firsthand and via reports from others. These underpinnings are reflected throughout this book.

My professional interests and career have long balanced on the cusp between marine science and policy. In 1988, after returning stateside from Belize, I took a marine policy fellowship and served as a staff member to the U.S. Senate's National Ocean Policy Study. Among my principal responsibilities in this position was staffing the reauthorization of the National Marine Sanctuary Act (NMSA). Like most Americans, I was previously unaware of the existence of the small National Marine Sanctuary Program (NMSP), one of the country's best kept secrets, buried deep within the Department of Commerce's National

Oceanic and Atmospheric Administration. Yet, I was excited and intrigued by the potential of this program to protect special ocean places. The successful reauthorization of the NMSA late in 1988 coincided with the approaching end of my Senate fellowship. As I looked to build on my experiences, I sought a position that would include the pursuit of highly protected ocean places and the possibility of unlocking the potential of the NMSP to create them.

Early in 1989, I accepted such a position with the Center for Marine Conversation (CMC, now The Ocean Conservancy) based on its reputation for sound science, policy expertise, and integrity and my experience with it as an outstanding source of information on marine sanctuaries, marine protected areas, and a host of other issues to me in my Senate position. From 1989–1994, I directed the organization's habitat, marine protected area, and ecosystem protection work, including major efforts to expand and strengthen the NMSP. These efforts were highly collaborative and phenomenally successful by many measures, including a doubling in the number of sanctuaries, a five-fold increase in sanctuary area, and a similar increase in funding for the program. They also reinvigorated the sanctuary program with improved public support, greater public recognition, and committed activists and supportive coalitions for many individual sites. Despite these successes, it remained clear that existing sanctuaries lacked comprehensive protection and any significant protection from fishing activities, and that only one site afforded short-term prospects for providing either.

From 1990–1995, the still new and budding Florida Keys National Marine Sanctuary (FKNMS) offered the only real opportunity to significantly address fishing impacts within the sanctuary system and among the best opportunities to develop marine reserve level protection within continental U.S. waters. The FKNMS emerged from the collaborative efforts to expand and strengthen the NMSP described above and was uniquely created via an act of Congress. Enacted in 1990, The Florida Keys National Marine Sanctuary and Protection Act (FKNMS&PA) created the nearly 3,000 nautical mile$^2$ sanctuary, provided a strong mandate for its protection, and established a framework for its management. The Act required the subsequent development of a comprehensive management plan for the area rather than providing one, but explicitly required consideration of geographical zoning (e.g., marine reserves) to protect its resources. By 1995, it was clear that successful implementation of marine reserves within the FKNMS was far from a done deal and would require substantial and expanded efforts and improved arguments and information regarding the efficacy and experience with marine reserves elsewhere. In recognition of this, I focused my efforts more fully on the marine reserve issue and initiated research on the global experience with marine reserves that eventually led to the development of this book.

In 1995, I organized an international workshop on the Global Experience and Efficacy of Marine Reserves, cosponsored by CMC and the Caribbean Marine Research Center (CMRC), and held at CMRC's Lee Stocking Island Field Station in the Bahamas. The workshop brought together leading marine reserve experts from around the globe with firsthand experience in the development and evaluation of marine reserves and included participants from six continents. This provided a wealth of information and diversity of perspectives on marine reserves, identified both commonalities and differences related to marine reserve development in different environments, and produced one of the first consensus statements on the benefits of marine reserves (see chapter 4). Many of the roots of this book can be traced back to this workshop and subsequent research, contacts, and experiences that grew out of it. The workshop, follow-up activities, and related research greatly expanded my horizons with respect to marine reserves and also contributed to the successful implementation of a limited system of marine reserves in the FKNMS in 1998, a much larger marine reserve off of Florida's Dry Tortugas, also within the FKNMS, in 2002, and a more extensive network of marine reserves in the Channel Islands NMS off California in 2003 (see chapter 8).

Initial plans to publish proceedings from the 1995 workshop were delayed and then cancelled due to rapidly evolving marine reserve developments and other related priorities. In less than a decade since that workshop, the number and size of marine reserves, the research related to them, and the evidence supporting their efficacy have all continued to expand dramatically. In the United States, development of marine reserves off the Florida Keys, including the larger Tortugas Ecological Reserve; the more extensive network of reserves off of California's Channel Islands; and of several remote island marine reserves in the Caribbean and Central Pacific are especially noteworthy and indicative of this trend. Among other more developed countries, the continued expansion of New Zealand's national network of marine reserves and the more recent and extensive advances in development of marine reserve networks in Australia (see chapter 11) likely lie at the leading edge of marine reserve progress. Similarly, developing national marine reserve networks in the Bahamas (see chapter 9) and in Belize (see chapter 10) are representative of the forefront of marine reserve progress among less developed countries.

In 1998, we initiated discussions with Island Press regarding the pressing need for a state-of-the-art book on marine reserves detailing the arguments and science behind, the evidence for, the global experience with, current trends, practices, and issues related to, and future prospects for them. This book is the result of those discussions and our attempt to fill this need. Probably the greatest challenge we have faced in putting it together has been trying to keep pace with the rapid and accelerating pace of progress with respect to both marine reserve development and science. A scientific colleague recently likened this effort to the labor of Sisyphus, accurately suggesting that just when we thought we were coming to closure on a piece of the book, new information or developments would surface that we felt compelled to include. At times, we certainly felt like Sisyphus!

As we look forward to the book's publication in 2004, we intend and believe it will provide a strong overview of the current state of the art with respect to marine reserve science, design, and use. In assembling the book, we recognized that it could not be fully comprehensive, all-inclusive, or completely up-to-date. Even if such a book were possible to compose, it would not remain so through publication. Instead, we strove to highlight major issues, critical arguments and information, representative and exceptional examples, key progress and trends, and likely future directions and prospects. Consequently, while some details may quickly become dated, we fully expect the major themes and ideas contained in the book will remain relevant for the foreseeable future. We further intend and expect it be relevant to a broad audience including non-expert scientists, students, managers, decision-makers, conservationists, stakeholders, and an increasingly educated and concerned segment of the lay public. The primary scientific literature on marine reserves has become so extensive at this point that no book could exhaustively review all of it. Rather, we have selected those examples we feel are most important for discussion, included a broad and representative cross section for additional depth, and synthesized both to provide a comprehensive overview. The lack of an existing easily accessible overview and synthesis on marine reserves was a primary motivation for writing this book and is the vacant niche we are attempting to fill with it. Readers interested in exploring the primary literature more fully should find a strong base here for such exploration.

The rapid and accelerating progress with respect to marine reserve development and science around the globe, across many and diverse areas and situations—driven by locally appropriate and variable approaches, but employing common themes—will likely continue. We avoid a one-size-fits-all approach to marine reserves, but rather highlight what has and hasn't proven successful in different situations, draw conclusions where appropriate based on these experiences, suggest general guidelines, and attempt to provide an accessible information base and foundation on which interested parties can draw and build approaches and solutions tailored to their individual needs.

Marine reserves remain controversial and contentious in many places and among some stakeholders, despite, and in some cases because of, the considerable progress made to date in many areas with the participation of many stakeholders. In the United States, there has been some backlash within certain user communities to the successful establishment of marine reserves in the Florida Keys and Channel Islands. Yet, in the long run, we believe that the resulting public debate on marine reserves will be a net benefit and that recent progress on marine reserve science, design, and use will continue and likely accelerate further. The spirited debate regarding marine reserves motivated us in preparing this book, not because we endeavor to end that debate, but because we strive to inform it. Attempts to stifle or avoid such debate often backfire and are unwise, though tools for keeping it civil, respectful, and constructive are warranted and discussed in the book (see especially chapter 6). A lively and vigorous discussion of marine reserve issues among many constituencies, across multiple public sectors, and at a variety of levels is highly desirable, much needed, and likely essential to their continued success as a key marine policy tool. Our experience and philosophy with respect to this is much in line with former U.S. President John F. Kennedy who voiced: "My experience in government is that when things are non-controversial and beautifully coordinated, there is not much going on."

Human alteration of marine ecosystems and their living inhabitants continues to accelerate and expand, but increased public awareness of such anthropogenic change and related changes in societal values and ethics offer some hope for the oceans' future. These two factors combined with the continued failure of other existing management tools to successfully address the former and adapt to or reflect the latter, fuel our belief that the use of marine reserves will continue to advance. Marine reserves and the debate about their use are at least as much about societal goals, values, and ethics related to marine resource use as their science and design, though debate over the latter often masks more fundamental disagreement over the former. Nonetheless, such discord will likely ameliorate somewhat as the needs of ecosystem protection and more traditional fisheries management increasingly converge. The first several chapters emphasize these themes.

Responsible stewardship and intergenerational equity are among the goals and benefits of marine reserves. At a personal level, furthering intergenerational responsibility was among the primary incentives for compiling this book. I feel privileged to have enjoyed a range of ocean experiences and retain vivid recollections of those from my formative years. I recall sport fishing with my father as a youth and his descriptions of the former ocean bounty from his youth and at times questioning their veracity. Several decades later, I am teaching my two young children to fish, enjoy, and protect the oceans and realize much to my amazement that I too have seen changes in ocean life of a similar or greater magnitude. Marine reserves afford a tool for preserving and restoring wild ocean ecosystems and their former abundance and diversity of marine life, so that our and future generations can continue to use and enjoy them. We hope this book helps to achieve that goal.

Jack A. Sobel
The Ocean Conservancy

# Acknowledgments

Completing this book was not easy and it did not happen overnight. Neither science nor policy writing is ever easy. The level of difficulty increases synergistically when you combine them. The fact that marine reserves were a hot issue scientifically and politically when we embarked on this journey, and have only become hotter since, increased the challenge further. Meeting this challenge required teamwork and support from many, all of whom we acknowledge and thank, a few of whom we specifically mention below.

Without the long-term support and commitment of The Ocean Conservancy and the Perry Institute for Marine Science's Caribbean Marine Research Center, the book likely would not have been completed. The authors' home institutions cosponsored the original international marine reserves workshop in 1995 that eventually gave rise to the book and supported their time in developing, writing, editing, and revising the book over the last several years. Current and former staff at both institutions also made key contributions to its completion, both directly and indirectly. Bob Cronan of Lucidity Information Design, L.L.C., deserves special thanks for many of the book's graphics.

The Ocean Conservancy thanks the following funding sources for their generous support of its ecosystem protection efforts that enabled the book's completion and production: Bernice Barbour Foundation, National Fish and Wildlife Foundation, Henry Luce Foundation, John D. and Catherine T. MacArthur Foundation, Moore Family Foundation, Panaphil Foundation, Surdna Foundation, Wiancko Charitable Foundation, and an anonymous Pennsylvania donor.

The authors would also like to thank the growing cadre of marine reserve and protected area experts who have advanced reserve science, design, and use, and directly or indirectly inspired, influenced, assisted, or provided the raw materials to the authors for creating this book. This includes a core group who participated in the 1995 workshop or otherwise contributed to the development of the ideas contained in the book. An incomplete list of these includes Jim Bohnsack, Gary Davis, Bill Ballantine, Tim McClanahan, Gary Russ, Callum Roberts, Mark Hixon, Jane Lubchenko, Bob Warner, Sylvia Earle, Billy Causey, and Mike Weber. Margaret Davidson of the National Ocean Service (NOS), the NOS National Office, and the National Oceanic and Atmospheric Administration Library also deserve a special thanks for logistical and research support.

Above all, we acknowledge the incalculable contributions of our families, without whose support and sacrifices the book would certainly not have been possible.

# I

# Principles and Concepts

ONE

# Our Oceans in Trouble

People who know the sea well know something is wrong. Children visit the sea and listen in disbelief to stories about the good old days. Then they grow up, have their own kids, tell their own stories, and understand something's missing, that their kids are being deprived of something that once brought them great pleasure. Sophisticated media coverage also increasingly highlights and documents these changes for us, but with each generation, the clock is reset and we forget what came before, minimizing the perceived change. Yet, one need only look at a map of the coast or walk around a coastal community to find names of places like Sheepshead Bay, where no one's caught a sheepshead in a generation; Halibut Cove, where no one may ever catch a halibut again; Jewfish Creek, where no one remembers the last jewfish; or Salmon Run, where the last run occurred before anyone alive today was born. In a very real way, we are losing our natural marine heritage and our biodiversity, and it matters.

Over a century ago, scientists first noted rapid changes occurring along the east coast of North America. According to the U.S. Commission of Fish and Fisheries, halibut from coastal New England had been nearly extirpated by 1878 (see the following quote). Dwindling cod stocks triggered a decline in landings from their historic peak in 1887, followed by other targeted groundfish species in the ensuing decades (Fig. 1.1; NMFS 1990). Natural oyster reef habitat had been virtually eliminated throughout the Chesapeake Bay and northeastern United States (Brooks 1996). Similar changes had already been observed in Europe. The following summary vividly encapsulates the changing sea state at that time.

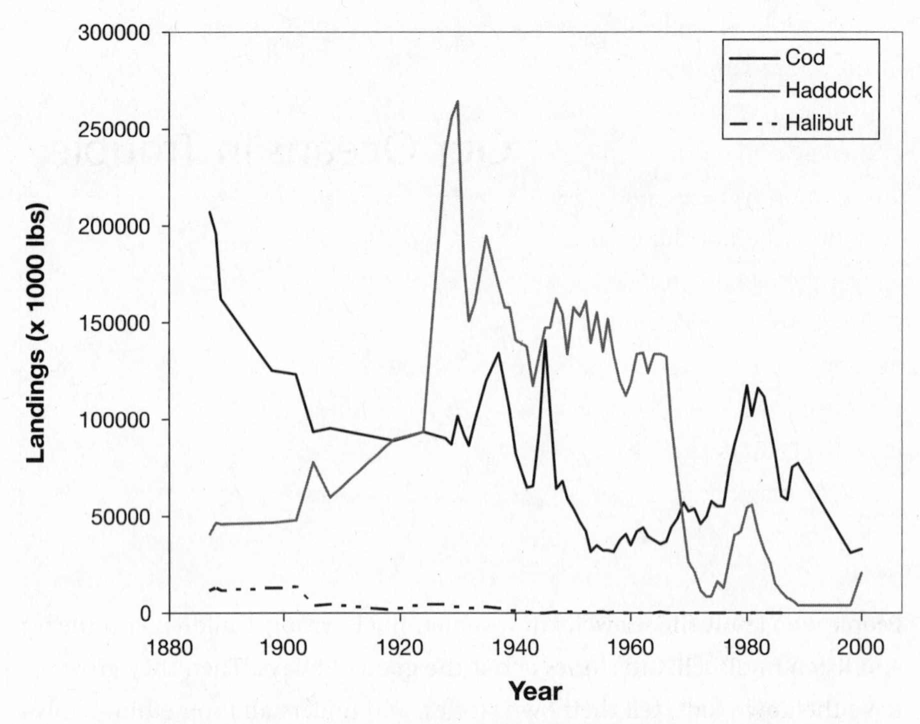


**FIG. 1.1 New England Cod, *Gadus morhua*, Landings 1887–2002.** Chronology indicates long-term decline in cod landings from a historic peak in the late 1890s, despite subsequent, temporary, and lower, interim peaks, and likely, increased effort. Latter peaks may reflect deployment or expanded use of new and more efficient gear, increased capacity or effort, geographic expansion, and stochastic changes in population(s). Other New England groundfish species show similar trends, though some show earlier (e.g., halibut, *Hippoglossus hypoglossus*) or later (e.g., haddock, *Melanogrammus aeglefinus*) peaks in landings. Source: Data from NMFS 1990 and 2002.

> Wherever . . . man plants his foot and the "civilization" is begun, the inhabitants of the air, the land, and the water, begin to disappear. . . . The fish, overwhelmingly numerous at first, . . . feel the fatal influence in even less time than the [terrestrial] classes. . . . The halibut, one of the best of our fishes, was so common along the New England coast as not to be considered worthy of capture. . . . It is only within [the last] few years that our people have come to learn their excellence and value, but they have already disappeared almost entirely from the inshores of New England, and have become exterminated in nearly all waters of less than five hundred feet in depth. (United States Commission of Fish and Fisheries 1880, p. xlv)

But this report represented a minority view. A century ago, the prevailing scientific and public views of the ocean's living resources remained closer to this Thomas Huxley (1883) vision presented in his inaugural address to the 1883 Fisheries Exhibition in London: "Probably all the great sea fisheries are inexhaustible." Despite this oft-quoted proclamation, Huxley did acknowledge in his address that some fisheries, even some sea fisheries, were in fact exhaustible. The scientist within him could not ignore the empirical evidence that some of the fisheries he researched, notably the European oyster and certain salmon fisheries, had already been largely depleted.

But this did not change the conventional wisdom, that (1) little threat of endangerment, extirpation, or extinction existed for most marine species or ecosystems; (2) the well-documented vulnerability of a few notable exceptions, including some marine mammals, sea turtles, sea birds, estuaries, and coral reefs, extended to little else, especially most marine fish and invertebrates; and (3) the main targets of the world's great fisheries, which still retained a cloak of inexorable, and even magical, invincibility, were somehow immune to such outcomes.

For most of the past century, this dogmatic view remained dominant. Although in recent decades, minority voices within and outside the scientific community started to question it, the assumption of many was that managers could protect any individual species through tools like catch limits, gear restrictions, and other traditional tools. Without compelling evidence to the contrary, sustainable fisheries management was seen as achievable and just around the corner, using these tools, though perhaps needing better information and more political will.

To be fair, for much of human history, the oceans did seem relatively resistant and resilient to our actions, capable of both maintaining themselves and supplying a continued stream of fish, shellfish, and other valuable commodities. Areas undiscovered by fishermen or too far from port, too deep, or too difficult to fish for other reasons, served as "natural refuges" from fishing and protected intact marine communities. This helped maintain healthy marine ecosystems, protect biodiversity, and support fisheries. However, new and improved gear and technology, increased capacity, shifting targets, and rising market prices have enabled exploitation of both previously unfished natural reserves and formerly nontargeted species. As a result, these natural reserves have largely disappeared; their ability to help protect biodiversity, maintain healthy ecosystems, and replenish other fished areas is greatly reduced; and both the magnitude and geographic scope of fishing impacts have been greatly increased.

Slowly, the tide of scientific and public opinion is turning. Within the past ten years it has accelerated, approaching bore velocity, and the prevailing views on this may now be amid a phase shift. A few years ago, we still lacked a strong article in a prestigious journal or a consensus statement from a respected independent group of prominent scientists on the true scope of marine endangerment. We now have several (e.g., Jackson et al. 2001).

## RISING TIDE OF MARINE ENDANGERMENT

While some questions related to the degree of extinction risk for marine fish and invertebrates remain lively topics, the same questions for marine mammals, sea birds, and sea turtles should have been resolved long ago. Human exploitation including fishing and other impacts clearly puts these animals at risk for extinction; the empirical evidence for their susceptibility is really beyond serious debate. Though less well known than their terrestrial counterparts, the rapid disappearance of Steller's sea cow (*Hydrodamalis gigas*), the Caribbean monk seal (*Monachus tropicalis*), and the great auk (*Pinguinis impennis*) (Roberts and Hawkins 1999) following brief contact with mobile human hunters in very different and geographically distinct ecosystems provides three of the most striking illustrations of their susceptibility. Steller's sea cow disappeared within just a few decades of contact with North Pacific whalers once seagoing whaling boats and technology arrived there. The Caribbean monk seal and great auk took slightly longer to succumb, but were still gone within a century or so of similar contact. The sea mink (*Mustela macrodon*) similarly disappeared from North Atlantic coastal waters by the close of the nineteenth century (COSEWIC 2002).

All of the great whales and sea turtles have teetered on the brink of extinction, but miraculously, none have thus far toppled from the precipice. Some have withstood extirpations, to which we lost the Atlantic gray whale (*Eschrichtius robustus*) (Mead and Mitchell 1984), along with the Atlantic walrus (*Odobenus rosmarus rosmarus*) (COSEWIC 2002). Others have seen dramatic declines such as those described for Caribbean sea turtles (Jackson 1997). All remain endangered or threatened and none have yet escaped extinction. Most have been given a respite through a complete or partial cessation of intentional, directed killing, but not all. Even some of those now fully protected from such directed take remain highly endangered. The northern right whale (*Balaena g. glacialis*) herd has been reduced to several hundred and continues to face threats from vessel strikes, entanglement in fishing gear, and minimum viable population size. Steller's sea lion (*Eumetopias jubatus*) likewise is still facing a suite of interlocking threats in the North Pacific.

Documented marine fish and invertebrate extinctions resulting from human impact are relatively few and less dramatic. Until recently, little attention has been paid to them. They remain more likely to go unnoticed, and threats to them are often more difficult to prove. However, there is a rapidly increasing suite of such organisms approaching the brink and a number that may already be extinct (Fig. 1.4). These include a diverse array of finfish, shellfish, and other invertebrates with a variety of life histories and distributions from around the globe. Perhaps most remarkable, the Canadian government recently listed two populations of the Atlantic cod, once seemingly ubiquitous across the North Atlantic, as endangered and threatened. Within the last three decades, the Newfoundland and Labrador cod population declined roughly 97 percent and the species virtually disappeared from some offshore areas (COSEWIC 2003).

The striking case of California's white abalone, *Haliotis sorenseni,* provides a clear and present example of the extinction risk posed to at least some marine species from targeted fisheries. This abalone occupied a relatively narrow depth range and small geographic range, but was fairly abundant in waters between 25 and 65 meters deep around California's Channel Islands until the early 1970s. At this time, a short-lived commercial fishery targeted this species, employing a handful of fishers for less than a decade. Within the span of just a few years, the fishery itself was extinct and the species was on the brink (Fig. 1.2). Commercial landings peaked at 65 tons in 1972, but plummeted to 0.15 tons in just four years. In the early 1990s, intensive searches in known habitats that once harbored densities of up to 10,000 abalone/hectare yielded only a few dozen. Abalone require minimum densities for successful fertilization and recruitment. There is no evidence of significant recruitment or landings in the last several decades. The white abalone appears to be approaching extinction, even though the brief, but intense, fishery that caused its initial collapse ended decades ago. Efforts are now being made to concentrate some of the few remaining adults in an attempt to facilitate successful reproduction, but it may be too late (Davis et al. 1996; Tegner et al. 1996).

The Nassau grouper (*Epinephelus striatus*), a large, long-lived species, formerly common throughout the Caribbean, provides another striking example of vulnerability to exploitation and associated extinction risk. Once an important apex predator, the dominant grouper on many Wider Caribbean coral reefs, and a species of considerable commercial importance, it is today absent or rare across much of the region. Where it still exists, it is much smaller and less numerous than it previously was. Despite its relatively broad distribution and once large numbers, it is exceptionally vulnerable to fishing. The Nassau grouper fears little, aggressively attacks baits, approaches divers, and eagerly enters traps. But the mating habits of the Nassau grouper may ultimately be its downfall. It is a protogynous (female first), hermaphroditic (sex-changing),

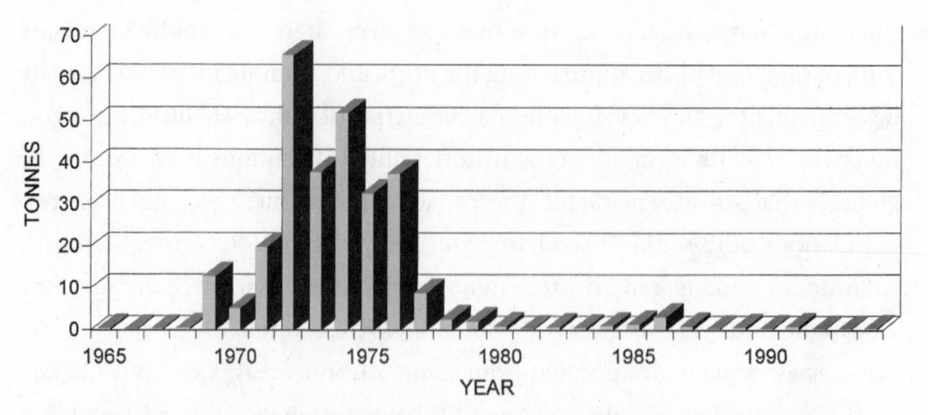


**FIG. 1.2 California White Abalone, *Haliotis sorenseni,* Landings 1965–1994.** Graph reflects reported commercial landings (metric tons). Landings for 1978–1994 include both miscellaneous abalone species and the white abalone. Period from 1969–1979 captures rapid rise and fall of short-lived commercial fishery executed by a handful of fishers for less than a decade that decimated population. Source: Davis et al., 1996; adapted with permission from the American Fisheries Society.

group spawner that aggregates in large numbers to spawn at specific sites for a short predictable time each year. These reproductive habits are a double whammy. First, the targeting of larger, older fish by fishermen means that the big males with greatest reproductive capacity are largely removed from the population. Nassau grouper are also very vulnerable to fishing while aggregating at their spawning sites, and fishers frequently target known spawning sites. Such aggregations once numbered in the tens of thousands of fish. At least a third of these once huge aggregations no longer exist. Despite closures to both spawning sites and targeted fishing, some aggregations and populations have not shown signs of recovery, possibly because measures came too late or because of continued bycatch. The Nassau grouper is currently listed as endangered on the International Union for Conservation of Nature and Natural Resources (IUCN) Red List and a candidate species for the U.S. Endangered Species Act (ESA) (Coleman et al. 2000; Sadovy and Eklund 1999).

Australia's unusual spotted handfish (*Brachionichthys hirsutus*) provides a third striking example of a marine fish recently brought perilously close to and now teetering on the brink of extinction. So named because of its somewhat peculiar habit of walking on its fins rather than swimming, this fascinating species was one of the first Australian fish discovered and could be among the world's first lost due to human activity. The handfish is restricted to a narrow range within a single Australian estuary and is capable of only limited movement. It lays a small number of benthic eggs that remain on the bottom and have limited dispersal capacity. The primary threat to its continued existence is predation on its benthic eggs by the exotic northern Pacific sea star. This alien sea star, likely introduced via ship ballast water, is not a natural predator of handfish eggs. Trawling, dredging, pollution, modification of freshwater flow, and other activities that could disturb its estuarine habitat are also potential threats, as is any targeted collecting that may result from its rarity or value (Pogonoski et al. 2002).

Despite the increased recognition of extinction risk for fish, a proliferation of petitions to list fish under the U.S. ESA, and a growing list of additions to its candidate species list, until recently there remained no exclusively marine domestic fish listed on it. Prior to 2003, the only marine fish listed under the U.S. ESA were anadromous species that spawn in fresh or estuarine water, with the possible exception of the totoaba (*Totoaba macdonaldi*), a species that spawns only in the northern Sea of Cortez near the mouth of the Colorado River in Mexico. However, the eventual listing of a domestic truly marine fish under the U.S. ESA was only a question of when and which species.

On April 1, 2003, the listing of the smalltooth sawfish (*Pristis pectinata*) as endangered under the U.S. ESA answered these questions. This majestic and charismatic species may grow to 25 feet (7 meters) in length and bears a large sawlike snout responsible for its name (Fig. 1.3). Once common in the United States from North Carolina to Texas, dramatic reductions in range and numbers now largely restrict it to the extreme southern tip of the Florida peninsula and a population size less than 1 percent of its historical abundance. The current distribution is focused around Everglades National Park and Florida Bay, where it was once abundant enough to be the target of a recreational bow and arrow fishery. Commercial fisheries landings and incidental take were primarily responsible for reducing this species and bringing it to the brink of extinction, but typical of many endangered species, a multitude of factors, including habitat loss, pollution, modified water flow, and continued bycatch now conspire to keep it there or finish it off (NMFS 2003). Perhaps it was foolish to wait so long.

Prior to the sawfish listing, the totoaba was likely the most marine fish listed under the U.S. ESA. It provides another excellent and interesting example of how human activities, often acting in concert, can rapidly endanger a marine fish. The largest member of the drum or croaker family (Scianidae), the totoaba was endemic to and abundant in the Sea of Cortez (Baja California), where it aggregated to spawn in the lower reaches of the Colorado River. The common names of this family stem from sounds produced by vibrating their swim bladders. Mexican fishers initially targeted this species in the 1920s, primarily for

**FIG. 1.3 Smallmouth Sawfish, *Pristus pectinata,* circa 1928.** Historic photograph shows a day's catch of smallmouth sawfish taken off southwest Florida. Recently listed as endangered under the U.S. Endangered Species Act, these large predators were once common from North Carolina to Texas, but are now rare and largely restricted to a small area off of southern Florida. Source: Photo courtesy of Al Stier and www.floridasawfish.com.

their swim bladders and the high price they fetched in Asian seafood markets. Nonetheless, the catch peaked by 1942 at close to 5 million pounds and declined precipitously thereafter. Although the directed catch was clearly responsible for the initial decline of this species, damming of the Colorado River and bycatch from an intensive shrimp fishery subsequently furthered its decline and continue to endanger it (Norse 1993).

The growing list of petitioned and candidate marine species under the U.S. ESA and elsewhere reflects the increased risk and recognition of that risk and the extension of human impacts further offshore (Fig. 1.4). Among the truly marine species recently listed or petitioned are the barndoor skate (*Raja laevis*) in the New England region due primarily to bycatch; several groupers in the southeastern United States and Caribbean, including the Nassau grouper, due to both directed take and bycatch; the boccacio rockfish (*Sebastes paucispinis*) on the U.S. Pacific coast due to directed take and bycatch; and a number of marine fish from Puget Sound in Washington state, reflecting the increasing concern about the status of marine fish species throughout the country.

## FROM SPECIES TO ECOSYSTEMS

> The diversity of life in the ocean is being dramatically altered by the rapidly increasing and potentially irreversible effects of activities associated with human population expansion. The most critical . . . contributors to changes in marine biodiversity are . . . fishing and removal of the ocean's invertebrate and plant stocks; . . . pollution; physical alterations to coastal habitat; invasions of exotic species; and global climate change. . . . These stresses have affected . . . life from the intertidal zone to the deep sea.
>
> —NRC 1995

In addition to its remarkable similarity to the observations of the U.S. Commissioner of Fish and Fisheries (1878) with which this chapter began, two important distinctions stand out in the NRC (1995) report, reflecting increases in both human understanding and impact in the last century. First, it defines *biodiversity* (a term unknown a century ago) to mean the variety or collection of life at three levels, genomes, species, and ecosystems, and recognizes that humans are now impacting all three. Fishing and other stresses are altering the genetic structure of some marine species, threatening or endangering the continued viability of others, and modifying complex marine ecosystems, including their associated species assemblages and physical environment. Equally significant, the NRC report recognizes the increasingly ubiquitous geographic scope of anthropogenic impact, which is no longer confined to nearshore, shallow water, or developed areas.

Because human populations contribute directly to all of the proximate stresses identified here, these stresses rarely occur in total isolation from one another and often result in cumulative impacts to species, genomes, and ecosystems. Depending on the nature of these cumulative interactions, the impacts

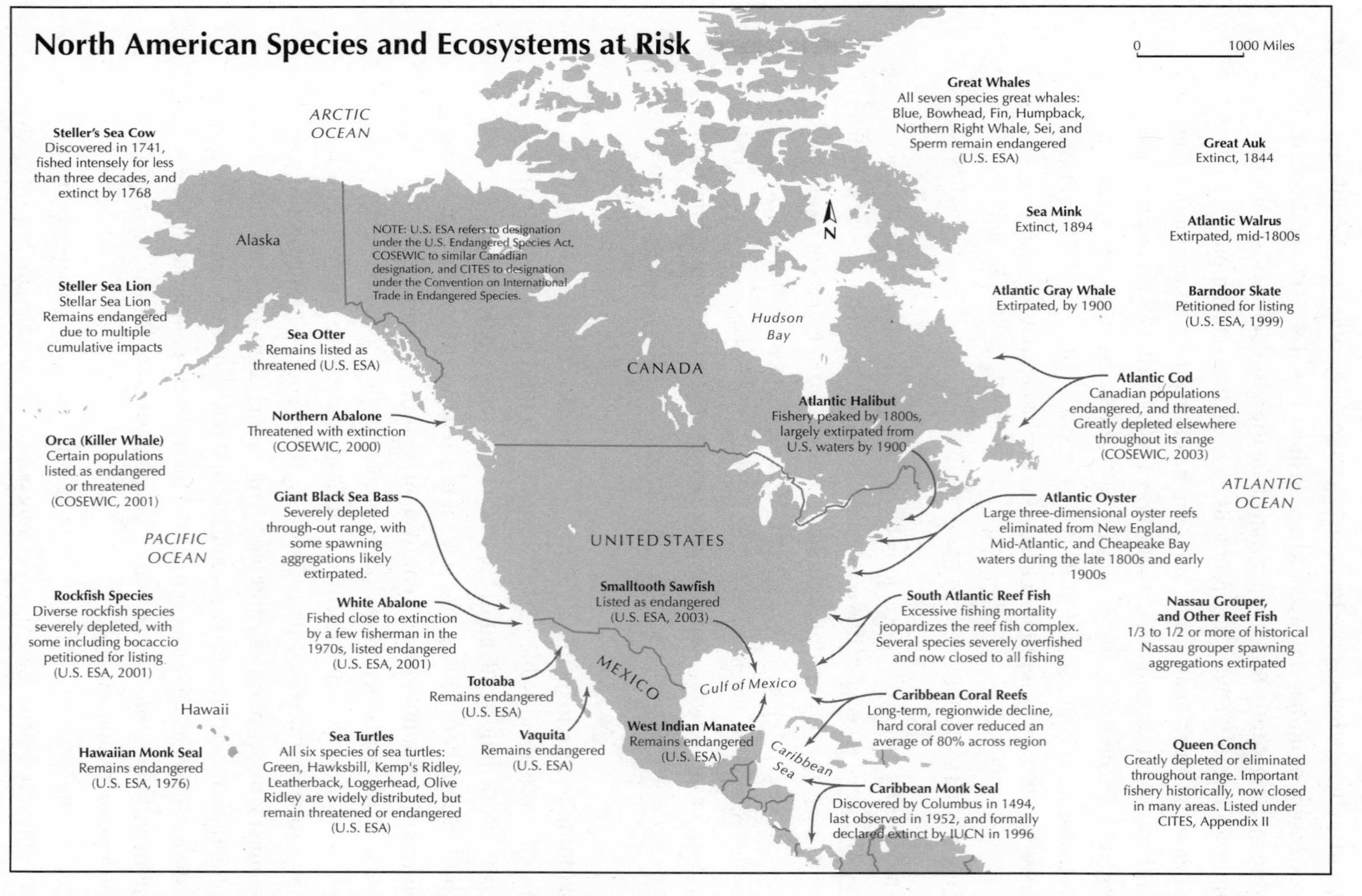


**FIG. 1.4 North American Marine Species and Ecosystems at Risk Map.** This figure highlights selected examples of species and ecosystems considered extinct, endangered, threatened, or at risk. Sources include COSEWIC 2002 and 2003; http://www.nmfs.noaa.gov/prot_res/; and Roberts and Hawkins, 1999.

are often synergistic. Fishing and related extractive activities are particularly widespread in the oceans and often remove critical components of ecosystems, making them more vulnerable or less resistant and resilient to other stresses, and are thus often implicated as key contributors to cumulative and synergistic impacts. Acting both independently and together these stresses have impacts ranging from the decline in important living marine resources, the loss of productive marine habitats, and reduced water quality to human health problems, mass mortality of fish and marine wildlife, and increasingly, population, species, community, or ecosystem endangerment or collapse. Furthermore, these anthropogenic stressors also interact with natural stressors, such as hurricanes, other storms, and climate variation (Jackson et al. 2001).

Despite the recognized gravity of these threats, a number of human and ecological attributes often frustrate our ability to fully understand, much less manage, them. These factors often heighten the threat to marine ecosystems and can include human ignorance and arrogance, scientific uncertainty, environmental variability, and biological complexity. Synergy, biocomplexity, ecological diversity, and redundancy in marine ecosystems can often delay, mask, or lead to their sudden collapse.

### Ignorance and Arrogance

World-renowned marine explorer, scientist, and diver Dr. Sylvia Earle often states, "The single most frightening and dangerous threat to the ocean is ignorance" (Carless 2001). We have much left to learn. But we do already clearly know (1) that specific human activities are causing profound changes to our oceans, (2) what some of these impacts are, and (3) how we can use existing information to better protect our ocean resources. Ignorance becomes a much more powerful threat when matched with arrogance. It is often not how little we know that gets us in trouble, but how much we think we know. Our frequent failure to recognize the limits of our knowledge and act accordingly (i.e., in a precautionary manner) causes the harm. We often seem to believe that if we only studied a little more and got a little more knowledge, we'd be able to overcome uncertainty and effectively manage nature. Yet the more we learn, the more elusive that goal seems to become.

### Uncertainty and Variability

Uncertainty and environmental variability are two similarly linked ecological attributes. Temporal and spatial environmental variability are among the most certain of ecological attributes. Even without human impact, both are nearly universally present and often of great enough magnitude to obscure otherwise reliable measures or indicators. An example of this is the frequent masking of the stock-recruitment relationship in fisheries management due to the high natural variability of key parameters that tend to mask the expected and widely held principle that there is a linkage between the size of an exploited stock and the level of recruitment. Although this widely held belief must be true at some level, the relationship can be difficult to detect or demonstrate due to the high signal to noise ratio. The high degree of natural environmental variability inevitably creates considerable uncertainty in many natural resource management decisions, especially those involving fisheries issues. Such uncertainty in the context of controversial management decisions nearly always leads to delay, inaction, or weak action, until it is too late to adequately protect the resource (Ludwig et al. 1993).

### Ecological Complexity and Interdependence

The ecosystem is the basic and functional unit of ecology because it includes both living organisms (biotic communities) and their associated nonliving (abiotic) environment, each influencing the properties of the other and both necessary for maintenance of life as we know it on the earth. Any unit that includes all of the living organisms in a defined geographic area interacting with their physical environment in such a way that energy flow leads to clearly defined trophic structure, biotic diversity, and material cycles constitutes an ecosystem. Living organisms and their surrounding physical environment are inseparably interrelated and interactive with each other. The notions of ecological complexity, interdependence, and integrity extend the ecosystem concept to recognize the holistic nature and complex relationships within and among natural systems. Ecological (or bio-) complexity is "the dynamic web of often surprising interrelationships that arise when components of the global ecosystem—biological, physical, chemical, and the human dimension—interact" (NSF 2001). Ecological interdependence simply refers to the many, varied, and complex ways in which ecosystem components interact, affect, relate to, and depend on one another for their health and survival. These ecological attributes often greatly amplify the impacts of perturbations to individual components throughout marine systems and consequently exacerbate threats to our oceans. The following case of the spiny lobster in the Marcus and Malgas Isles illustrates this well.

*Spiny Lobsters*

Various members of the diverse, fascinating, and widespread spiny or rock lobster family (Palinuridae) play critically important ecological roles in both tropical coral reef and temperate kelp systems. They are among the most valuable and prized fisheries resources in the world and are the targets of intense commercial and recreational fisheries around the world. Spiny lobsters are technically considered omnivores, but frequently function primarily as mid-level carnivores within their marine benthic communities. As such, they govern the number and sizes of invertebrates like sea urchins, mussels, and gastropods (Lipcius and Cobb 1994).

Many aspects of the spiny lobster's biology are remarkable. For example, its early life history includes an incredibly long larval period that can last up to two years, cover hundreds or thousands of miles, and involve a dozen form changes. But no aspect is more fascinating than that revealed by its dramatically different fates off Malgas and Marcus Islands in South Africa. When studied in the 1980s, these adjacent "twin" islands shared similar physical characteristics, but dramatically different biological communities and food-web relationships that were stable and persisted over time, apparently representing alternative stable states (Barkai and McQuaid 1988).

On Malgas Island, the rocky habitat was dominated by seaweeds and by superabundant rock lobsters (*Jasus lalandii*), which constituted an extraordinary 70 percent of the total benthic biomass. Rock lobsters there consumed settling mussels and prevented the establishment of mussel beds. They also preyed on several species of whelks, though one whelk species was partially protected from lobster predation by a commensal seaweed that covered its shell, and other larger whelks were too large for the lobsters to eat. By contrast, Marcus Island maintained extensive mussel beds and large populations of sea urchins, sea cucumbers, and especially whelks, but lobsters were conspicuously absent. Whelks were less abundant, less diverse, and dominated by different species on Malgas. At some point, the lobster was lost as the key predator from Marcus Island and something prevented its return (Barkai and McQuaid 1988). But what?

An elegant series of experiments partially unraveled the mystery of Malgas and Marcus. Initial studies showed that selective predation pressure from lobsters on different whelk species on Malgas limited their abundance there. In contrast, the absence of lobsters on Marcus allowed for far greater abundance, diversity, and composition of whelk there. When lobsters were transplanted from Malgas to Marcus Island in cages and control lobsters were kept in cages on Malgas as well, all caged animals on both islands survived until their release over nine months later. In other words, there was nothing about the water quality, temperature, or currents that controlled their survival.

The next experimental step was to install artificial shelters on Marcus to ensure suitable habitat for lobsters there and then transfer and release a thousand tagged lobsters to them. This release produced sudden, dramatic, and remarkable results. The usually predatory and apparently healthy lobsters were engulfed, overwhelmed, and consumed immediately by hordes of whelk, their normal prey. Hundreds were attacked instantaneously and within days no live lobsters could be found on Marcus. To ensure that the lobsters hadn't been injured or otherwise made susceptible to attack in the tagging and transfer process, the experiment was repeated five additional times with smaller numbers of unmarked lobsters with similarly gruesome results. The results were all the more spectacular considering the lobsters' strong tail, rapid reverse swimming escape response, and armored shell. Close observations revealed that the lobsters initially escaped by flicking their tails, but each time they contacted the substrate, more whelk attached to them until their weight prevented escape. Lobsters were mobbed to death on average within fifteen minutes by more than 300 whelks that stripped them of all flesh within an hour (Barkai and McQuaid 1988).

According to local fishermen, lobster populations on the two islands had been similar to each other just two decades earlier. The cause of the initial disappearance of lobsters from Marcus Island remains a mystery and may have been due to either overfishing or pollution. Regardless of the initial cause, the whelk have proven capable of reversing their normal role as rock lobster prey and excluding them by aggressively preying on them when they return. Unlike lobsters, whelk can only prey on damaged mussels, not on healthy ones, and mussels are filter feeders obtaining their nutrition from the water column, enabling them to coexist at high densities when lobsters are excluded (Barkai and McQuaid 1988).

The Malgas–Marcus saga points out the complex, variable, and often hard to predict or see predator–prey, competitive, and other relationships that frequently exist among members of biological communities and between them and their physical environment. Certainly, as already noted here and as is well documented, spiny or rock lobsters often play critical roles in structuring a range of tropical and temperate benthic communities. For example, research on the role of fishing impacts on another spiny lobster (*Panularis interuptus*) in the dynamics of southern California kelp communities concluded that heavy fishing pressure on the lobster likely contributed to the release of sea urchin populations and the episodic destructive urchin grazing observed since the 1950s, with the associated urchin barrens and reduced kelp forest cover. Such fishing pressure reduces the number and density of lobster populations, and it also reduces lobster size (Fig. 1.5), which means that larger urchins escape being eaten by lobsters (Tegner et al. 1996; Tegner and Levin 1983).

Earlier, we mentioned the extreme intensity of many commercial and recreational lobster fisheries. To provide an idea of this intensity level for one fishery, the industrious, hardworking Florida Keys fishermen remove nearly all of the adult spiny lobster (primarily *Panularis argus*) from shallow water areas in less than half a year of fishing (Hunt 1994). Recreational lobster fishing is an important factor in the intensity of this fishery and can rival or even exceed the commercial catch in some areas (Blonder et al. 1988; Davis 1977; Eggleston and Dahlgren 2001). Scientists, fishers, conservationists, and others may continue to debate whether such current intense harvest levels are sustainable from a lobster fishing perspective, but given what we already know there should be little doubt that this level of lobster removal is likely having profound impacts on ecosystem dynamics, community structure, and nonconsumptive recreational activities. We may not know exactly what these are, what their extent is, or when they will manifest into a demonstrable collapse or problem, but we should be concerned.

In 1997, commercial lobstermen in the Florida Keys discovered an unusual and unprecedented sea urchin–sea grass overgrazing event (Fig. 1.6). Urchins were piled on top of one another in a mounded front 2 miles long at densities up to 364 per square meter. From September 1997 to May 1998 the urchin front consumed and denuded the valuable sea grass habitat within its path. The front eventually receded, but the damaged areas have still not recovered and there have been some less intense recurrences (Hunt 2001; Rose et al. 1999; Sharp 2000). We may never know whether this event was related to the lobster fishery or another anthropogenic or natural stress, but it is certainly plausible. What is clear is that the multiple interactions among species and stressors demonstrate that the largely single species approaches of the past, aimed at controlling one variable, such as fishing gear, or protecting one species, typically one of commercial value, are simply inadequate.

## MARINE RESERVES: A HOLISTIC, ECOSYSTEM-BASED APPROACH

Long before species are threatened with extinction, the critical roles they play in maintaining healthy ecosystems are often impaired. While the number of

**FIG. 1.5 Channel Island Spiny Lobsters from the "Good Old Days," circa 1960.** Historic photograph shows size of lobsters caught during early days of scuba-diving off the Channel Islands. Source: Photograph courtesy of Dick Holt.

marine species facing biological extinction continues to grow, many more are becoming ecologically irrelevant or subject to ecological extinction. Still more remain ecologically relevant, but function in a reduced or altered capacity from an ecosystem perspective. Spiny lobsters do not yet face biological, ecological, or even commercial extinction, but have been severely depleted in many places. As a result, the ecosystem functions they once provided have been profoundly altered or lost, sometimes with dramatic consequences. In most ma-

**FIG. 1.6 Florida Keys Urchin Overgrazing Event, circa 1997–1998.** Photograph captures unusual and unprecedented event during which sea urchins were piled up in a mounded front over two miles long with densities up to 364 urchins/meter$^2$. The advancing urchin front destroyed valuable seagrass habitat in its path that has yet to recover. The cause of the outbreak remains a mystery. Source: Photograph courtesy of the Florida Fish and Wildlife Conservation Commission.

rine ecosystems, a myriad of species have been similarly depleted and their functional contributions lost or reduced. Globally, large predatory fish biomass has been reduced over 90 percent (Myers and Worm 2003). Depletion of a single predator can reverberate and impact a host of other species, yet we continue attempting to manage most species independently. What happens to species and ecosystems when you deplete an entire class of predators? As each thread of life is removed, the fabric that once stitched together marine ecosystems frays. The tearing apart and impairment of these living systems places even greater stress back on already depleted species. The resulting downward spiral often leads to collapse of species and ecosystems and must be reversed.

Marine "no-take" reserves, areas closed to fishing and all other extractive activities, are among the most essential tools required to protect and restore the health of our oceans from multiple stressors. Why? They are uniquely tailored to help prevent and reverse the downward spiral that results from removing critical components of living marine systems in areas subject to exploitation. At the most basic level, they are the only approach to marine resource management specifically designed to protect the integrity of marine ecosystems and preserve intact portions and examples of them. In fact, this is their primary purpose.

Protection of ecosystem integrity encompasses three components: ecosystem health, resilience, and potential for continued self-organization. *Ecosystem health* refers to its current state or condition at a point in time. *Resilience* refers to the ecosystem's ability to respond to additional stress caused by external influences. The final component refers to the ecosystem's capacity for development, regeneration, and evolution under normal circumstances (Kay and Regier 1999). Marine reserves contribute to the protection of ecosystem integrity by greatly strengthening and supporting each of the component parts of that integrity.

Marine reserves can function as a preventative or as insurance to maintain, protect, and restore healthy marine ecosystems. They can also serve as a cure: They can help restore and recover an ill or injured ecosystem. They can help stop the bleeding, stabilize and right the patient, protect the vital organs, and set them on a path to recovery. They also allow us to monitor them to provide critical data needed for decisions about additional therapy. For healthy systems, they may be employed as part of a comprehensive, preventive health maintenance program. For both, they can provide insurance via increased resilience to speed recovery following an unexpected or catastrophic event. Thus, increased stress only enhances their value. Ecosystems are inherently dynamic and the long-term goal should not be to freeze them at a point in time, but rather to maximize their options for continued evolution. Over longer time scales, the role of marine reserves in protecting ecosystems' capacity for continued natural evolution will likely grow and is among their most important.

Marine no-take reserves do not seek to protect just one species, control a single variable, or eliminate natural change. Rather, they aim to avoid anthropogenic perturbations to individual species and across whole ecosystems while supporting natural diversity, variability, and evolution. They preclude extractive or consumptive use within their boundaries but are essential to responsible, sustainable use outside their boundaries. Marine reserves do not allow fishing within their boundaries, but effective networks of them may be critical to optimizing and preserving fishing opportunities outside of them. Marine reserves need not ban nonconsumptive recreational activities and can enhance opportunities for such activities. Effective marine reserves do not require complete knowledge about individual species, their natural interactions, or the complex ecosystems of which they are integral components, but are essential to improving our understanding of each.

Marine reserves are likely the only tool for achieving certain conservation objectives, the best at achieving others, and a key contributor to achieving still others. Although necessary, they are not a magic pill or cure-all and not sufficient by themselves to protect the oceans from all human abuse independent of other measures.

## MARINE PROTECTED AREA NOMENCLATURE: SPEAKING A COMMON LANGUAGE?

As highly regarded a management tool as marine reserves have become, there remains a great deal of confusion as to exactly what they are. Much of this relates to broader terminology issues because marine reserves are part of a spectrum of management mechanisms under the umbrella term *marine protected area* (MPA). Among the first things anyone new to MPAs notices is that there are a plethora of terms to describe them, the terms are not used consistently, and their names often bear no resemblance to reality or the level of protection they afford. Protected areas are often left unprotected. Sanctuaries frequently provide limited or no sanctuary for their inhabitants. Reserves are rarely held in reserve. What is called a sanctuary in one place is referred to as a reserve somewhere else. What is called an MPA by someone in one place might be referred to as a marine managed area (MMA) or marine conservation area (MCA) by someone else or in some other place. MPA to some people in some places is a very general term that covers a range of place-based protections, whereas to other people in other places it is a very specific term that connotes a very high level of protection. There is little consistency from place to place or within a place. There is little consistency or standardization among or within stakeholder groups. In short, confusion abounds.

## MARINE RESERVE AND MPA TERMINOLOGY

**Definition:** *Marine reserve = marine "no-take" reserve = an area of the sea in which all consumptive or extractive uses, including fishing, are effectively prohibited and other human interference is minimized to the extent practicable.*

The focus of this book is marine no-take reserves, or marine reserves for short. In the context of this book, we intend either of these terms to define an area of the sea in which all consumptive or extractive uses, including fishing, are effectively prohibited and other human interference is minimized to the extent practicable. Both within and beyond the scope of this book, we believe there is an advantage to and recommend using the longer term, *marine no-take reserve,* whenever possible, to make the meaning more explicit, given the variable usage of these terms and the resulting confusion. Of course, one must explicitly define these terms when they are used in new contexts or with new audiences for the same reason. This book discusses some of the nuances of this definition but will not dwell on or overemphasize them unnecessarily. Given the variable usage of and confusion regarding MPA terminology, there is no ideal term for the book's subject, but we selected the preceding term(s) and definition to be as clear, accurate, and consistent with other accepted terminology as possible, and to avoid unnecessary confusion.

For the purpose of this book, we define MPAs as distinct from marine reserves:

**Definition:** *Marine protected area (MPA) = "Any area of intertidal or subtidal terrain, together with its overlying water and associated flora, fauna, historical and cultural features, which has been reserved by law or other effective means to protect part or all of the enclosed environment."* (Kelleher 1999)

The above definition, originally adopted by the World Conservation Union (IUCN) nearly fifteen years ago, is among the most widely used and accepted. Given the large and broad governmental and nongovernmental organization (NGO) membership of IUCN, this definition may be considered an international standard. The U.S. government similarly defined MPA as "any area of the marine environment that has been reserved by Federal, State, territorial, or local laws or regulations to provide lasting protection for part or all of the natural and cultural resources within them" in MPA Executive Order #13158 (The White House 2000). Both the nearly identical IUCN and U.S. definitions comport with the idea that MPAs include a broad spectrum of protective management regimes with variable levels of protection for ocean waters.

One more definition is important in understanding marine reserves:

**Definition:** *Marine Wilderness = "An area of the sea . . . along with coastal land where appropriate, that has been protected to preserve or restore its natural character, condition, vistas, living communities, and habitats for present and future generations to enjoy, experience, explore, and study, but leave unaltered. Ocean wilderness areas are large, generally at least 100 square miles, closed to all extractive activities, including all forms of fishing, and to other damaging human activities as needed to ensure the natural communities within flourish, as much as possible unaffected by human activities."* (The Ocean Conservancy 2001)

The concept of marine wilderness has been discussed for at least four decades, following passage of the U.S. Wilderness Act of 1964. The intensity of that conversation has increased greatly in recent years due to recognition of the profound alteration of marine ecosystems discussed briefly earlier in this chapter and detailed in later ones. Nonetheless, progress in protecting ocean wilderness remains slow. The preceding definition, adopted by The Ocean Conservancy in 2001, is similar to that currently under consideration by a broader consortium of conservation interests. Though consistent with and similar to the marine reserve definition provided earlier here, it builds on and expands it in several ways. The most critical differences are that it suggests a minimum size, explicitly states as a goal preserving or restoring the natural character, and raises the bar of protection with respect to other human activities.

## EVOLUTION OF TERRESTRIAL AND MARINE PROTECTED AREAS AND CONSERVATION ETHICS

> Imagine that within a national park one of the large carnivores, say wolves, were allowed to be hunted and killed. Or, if not, chased until exhausted and then released. . . . Yet this situation is the status quo for freshwater and marine fish [even] within our national parks. . . . The ordainment of fishing but not hunting indicates an unjustified dichotomy between aquatic and terrestrial species and ecosystems. The results of this . . . management . . . may be causing a[n] . . . erosion of . . . park . . . resources. Research suggests that . . . fish, particularly top predators, play important ecological roles within aquatic ecosystems. . . . In coral reef ecosystems . . . urchins are controlled by a few edible triggerfish and changes in these top predators can have unexpected consequences that can affect the entire ecosystem. (McClanahan, *Bioscience,* 1990)

A little over a century ago, under the leadership of President Teddy Roosevelt and others, the United States began to safeguard great pieces of America's landscape as national parks and wildlife refuges and to espouse a national conservation ethic. Without their great foresight, Yellowstone, Yosemite, and the other great terrestrial landmarks that are part of our national legacy might not exist. Nor would they be safeguarded today had we simply drawn lines on a map back then and failed to develop a stronger terrestrial conservation ethic, invest in their future, or strengthen their protection. The resulting extensive and diverse system of U.S. terrestrial protected areas has served to protect a part of America's natural heritage and to provide a model, which other nations have modified to protect their own natural landscapes. The first of these, Yellowstone National Park, was created in 1872, just about the time the U.S. Fish

Commission was making its observations about the decline of ocean resources generally and the disappearance of fish off the New England coast specifically.

In 1966, nearly a century later, another distinguished group of scientists raised remarkably similar concerns to President Johnson in Effective Use of the Sea, Report of the Panel on Oceanography, President's Science Advisory Committee. They recognized that (1) the near-shore environment was critically important; (2) it was undergoing rapid modification due to human activity, the details of which were unknown, but were broadly undesirable; and (3) this problem was urgent and its dangers had not been adequately recognized (The White House 1966). Among the report's major recommendations to meet its long-range goals of increasing marine food resources and preserving the near-shore environment was the establishment of a national system of marine wilderness preserves and the extension to marine environments of the basic principles and policies of the Wilderness Act of 1964 (Public Law 88-577) to secure for the American people of present and future generations the benefits of an enduring resource of wilderness. Specific purposes identified for such a system included (1) provision of ecological baselines against which to compare modified areas, (2) preservation of major types of unmodified habitats for research and education in marine sciences, and (3) provision of continuing opportunities for marine wilderness recreation.

Diverse national systems of MPAs generally and marine reserves specifically offer tremendous potential to protect our marine legacy and save, study, and sustainably utilize the world's marine biological diversity. Despite this potential, the development of MPAs, their conceptual framework, and the underlying ocean conservation ethic necessary to support them have trailed their terrestrial counterparts by nearly a century. The dumping of nerve gas off Florida and the Santa Barbara oil spill triggered public outrage that contributed to congressional consideration of eleven bills in 1968 to establish sanctuaries and oil drilling moratoria off the coasts of California, Massachusetts, and New Hampshire (CNA 1977). In 1970, a report from the President's Council on Environmental Quality (CEQ) rekindled interest in ocean dumping and sanctuary legislation. In 1972, exactly one hundred years after Yellowstone was created as the country's first national park, the U.S. Congress recognized the lag in development of MPAs, finding that "this Nation has recognized the importance of protecting special areas of its public domain, but these efforts have been directed almost exclusively to land areas above the high water mark." As a result, Congress created the National Marine Sanctuary (NMS) Program to "identify areas of the marine environment of special national significance due to their resource or human-use values" and "provide authority for comprehensive and coordinated conservation and management of these areas" consistent with "the primary purpose of resource protection" (NMSA 2000).

A couple of years into this new millennium, our U.S. MPAs remain near where our terrestrial ones stood at the start of the last century. We have begun to identify and designate some of the cornerstone marine landmarks, areas like the Florida Keys, Monterey Bay, and the Northwestern Hawaiian Islands, as worthy of national marine sanctuary or similar protected status, but we have not yet developed strong enough management plans for them or properly invested in their future, and there remain geographic holes in the system that require filling. To date, less than 1 percent of U.S. marine waters have received national marine sanctuary or similar MPA status, and less than 1 percent of national marine sanctuary waters have received stronger protection as marine no-take reserves (The Ocean Conservancy 2002). McCardle (1997) calculated that the State of California, with over 100 MPAs, fully protects less than .2 percent of its waters from all fishing activities, as marine no-take reserves, and provides effective enforcement for only a fraction of these areas. Similarly, as of 1997, 72 out of Canada's 110 MPAs provided no protection to either marine species or habitats. North America is far from unique with respect to this paucity of strong protection for marine areas. Rather, this condition with relatively few MPAs, comprising a limited area, and affording little protection even where they do exist "characterizes the situation worldwide, in countries rich and poor, in waters warm and cold" (Fig. 1.7); (Roberts and Hawkins 2000).

> As humanity increases its influence on the world's environment . . . marine ecosystems are experiencing pressures that will soon equal terrestrial counterparts. One of the future roles of our parks will be to maintain pristine ecosystems that can be compared to other managed and mismanaged ecosystems. This preservation will be difficult with the parks' many external influences, but will be impossible if internal management allows recreation and resource use to supercede preservation. The subjectivity of the fishing–hunting dichotomy must be relinquished to a more objective management plan that preserves aquatic in the same manner as terrestrial species. (McClanahan 1990)

Since at least the late 1800s, the United States has clearly been a global leader in the effort to conserve terrestrial areas through parks and protected areas. Though far from perfect, the United States together with much of the world has developed a system of terrestrial managed areas that includes a full spectrum of protection levels. The IUCN provides a classification system de-

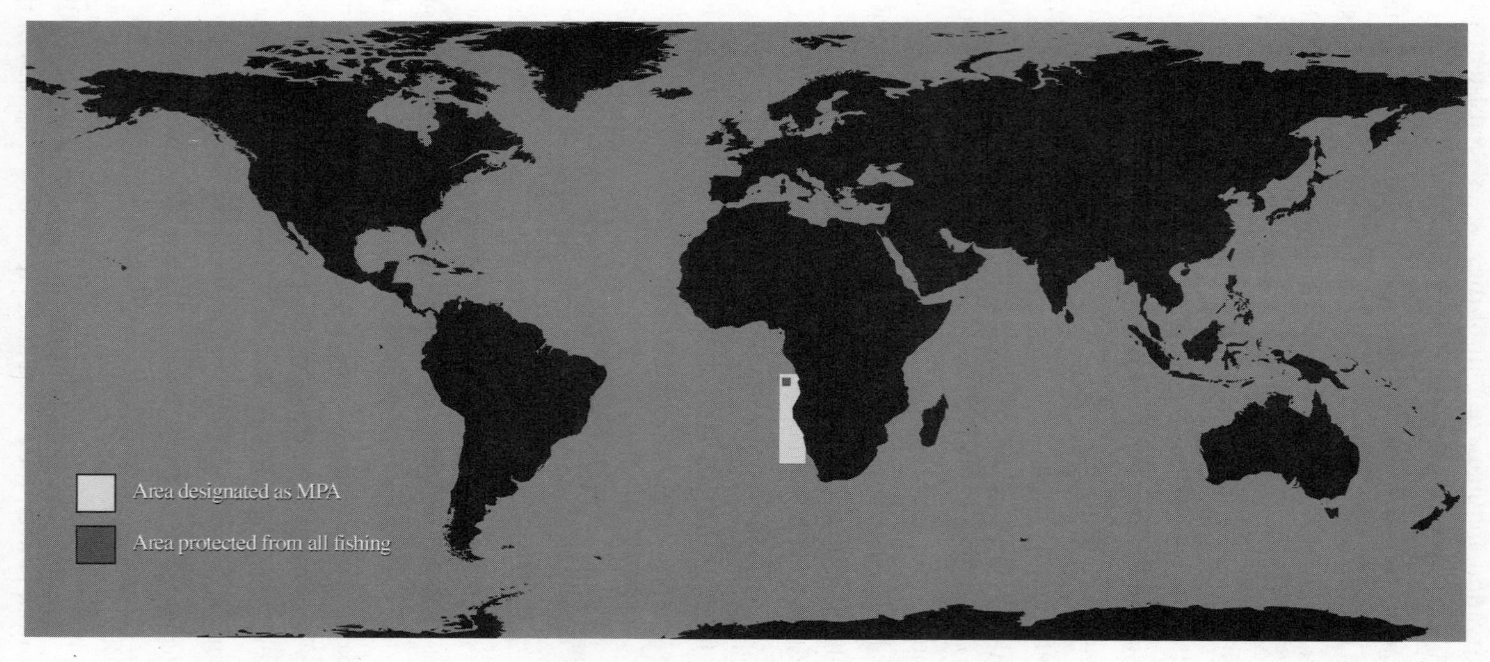


**FIG. 1.7 Fraction of Ocean Protected as Marine Reserves and Marine Protected Areas (MPAs).** This map shows the small portion of the ocean included in marine protected areas of any type, < 0.5%, and the even smaller portion included in marine reserves protected from all forms of fishing, < 0.0001 percent. The majority of existing MPAs are not yet adequately managed. Source: Figure courtesy of Callum Roberts and Julie Hawkins, adapted from Roberts and Hawkins (2000).

scribing a full range of protected areas applicable to both marine and terrestrial areas (Box 1.1) plus some other public managed area categories not considered to be protected areas under the IUCN system. Collectively, this diverse set of terrestrial protected and managed areas, with variable levels of protection, contributes greatly to conserving the world's biodiversity. There is no similar comprehensive system of MPAs yet in place to protect our marine biodiversity. Some MPAs do exist and make a contribution, but they cover a much smaller portion of the marine environment than terrestrial areas, are greatly skewed to the less protected end, are often poorly funded and weakly implemented, and consequently provide far less protection. The range of MPAs is truncated, with IUCN categories 1 and 2 (Box 1.1) nearly completely lopped off. Neither the United States nor the world has done well in this regard. Unlike the terrestrial case, the United States has not provided strong global leadership on this and, if anything, has probably learned much from other countries, especially island and other nations with extensive coastlines, which have more clearly recognized the value of their coastal assets.

In the mid 1900s, Aldo Leopold recognized the need for humanity to develop a land ethic to advance terrestrial conservation. If he were still alive today, he might argue that this is still lacking and we are still losing the war, but on the terrestrial side some progress has clearly been made and some battles have been won. Today, development of a similar "ocean ethic" must go hand-in-hand with advancing ocean conservation and marine no-take reserves. Some reserve opponents cast their argument from a right to fish perspective. The appropriate question to ask is not, To fish or not to fish? but rather, Must we fish everywhere? On land, we have already decided that hunting should not occur everywhere. Science strongly supports the immediate need for a similar ocean ethic and a societal decision that fishing should not occur everywhere, if we want to stem the tide of changes described in chapters 1–3 of this book, protect ocean ecosystems and species, and halt or prevent the loss of marine biodiversity. However, society, not science alone, will ultimately need to decide whether or not these are things we want to do.

> The last word in ignorance is the man who says of an animal or plant: "What good is it?" If the land mechanism as a whole is good, then every part is good, whether we understand it or not. If the biota in the course of eons, has built something we like but do not understand, then who but a fool would discard seemingly useless parts? To keep every cog and wheel is the first precaution of intelligent tinkering. (Leopold 1953, 146–147)

**Box 1.1 IUCN Categories of Protected Areas**

| Category | Purpose |
|---|---|
| Ia | **Strict nature reserve/wilderness protection area:** managed mainly for science or wilderness protection—an area of land and/or sea possessing some outstanding or representative ecosystems, geological or physiological features, and/or species, available primarily for scientific research and/or environmental monitoring |
| Ib | **Wilderness area:** protected area managed mainly for wilderness protection—large area of unmodified or slightly modified land and/or sea, retaining its natural characteristics and influence, without permanent or significant habitation, which is protected and managed to preserve its natural condition |
| II | **National park:** protected area managed mainly for ecosystem protection and recreation—natural area of land and/or sea designated to (a) protect the ecological integrity of one or more ecosystems for present and future generations, (b) exclude exploitation or occupation inimical to the purposes of designation of the area, and (c) provide a foundation for spiritual, scientific, educational, recreational, and visitor opportunities, all of which must be environmentally and culturally compatible |
| III | **Natural monument:** protected area managed mainly for conservation of specific natural features—area containing specific natural or natural/cultural feature(s) of outstanding or unique value because of their inherent rarity, representativeness, or aesthetic qualities or cultural significance |
| IV | **Habitat/species management area:** protected area managed mainly for conservation through management intervention—area of land and/or sea subject to active intervention for management purposes so as to ensure the maintenance of habitats to meet the requirements of specific species |
| V | **Protected landscape/seascape:** protected area managed mainly for landscape/seascape conservation or recreation—area of land, with coast or sea as appropriate, where the interaction of people and nature over time has produced an area of distinct character with significant aesthetic, ecological, and/or cultural value, and often with high biological diversity. Safeguarding the integrity of this traditional interaction is vital to the protection, maintenance, and evolution of such an area. |
| VI | **Managed resource protected area:** protected area managed mainly for the sustainable use of natural resources—area containing predominantly unmodified natural systems, managed to ensure long-term protection and maintenance of biological diversity, while also providing a sustainable flow of natural products and services to meet community needs. |

Cardiff University and IUCN 2002 Speaking a Common Language Information sheet #3. July 2002. Available online: http://www.cardiff.ac.uk/cplan/sacl.

The visionary conservationist Aldo Leopold penned these words, his "first principle of conservation," in an essay entitled Conservation, driven by the vast destruction and degradation of terrestrial and freshwater ecosystems he observed prior to the mid-1900s. A half century later, as we increasingly recognize and document similar catastrophic changes to marine systems (e.g., Jackson et al. 2001; NRC 1995), it is clear that these and other concepts he developed back then apply equally well to marine systems now.

Particularly worthy of extension to the marine environment is his recognition, fifty years before the term *biocomplexity* was coined, that the outstanding scientific discovery of the twentieth century was not a technical device, such as a television, radio, or, by extension, a computer, space station, or submarine, but rather the complexity of living organisms, their interactions, and their interdependencies. Similarly, his recognition that only those who know the most about these organisms and their relationships can also fully appreciate how little we know about them, and why our aim must be to preserve all of the parts and connections, even though the best scientist cannot recognize all of them, remains as applicable, if not more so, for marine systems than for terrestrial ones. Furthermore, his writings and the examples he provided clearly recognized the need to protect entire ecosystems and retain their integrity, even though his "land ethic" predated the terms *ecosystem management* and *ecological integrity,* let alone their application to the marine realm (Leopold 1953).

## CONCLUSION: URGENT NEED FOR NEW APPROACHES, INCLUDING MARINE RESERVE NETWORKS

> The world we have created today as a result of our thinking thus far has problems which cannot be solved by thinking the way we thought when we created them.
>
> —Albert Einstein

New ways of thinking and acting, changes to the status quo, and rapid action are urgently needed to restore some of what's been lost, hold on to some of what's left, protect our options, understand our choices, improve our decisions, and preserve some of the oceans' wilderness for current and future generations to enjoy. A more precautionary, ecosystem-based approach is necessary to do so. As Leopold suggested, we can't afford to lose the pieces. Marine reserves, areas in which all marine life is protected from all forms of fishing and other extraction, and the subject of this book, provide a safety net for them and are an essential part of such an approach, but not the whole answer. Nearly every scientific and government study, report, and reference discussed in this opening chapter and many elsewhere reached this same conclusion.

Chapters 2 and 3 focus more closely on the state of marine ecosystems and fisheries, the impacts of fishing on all levels of biodiversity, and why marine reserves are needed to address some of these impacts. Chapter 4 develops the marine reserve concept more fully and describes what they can do that others tools can't. It also explores the scientific underpinnings of marine reserves, including some of the existing evidence for their efficacy with respect to these. Chapter 5 lays out critical marine reserve design issues and approaches. Chapter 6 examines the equally important human dimensions of reserves. Chapter 7 explores some of the current priorities in scientific and research issues related to reserves and some of the traditional as well as exciting new tools for addressing these. The second part of the book, chapters 8 through 11, tours and reviews the global experience with marine reserves, extracting specific points from a host of sites and providing more detailed case studies for a smaller number. The final chapter also draws on the entire book to provide conclusions and recommendations regarding the use of marine reserves.

## REFERENCES

Barkai, A., and C. McQuaid. 1988. Predator–prey role reversal in a marine benthic ecosystem. *Science* 247 (4875):62–64.

Blonder, B. I., J. H. Hunt, D. Forcucci, and W. G. Lyons. 1988. Effects of recreational and commercial fishing on spiny lobster abundance at Looe Key National Marine Sanctuary. In *Proceedings of the 41st Annual Meeting of the Gulf and Caribbean Fisheries Institute,* Charleston, SC: GCFI, 487–491.

Brooks, W. K. 1996. *The Oyster.* Baltimore, MD: Johns Hopkins University Press.

Carless, J. 2001. Ocean exploration: Unlocking the mysteries of the deep to protect the future. *Blue Planet Quarterly* 1(2):18–25.

Center for Natural Areas (CNA). 1977. *Assessment of the Need for a National Marine Sanctuary Program.* Washington, DC: Office of Coastal Zone Management (OCZM). CNA/OCZM 7-35118.

Coleman, F. C., et al. 2000. Long-lived reef fishes: The grouper–snapper complex (AFS policy statement). *Fisheries* 25(3):14–20.

Committee on the Status of Endangered Wildlife in Canada (COSEWIC). 2002. *Canadian Species at Risk.* Ottawa, ON: Canadian Wildlife Service, Environment Canada.

———. 2003. *Canadian Species at Risk.* Ottawa, ON: Canadian Wildlife Service, Environment Canada.

Davis, G. E. 1977. Effects of a Recreational Harvest on a Spiny Lobster, *Panulirus argus,* Population. *Bulletin of Marine Science,* 27(2):233–236.

Davis, G. E., P. L. Haaker, and D. V. Richards. 1996. Status and trends of white abalone at the California Channel Islands. *Transactions of the American Fisheries Society* 125:42–48.

Eggleston, D. B. and C. P. Dahlgren. 2001. Distribution and abundance of Caribbean spiny lobsters in the Key West National Wildlife Refuge: relationship to habitat features and impact of an intensive recreational fishery. *The Journal of Marine and Freshwater Research,* 52: 1567–1576.

Hunt, J. H. 1994. Status of the fishery for *Panulirus argus* in Florida. In *Spiny Lobster Management,* 144–157. Cambridge, UK: Blackwell Scientific.

Huxley, T. H. 1883. Inaugural Address Fisheries Exhibition, London. In *The Fisheries Exhibition Literature,* 1–11. London, 1885.

International Union for Conservation of Nature and Natural Resources (IUCN). 1994. *Guidelines for Protected Area Management Categories.* Gland, Switzerland: Commission on National Parks and Protected Areas.

Jackson, J. B. C. 1997. Reefs since Columbus. *Coral Reefs* 16:23–32.

Jackson, J. B. C., et al. 2001. Historical overfishing and the recent collapse of coastal ecosystems. *Science* 293(27 July):629–637.

Kay, J. J., and H. Regier. 1999. Uncertainty, complexity, and ecological integrity: Insights from an ecosystem approach. In *Implementing Ecological Integrity: Restoring Regional and Global Environmental and Human Health,*121–156. NATO.

Kelleher, G. 1999. *Guidelines for Marine Protected Areas.* Gland, Switzerland and Cambridge, UK: IUCN.

Leopold, A. 1953. Round River. In *The Journals of Aldo Leopold.* New York: Oxford University Press.

Lipcius, R. N., and J. S. Cobb. 1994. Introduction: Ecology and fishery biology of spiny lobsters. In *Spiny Lobster Management,* 1–24. Cambridge, UK: Blackwell Scientific.

Ludwig, D., R. Hilborn, and C. Walters. 1993. Uncertainty, resource exploitation, and conservation. *Science* 260:17–18.

McArdle, D. A., ed. 1997. *California Marine Protected Areas.* La Jolla: California Sea Grant College System.

McClanahan, T. R. 1990. Are conservationists fish bigots? *Bioscience* 40(1):2.

Mead, J. G., and E. D. Mitchell. 1984. *Atlantic Gray Whales. The Gray Whale,* Eschrichtius robustus. Orlando, FL: Academic Press, 33–53.

Myers, R. A., and B. Worm. 2003. Rapid worldwide depletion of predatory fish communities. *Nature* 423(15 May):280–283.

National Marine Fisheries Service (NMFS). 1990. *Historic Fishery Statistics Atlantic and Gulf States 1879–1989.* Silver Spring, MD: NMFS.

———. 2002. Commercial Fisheries and Marine Recreational Fisheries Statistics Survey. Available online: http://www.st.nmfs.gov/st1/ (last accessed Dec. 2003). (NMFS Fisheries Statistics and Economics Web site).

———. 2003. *Endangered and Threatened Species: Final Endangered Status for a Distinct Population Segment of Smalltooth Sawfish* (Pristis pectinata) in the U.S. Federal Register 68: 15674–15680, Washington, D.C.

Norse, E. A. 1993. *Global Marine Biological Diversity: A Strategy for Building Conservation into Decision-Making.* Washington, DC: Island Press.

National Research Council (NRC). 1995. *Understanding Marine Biodiversity: A Research Agenda for the Nation.* Washington, DC: National Academy.

National Science Foundation (NSF). 2001. Biocomplexity in the environment. Available online: http://www.nsf.gov/geo/ere/ereweb/index.cfm.

Pogonoski, J. J., D. A. Pollard, and J. R. Paxton. 2002. *Conservation Overview and Action Plan for Australian Threatened and Potentially Threatened Marine and Estuarine Fishes.* Canberra: Environment Australia.

Roberts, C. M., and J. P. Hawkins. 1999. Extinction risk in the sea. *Trends in Ecology and Evolution* 14:241–246.

———. 2000. *Fully Protected Marine Reserves: A Guide.* Washington, DC: World Wildlife Fund Endangered Seas Campaign.

Rose, C. D., W. C. Sharp, W. J. Kenworthy, J. H. Hunt, W. G. Lyons, E. J. Prager, J. F. Valentine, M. O. Hall, P. E. Whitfield, and J. W. Fourqueran. 1999. Overgrazing of a large sea grass bed by the sea urchin, *Lytechinus variegatus,* in outer Florida Bay. *Marine Ecological Progress Series* 190:211–222.

Sadovy, Y., and A. M. Eklund. 1999. *Synopsis of Biological Information on* Epinephelus striatus *(Bloch, 1972), the Nassau grouper, and* E. itajara *(Lichtenstein, 1822) the jewfish.* Washington, DC: U.S. Dept. of Commerce.

Sharp, W. C. 2000. *Destructive Urchin Grazing in a Sea grass Bed in Western Florida Bay: When Should Resource Managers Intervene?* Key West, FL: The Nature Conservancy.

Tegner, M. J., and L. A. Levin. 1983. Spiny lobsters and sea urchins: Analysis of a predator–prey interaction. *Journal of Experimental Marine Biology and Ecology* 73:125–150.

Tegner, M. J., L. V. Basch, and P. K. Dayton. 1996. Near extinction of an exploited marine invertebrate. *Trends in Ecology and Evolution* 11:278–280.

The Ocean Conservancy (TOC). 2001. *Marine Wilderness Campaign Launch Press Materials.* Washington, DC: The Ocean Conservancy.

———. 2002. *Health of the Oceans 2002 Report.* Washington, DC: The Ocean Conservancy.

United States Commission of Fish and Fisheries. 1880. *Report of the Commissioner of Fish and Fisheries for 1878.* Washington, DC: Government Printing Office.

The White House. 1966. *Effective Use of the Sea: Report of the Panel on Oceanography, President's Science Advisory Committee.* Washington, DC: The White House.

———. 2000. Executive Order 13158 of May 26, 2000: Marine Protected Areas. Federal Register vol. 65, No. 105 Wednesday, May 31, 2000, Presidential Documents, P34909–34911.

TWO

# The State of Marine Ecosystems and Fisheries

During World War II, when U.S. General Douglas MacArthur needed to choose between a single-engine and a two-engine fighter design, he inquired not which flew faster or farther or better, but rather whether the two-engine fighter could fly on only one engine. He reasoned that the more complex two-engine plane would be less vulnerable to the natural stresses of flying and enemy attacks only if it could fly with just one engine, but more vulnerable, if it required both to fly. The correct answer to MacArthur's question might have been: It depends. Under ideal and stable meteorological conditions and in the absence of enemy fire, a two-engine fighter may be able to fly with a single engine, but is that likely to be the full range of conditions it faces? Most likely not.

Similarly, a hypothetical marine ecosystem may be able to survive for a while, missing some of its seemingly redundant or excess parts, under perfect and stable environmental conditions, and in the absence of human impacts; but in reality, even the most remote marine ecosystems never experience such constancy. Environmental variability is universal among marine ecosystems and all or virtually all of such systems are facing increasing human impacts. The need to protect all of a complex, diverse, and interdependent natural system's cogs and wheels to preserve its ecological integrity, resistance, and resilience in the face of multiple natural and anthropogenic stresses makes real living marine ecoystems especially vulnerable. Keeping all of their pieces, even the seemingly useless ones, maximizes the probability that healthy, natural marine ecosystems will persist and thrive through time. Paramount among the growing threats to such systems is fishing and related extractive activities.

The spreading geographic scope of marine fisheries is now nearly global. These fisheries have increased in intensity, slowly for much of the past five hundred years, and more dramatically over the last century. The recent acceleration is at least in part attributable to the rapid development and increased availability of fishery technologies. For example, intense North Atlantic ground fisheries, initially focused on the cod family (*Gadidae*), originated in Europe, then spread outward from a limited number of European ports to cover all nearshore European waters and westward to North America by the 1500s. The introduction of steam and other motor-powered vessels and the development of modern trawl gear in the 1800s intensified and expanded this fishery into adjacent, offshore waters. This offshore expansion was further accelerated by the development of large factory trawlers and fleets by the 1960s. More recent advances in trawl gear including "rock hopper" trawls enable access to rocky areas, previously unfishable with earlier gear. Most recently, new trawling gear and techniques continue to lower the depth limits that can be effectively trawled to more than 5000 feet and have opened up previously unexploited or lightly exploited highly diverse, fragile, and vulnerable deep ocean habitats, including seamounts (Broad 1998; Koslow et al. 2001; Kurlansky 1997; Moore 1999; Moore and Mace 1999).

The pervasiveness of the current damage to marine systems rivals that which Leopold (1953) predicted on land more than half a century ago. Human impacts have now been documented to extend throughout the oceans, even to deep-sea environments (Jackson et al. 2001; Koslow et al. 2001; NRC 1995). Jackson et al.'s (2001) innovative and retrospective work extends our understanding of how long such impact has been occurring, how pervasive it is, and what it portends for our future. Their work also provides practical and achievable mechanisms, including marine reserves, for restoring and protecting ocean environments. Myers and Worm (2003), focusing on more recent high-tech, "industrialized fisheries," demonstrated their global extent and documented their rapid and dramatic depletion of worldwide predatory fish communities. They found that such fisheries typically reduce predator community biomass by 80 percent in less than fifteen years and estimated that large predatory fish biomass today is at levels less than 10 percent of pre-industrial levels. They concluded that large predatory fish declines in coastal waters extended throughout the global ocean with potentially serious impacts to coastal ecosystems. The following sections describe some of the ways that human activities affect key marine ecosystems, especially through fishing activities. Then we return to the protection and restoration of marine ecosystems, the primary focus of this book.

## OYSTER REEFS

> Progress, far from consisting in change, depends on retentiveness. Those who do not remember the past are condemned to repeat it.
>
> —George Santayana

As we enter the twenty-first century, considerable attention, concern, and discussion are focused on the current declining state, prior loss, continued degradation, and threatened future of valuable coral reefs around the world (see next section). Yet a century ago, similar attention, concern, and discussion about then valuable oyster reefs failed to avert their demise. The wholesale destruction of oyster reefs and concurrent decline of oyster populations in many parts of the world provides a stunning portrait of severe ecosystem-level damage due to intensive fishing focused on critical ecosystem components and offers lessons for the future. Clearly, the removal of such key, stationary, structure-producing species that both provide important habitat and protect water quality via filtration can be expected to have wide-ranging ecosystem impacts. Can we learn the lessons of the past?

Over a century ago, a remarkable man, scientist, and former Chesapeake Bay oyster commissioner, W. K. Brooks, wrote a remarkable book, *The Oyster,* about this remarkable animal (Brooks 1996). In 1891, when the book was first published, a consistent, empirical record of repetitive oyster collapses already existed for intensively fished populations throughout Europe and North America. The insightful and prescient Brooks reviewed this record, documented the already declining status of Chesapeake Bay oysters, detailed the oyster's critical roles in bay and estuarine ecology, and predicted the pending collapse of the oyster resource and its consequences, but to no avail. Between the late 1880s and late 1990s, Chesapeake Bay's oyster population followed the historical pattern of oyster collapses elsewhere with production declining from peaks of over 15 million bushels to lows of less than 100,000 bushels by 1993 (Fig. 2.1). Scientists believe this reflects a remnant population of less then 1 percent of what existed in the early 1900s and a near complete loss of the large, high-relief oyster reefs that once dominated the bay (Brooks 1996; Horton and Eichbaum 1991; NMFS 1990; NRC 1995).

The impact of oyster population collapse and oyster reef disappearance in Chesapeake Bay and throughout their range extends far beyond the loss of the oysters. Oysters and oyster reefs play critical roles in bay and estuarine ecology, among which are filtering particulates to maintain water quality, providing

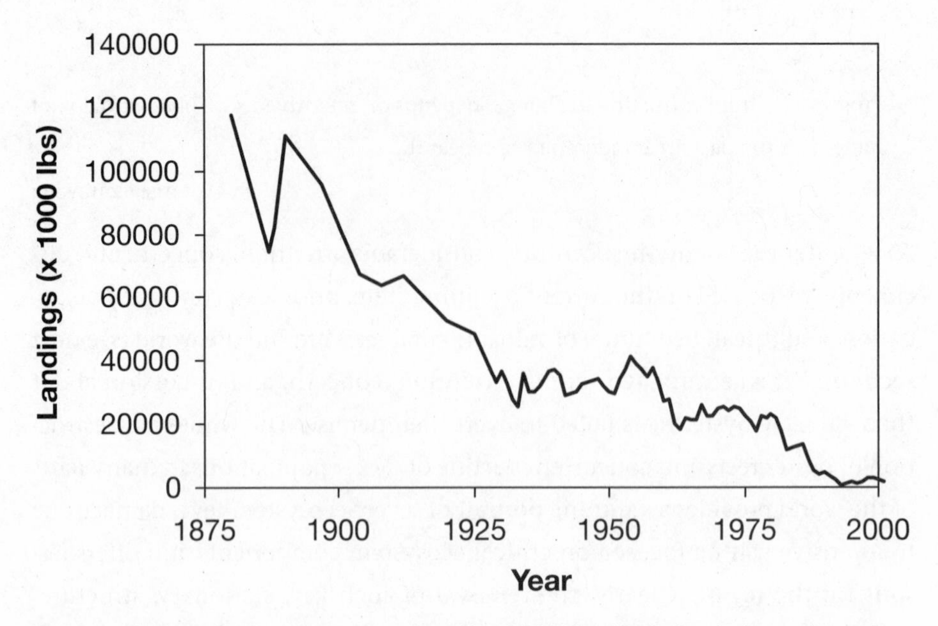


**FIG. 2.1 Chesapeake Bay Oyster Landings from 1880 to 2001.** Graph shows dramatic long-term decline in Chesapeake Bay oyster landings from peak catches well over 100 million pounds in the 1880s to near zero by the mid-1990s. This drop in landings probably reflects a remnant oyster population less than 1 percent of that existing in the late 1800s. Much of this sharp decline is clearly linked to fishing impacts rather than pollution or disease impacts that occurred much later. Similar collapses occurred elsewhere around the globe where large-scale commercial oyster fisheries developed. Source: Data from NMFS 1990 and 2002.

hard physical structure that provides essential habitat for many species, and mediating energy flow. Their demise results in lost filtering capacity, degraded water quality, eutrophication, reduced areal extent and spatial complexity of hard bottom and sea grass habitats, loss or reduction of many other species, and altered energy flow. A century ago, historically abundant oysters filtered the Chesapeake's volume in less than a week, but today's remaining few require more than a year to achieve the same feat (Newell 1988). Consequently, algae, sediments, and nutrients are less effectively removed from the water column, which leads to degraded water quality, reduced water clarity, and increased nutrients, all of which adversely affect sea grass and other submerged aquatic vegetation (SAV). Thus, two critical habitats, oyster reef (hard bottom) and sea grass (SAV) have been greatly diminished qualitatively and quantitatively as a result of oyster removal, the former more directly and the latter indirectly. In addition to the direct loss of oysters and oyster reef ecosystems, such changes alter, destabilize, and reduce the resilience of the larger bay and estuarine ecosystems of which they are but a piece (Brooks 1996; Horton and Eichbaum 1991; NRC 1995).

Recent studies discuss the broad impact of the oyster's ecological or functional extinction in more detail than is possible here, but we will likely never understand the full extent of this loss. The extensive impacts of oyster extraction to the Chesapeake are among the best studied, but similar ecosystem-level alteration is shared across the broad geographic range in which oysters are ecologically important. The story was much the same across the North Atlantic from France to England and from the Canadian Maritimes to New England and the Mid-Atlantic, and around the world in Australia and New Zealand (Brooks 1996; Jackson et al. 2001). Nonetheless, as recently as 1991, Larry Simms, a prominent Maryland waterman, claimed that "[o]verharvesting is a lie. All we need is one good spat set" (Brooks 1996).

## CORAL REEFS

Even as we struggle to fully understand and appreciate the ecological impact of the historical demise of our oyster reefs, the world faces the continued decline and potential loss of even more extensive, valuable, complex, and fragile natural reef systems—our spectacular coral reefs. For several decades, an increasingly vocal chorus of concern has risen from an increasingly large group of coral reef scientists, initially about the decline of the specific reef systems they knew best. In the past decade, this chorus of concern has reached a crescendo, broadened from focusing on specific, isolated reef systems to reefs generally, and spread to include a much wider cross section of the public, conservationists, fishers, divers, local community activists and leaders, decision makers, and even high-level elected leaders.

Despite the widespread concern, science cannot precisely tell us the turning point when an endangered coral reef will be lost or damaged forever or even how much has already been irretrievably lost. In 1992, a multinational group of coral reef scientists took a first stab at this. Scientists extensively reviewed, debated, and revised their initial estimates from 1992 to 1996 and reported them in the *Status of the Coral Reefs of the World: 1998* (Wilkinson 1998). They concluded that 10 percent of the world's reefs had already been irretrievably destroyed, 30 percent more were critically endangered within two to ten years, and another 30 percent were similarly threatened with destruction on a ten- to thirty-year time scale. *Status of the Coral Reefs of the World: 2000* reviewed, updated, and largely confirmed these dire predictions, but made some key changes (Wilkinson, 2000). It also found that about two-thirds of the world's reefs have already been destroyed or will be within thirty years unless strong action is taken to address anthropogenic stresses quickly (USCRTF 2000; ICRI 1996; Wilkinson 2000, 363; 1998, 184). A recent comprehensive review of Caribbean coral reef research studies concluded that live coral cover had already declined an average of 80 percent across this region's valuable and vulnerable reefs (see Fig. 2.2; Gardner et al. 2003).

> The world's coral reefs and associated sea grass and mangrove habitats are in serious jeopardy, threatened by an increasing array of overexploitation (i.e., fishing), pollution, habitat destruction, invasive species, disease, bleaching and global climate change. The rapid decline of these ancient, complex, and biologically-diverse marine ecosystems has significant social, economic and environmental impacts here in the U.S. and around the world. (USCRTF 2000)

At both national and international levels, there is an increasingly broad consensus that multiple human stresses have contributed to the overall decline of reefs and that no single stress is solely responsible, but the agreement breaks down some with respect to which stresses are most important. This growing consensus is remarkable given the extraordinary complexities and synergies of reefs and the threats to them, the geographic and temporal variability of both, and the varied values, beliefs, and interests of different stakeholders. Nonetheless, there is strong and diverse agreement that fishing and pollution, especially eutrophication and sedimentation, are among the key drivers in reef decline across broad geographic areas, that they interact synergistically, and that global warming threatens reefs on an even larger scale (USCRTF 2000; ICRI 1996; Wilkinson 2000, 363; 1998, 184).

Most recently, *Status of the Coral Reefs of the World: 2002* reported successes demonstrating our ability to reverse coral reef decline through the use of marine reserves, marine protected areas (MPAs), and other initiatives that address multiples stresses, including fishing (Wilkinson, 2002). However, it also indicated that such efforts were likely not keeping pace with concurrent increases in direct human-induced stress to reefs combined with the growing indirect impacts associated with global climate change. Particularly troubling were indications that even apparently healthy reef systems not yet experiencing loss of live coral cover may nonetheless be at high risk due to potentially catastrophic collapse of coral recruitment linked to loss of crustose coralline algae, a key settlement habitat, and corresponding major increases in turf and macroalgae. Equally disturbing was a summary of results from the global Reef Check mon-

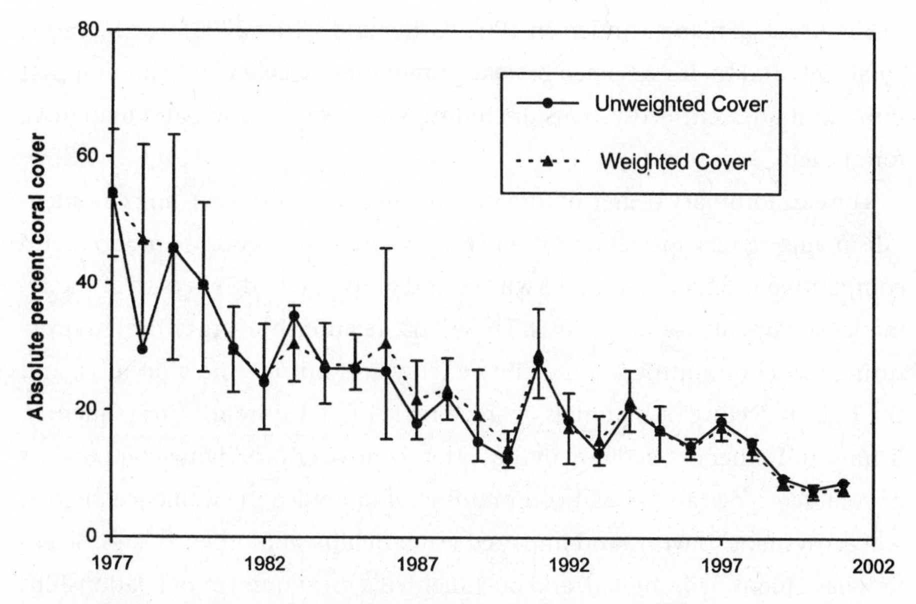


**FIG. 2.2 Caribbean Coral Reef Decline since 1977.** Graph shows results of comprehensive review of Caribbean coral reef studies indicating an average loss of 80 percent of live coral cover across the region's valuable reefs in less than three decades. Source: Adapted from Gardner et al. 2003 with permission from *Science*.

itoring program (1997–2001) suggesting that spiny lobster were being severely overfished at a global scale; several large species of reef fish were now in critical condition and other key reef species showed significant declines, also due to overfishing; and *Diadema* sea urchin abundances had declined significantly in the Pacific and were approaching levels found in the Caribbean, which reflected possible ecological destabilization (see following discussion; Wilkinson 2002).

Coral reefs are much more than just corals. They are a myriad of interwoven and interdependent habitats and associated organisms (Fig. 2.2). Some scientists have even suggested that *coral reef* is a misnomer due to its overemphasis on corals and a failure to recognize other equally important components, but their suggested alternatives have lacked popular appeal. Coral reefs are among the most diverse and complex ecosystems on Earth, providing homes to nearly 100,000 known organisms and likely a million or more yet to be discovered (Jackson 2001; Knowlton 2001; Reaka-Kudla 1997). The already known portion includes more than 4000 species of fish (McAllister 1991; McAllister et al. 1994), and extraordinary plant, or more accurately algal, diversity (Littler and Littler 2000). Coral reefs and marine systems in general are much more diverse at higher taxonomic levels (Sepkoski 1995) and consequently harbor much more of the world's genetic diversity than terrestrial systems. For example, the diversity of photosynthetic machinery in green algae alone dwarfs that found in all terrestrial plants (Anderson 1992; Littler and Littler 2000). Furthermore, coral reefs and their associated biota are often inextricably tied to and support important adjacent ecosystems, including sea grass meadows and mangrove forests, and their biota.

The extraordinary degree of interdependence, specialization, and commensalism among reef species and the fierce predator–prey, grazer–producer, and competitive interactions found within and among reef dwellers rival the remarkable diversity found on reefs. These close relationships are critical to structuring reef communities, controlling energy and nutrient flow on reefs, and the tight recycling of materials characteristic of reef systems. Consequently, fishing and other extractive activities often remove critical living components of coral reefs, destabilize reef communities, and reduce the resilience of coral reef ecosystems to withstand impaired water quality and other stresses. Scientists have identified, studied, and documented a great number of relationships and interactions among coral reef species and the related impacts of fishing activities to reefs, but have thus far only scratched the surface of these vast topics.

> Herbivorous fishes have a profound impact on the distribution, abundance and evolution of reef seaweeds. On shallow fore reefs, fishes are often estimated to consume from 50%–100% of total algal production and to take as many as 40,000–156,000 bites/m$^2$/day. Grazing on these portions of coral reefs equals or exceeds grazing rates for any other habitat, either terrestrial or marine. (Mark Hay, *The Ecology of Fishes on Coral Reefs,* 1991)

What is the nature of the relationship between fish and reef habitats and what happens to coral reefs when such extraordinary grazing pressure is removed, diminished, or altered? Science and experience provide partial answers to these questions, though our knowledge remains imperfect due to the complexity of reef systems. Fish–habitat interactions on reefs fall into three major classes: (1) the direct relationship between reef structure and shelter; (2) the feeding interaction involving reef fishes proper and the sessile biota, including algae, that gives rise to critical secondary effects, such as the mediation of competition and other interactions between coral and algae; and (3) the role of reef structure and the feeding patterns of reef planktivores and carnivores and the resulting link between their feeding activities and the recycling of nutrients and energy between reefs and adjacent habitats (Hay 1991). Upsetting the fragile balance among these interactions, especially grazing pressure, can cause catastrophic change to reef systems that can have delayed impact and is not always predictable.

Reef fish are not the only key reef macroherbivores (large grazers) nor do they act in isolation, but rather interact with invertebrate grazers such as sea urchins in structuring coral–algae communities. A pioneering urchin exclusion experiment, in which 7000 *Diadema* sea urchins were removed from a small patch reef in the U.S. Virgin Islands, produced rapid and remarkable results and clearly demonstrated the key role of urchins. Dramatic changes in the reef community occurred almost immediately. Within about six months a thick cover of algae obscured the reef surface and overgrew and killed some of the coral colonies, and the characteristic white halo zone surrounding the reef had completely disappeared (Ogden and Lobel 1978; Sammarco et al. 1974). Unbeknownst at the time, this experiment foreshadowed equally dramatic and rapid changes that would occur nearly a decade later on a much grander scale throughout the Caribbean (see below).

The massive Caribbean-wide epidemic that killed off over 95 percent of the previously ubiquitous black-spined sea urchin, *Diadema antillarum* (Fig. 2.3; Lessios et al. 1983, 1984) illustrates the key role that coral reef grazers play. Intense grazing on intact shallow water reefs removes nearly all algal growth leaving behind only those with strong physical, structural, or chemical defenses and the small rapidly growing filamentous algae that are maintained as low-lying algal turfs by near constant grazing. Despite the limited biomass of reef algae, their rapid growth and consumption by local herbivores makes such reefs among the most productive habitats on Earth and the herbivore–algal interactions among the most critical in structuring reef communities. Consequently, reef fishes exert tremendous influence on the distribution, abundance, and diversity of reef algae. Maintaining the spatial heterogeneity and mosaic nature of coral reefs is one way herbivory contributes to this (Hay 1991). Steneck (1988) hypothesized that macroalgae are competitively dominant in the absence of herbivores, and modern shallow reefs could not have existed prior to the evolution of herbivorous fishes.

The various interactions among fish and invertebrate grazers demonstrate that although there is redundancy of certain ecological functions on reefs within guilds, the roles played by individual members are not identical or interchangeable (Choat 1991). Put another way, all grazers are not alike, all predators are not alike, and so forth. The complex relationships among members of these functional groups and among the groups reflect the inherent complexity, diversity, and interdependencies of intact natural coral reef systems. Protecting the ecological integrity, stability, and resilience of reefs will likely require keeping all the cogs and wheels, as Leopold stated, rather than repre-

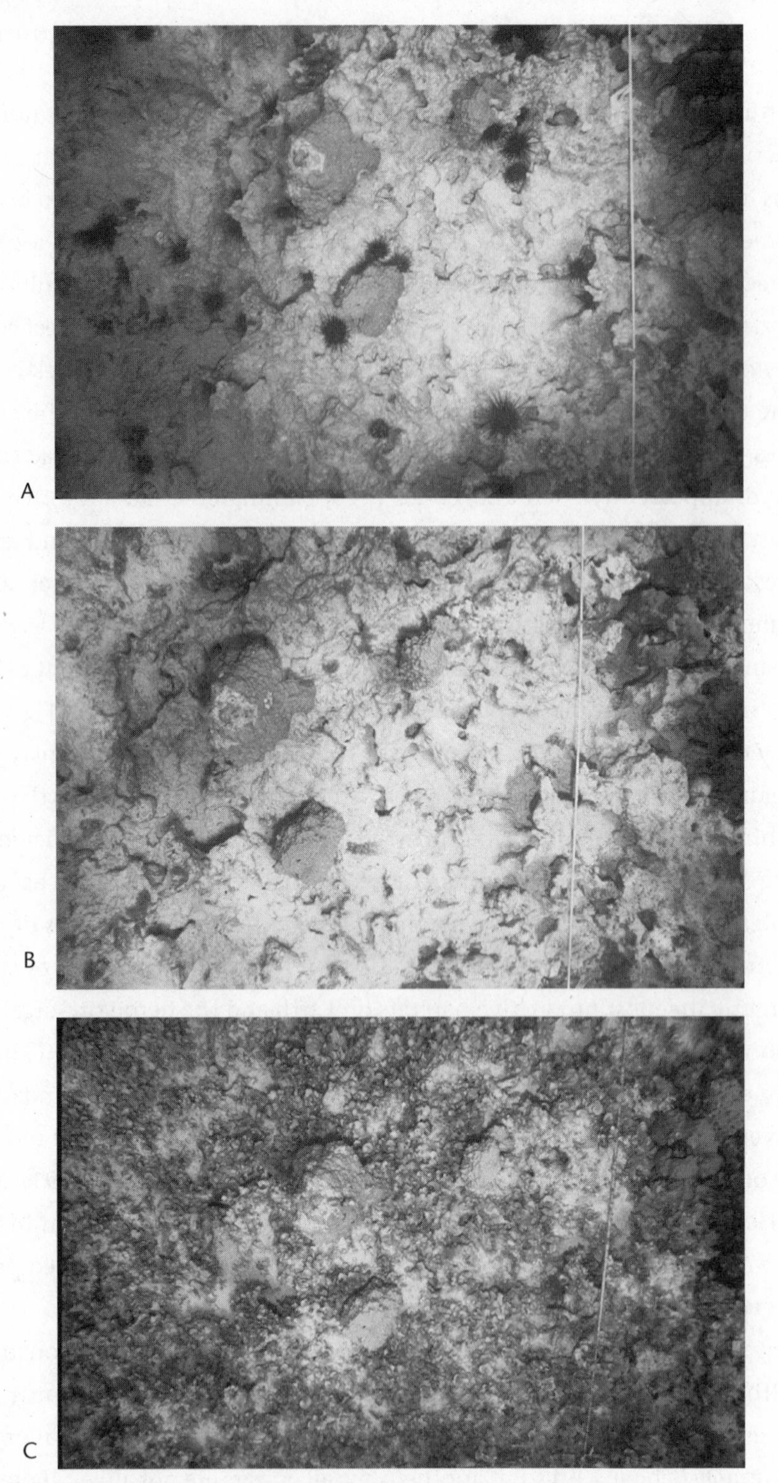


**FIG. 2.3 Caribbean Sea Urchin, *Diadema antillarum,* Die-off and Impact of Reduced Grazing on Reefs.** Time series of three photos taken on same one square meter coral reef plot showing (A) numerous urchins present on 2-6-84 during die-off; (B) the complete absence of urchins just a little over a month later on 3-11-84; and (C) the eruption of algae overgrowing the reef just three more months later on 6-23-84 following loss of urchin grazing. Additional time series photos (not shown) show eventual slow mortality of coral colonies two years later. Source: Photos and information courtesy of John Ogden.

sentatives of each group. This is especially true as reefs are exposed to variable natural and anthropogenic stresses acting individually and together.

## THE CAUTIONARY TALE OF CARIBBEAN REEF DECLINE

The coral reefs of the Wider Caribbean remain the most diverse shallow water marine communities in the Atlantic Ocean and are among the most important, diverse, and endangered biological systems in the world. These fragile systems have experienced gradual, but accelerating, degradation over the past quarter century. Sharp declines have been well documented in some areas, especially near human population centers, but comprehensive, long-term change information is lacking for most areas. During this period and especially since 1982, a combination of anthropogenic impacts and natural events has led to dramatic declines in coral cover and algal overgrowth of corals. The primary human impacts include reduced grazing due to overharvesting of living marine resources, especially herbivores, and water quality degradation due to coastal development, especially eutrophication. Reduced grazing capacity and water quality degradation due to human activities have acted in concert to shift the fragile balance of reef communities from coral- to algae-dominated systems. This has shifted the fragile balance on the reef from coral to algal domination and has been greatly exacerbated by natural factors, including a catastrophic pan-Caribbean die-off of the herbivorous black-spined sea urchin, *Diadema antillarum,* and several severe hurricanes.

After roughly two decades of reported western Atlantic reef decline and a slow transition from coral-dominated to algal-dominated reef systems in many areas, then most often attributed to water quality degradation, an unforeseen and widespread massive sea urchin die-off abruptly and heavily impacted reefs across the region. The catastrophic nature of the urchin die-off, the most extensive reported for any marine animal, greatly exacerbated this shift and provides an interesting and unique window through which to view the synergy of multiple stressors acting in concert on a single ecosystem. In particular, the urchin, *D. antillarum,* had been ubiquitous and abundant in coral reef systems throughout the region and was a principal herbivore and reef bio-eroder. In January 1983 a die-off of *D. antillarum* was first observed in Panama. It spread throughout the Wider Caribbean over the next thirteen months. Percent mortality during the die-off exceeded 90 percent throughout the region, and 99 percent in some locations. In St. Croix, shallow water algal biomass increased by over 25 percent within five days of the die-off and by over 300 percent over the next several years (Fig. 2.2; Carpenter 1988; Lessios 1988 a & b). In St. John, algal biomass increased thirty-fold in the first year before dropping back to five times the initial level (Lessios 1988a; Levitan 1988a). On some Jamaican reefs, percent cover for most algae increased from 1 to 95 percent within two years (Hughes 1994; Lessios 1988a). These changes were associated with changes in algal community structure and shifts from crustose coralline algae to fleshy and filamentous forms. Declines in both coral cover and coral recruitment were associated with the die-off and algal increases. The extent of changes in less populated areas and less fished reefs was less remarkable.

> The greatest triumph of a scientist is the crucial experiment that shatters certainties of the past and opens up new pastures of ignorance. (William D. Ruckelshaus)

Well into the 1990s, many intelligent people, including some prominent coral reef scientists, conservationists, managers, and resource users continued to share the widespread belief that water quality degradation, especially nutrient pollution, remained the primary or even sole threat to Caribbean (and other) coral reefs, despite considerable and mounting evidence that reduced and altered herbivory resulting from a combination of fishing activities and natural stresses was responsible for many observed changes, at least as important as pollution, and acted synergistically with increased pollution, in many places. Although some continue to hold this view religiously, a landmark study entitled *Catastrophes, Phase Shifts, and Large-Scale Degradation of a Caribbean Coral Reef* (Hughes 1994) and other studies began shifting the tide of many people's thinking. Hughes drew on a wealth of historical information to identify and document the key factors that ultimately contributed to the dramatic decline of Jamaica's storied and valuable reefs. The quality of the available information provided a compelling case that the cumulative impact of chronic overfishing combined with the *Diadema* die-off and two hurricanes was primarily responsible for the collapse of Jamaica's storied and valuable coral reefs and a phase shift to algal-dominated reefs.

Jamaica's population had been growing exponentially for over a century and its growth remains high. Chronic and increasingly severe and widespread overfishing occurred, first on Jamaica's narrow north coast, and then spread across its more extensive south coastal shelf, "increasingly accessible to a modernizing fishing fleet" during the 1970s to mid-1980s. The overfishing pattern had removed up to 80 percent of the fish biomass on the north coast's fringing reefs by the late 1960s, was operating at two to three times sustainable levels by 1973, and over decades had reduced adult stocks to the point where most fish species were being harvested prior to minimum reproductive sizes. Large predatory fish, sea turtles, and manatees had virtually disappeared along with adult members of many other species. Remaining species, including many herbivores, were greatly reduced in both size and number. Local replenishment of fish was virtually nonexistent and depended largely on recruitment from elsewhere. The overfishing pattern was repeated along the south coast between 1970 and the mid-1980s during which time fishing effort doubled, catch per unit effort (CPUE) dropped in half, and species composition changed markedly, indicative of severe nationwide overfishing.

What was the impact of such rampant overfishing? Interestingly, there was little apparent change initially in either coral cover or benthic diversity, which remained high from the 1950s through the 1970s. There were relatively few macroalgae during this period despite the paucity of herbivorous fish due to increased grazing by unusually high levels of *Diadema* resulting from removal of their predators and competitors via fishing. *Diadema* abundance was directly related to fishing pressure throughout the Caribbean prior to 1983. Extensive damage was inflicted on Jamaica's reefs by megahurricane Allen in 1980. A major algal bloom and coral loss occurred immediately afterward but was followed by a partial recovery that continued into 1983. At that point, it still looked like Jamaica's reefs might be salvaged.

In 1983, the unprecedented *Diadema* die-off occurred throughout the Caribbean and struck Jamaica's already weakened reefs like a death knell. Although this pandemic wreaked havoc throughout the region, it was exacerbated in places like Jamaica by the decades-long severe overfishing of herbivorous and predatory fish. With no check on its growth, a spectacular systemwide algal bloom reignited in 1983 and was still impacting Jamaican reefs more than a decade later. It resulted in a complete community phase shift from coral- to algal-dominated reefs including a remarkable decline in mean coral cover from 52 to 3 percent and an even more remarkable and corresponding increase in fleshy algal cover from 4 to 92 percent over a time span from just prior to the die-off to about a decade later. Now, two decades later, Jamaica's coral reefs have not yet recovered (Fig. 2.4).

More recently, in a paper entitled "Reefs since Columbus" (1997), Jeremy Jackson showed that Caribbean reefs had been decimated by fishing and other human impacts long before the time span considered by Hughes. He established that fishing impacts had begun several centuries earlier, that major ecosystem level impacts had already occurred prior to 1900 in Jamaica and elsewhere, and that modern ecology and reef observation began after major com-

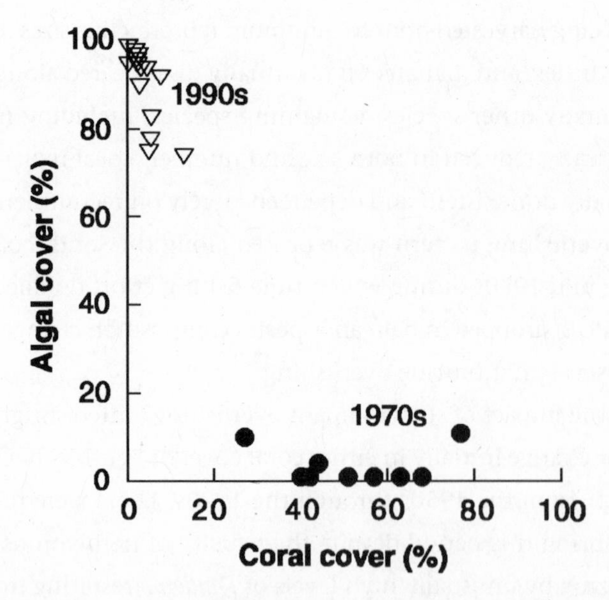


**FIG. 2.4 Jamaican Coral Reef Collapse, Phase-Shift, and Failure to Recover.** Changes in algal cover versus coral cover on Jamaican reefs from 1970s to 1990s showing coral collapse, phase-shift from coral-dominated to algal-dominated reefs, and failure to recover two decades later. Source: Reprinted from Hughes 1994 with permission from *Science.*

ponents of these systems had already been removed and/or altered. Jackson estimates that green turtles numbered between 6.5 and 660 million in the Caribbean shortly after the arrival of Columbus and suggests that other large vertebrates, including the other sea turtles, West Indian manatees, the now extinct Caribbean monk seal, and large fishes, were similarly abundant but had already been greatly reduced in Jamaica and many areas by 1900. Jackson levels an important criticism of coral reef research as well as marine research more generally:

> Studying grazing and predation on reefs today is like trying to understand the ecology of the Serengeti by studying the termites and the locusts while ignoring the elephants and the wildebeests. . . . [C]oral reef ecologists have been so devoted to dissecting small-scale processes that they have not seen the reefs for the corals. (Jackson 1997)

Most recently, Jackson et al. (2001) further elaborate on the connection between historical overfishing and the collapse of coral reef systems in the Wider Caribbean and elsewhere. Such changes are most apparent in the Wider Caribbean, but also prevalent elsewhere, including Australia's Great Barrier Reef, despite extensive protection there for several decades. During the 1980s, large branching corals (*Acropora* spp.) that had dominated Caribbean reefs for more than 500,000 years largely disappeared; Caribbean reef corals suffered rapid, catastrophic decline after being overgrown by algae, following the mass mortality of the sea urchin, *Diadema,* which removed the last major herbivore, long after fishing had greatly reduced or eliminated others; and coral community structure and dominance hierarchy, both of which had been stable and predictable based on a strong baseline, were greatly altered. Although evidence suggests that large vertebrate herbivores, including sea turtles, manatees, and fish had been greatly reduced in many areas before the turn of the century, these gross changes appeared or were noticed only after the *Diadema* was lost to disease (see Fig. 2.4).

The unfortunate truth is that the fate of Caribbean reefs is not unique. Halfway around the world and two decades ago, when fishing pressure was much less intense than it is today, a researcher with foresight named Gerald Goeden (1982) published a paper, entitled "Intensive Fishing and a 'Keystone' Predator Species: Ingredients for Community Instability." The paper demonstrated fishing-induced reductions to coral trout, a heavily sought grouper, and associated changes in community structure, which were unpredictable but quantifiable. He went on to suggest that continued fishing could create community instability and lead to an irreversible altered state, and he drew the following conclusion:

> Should this be the case then the concept of renewable resources may not be broadly applicable to the coral reef and the relationship between keystone species abundance and $F_S$ (instantaneous fishing mortality) may be less a management tool than a picture of the demise of the reef fish community as we know it. (Goeden 1982)

This statement may be worth reconsidering in light of the increasingly frequent, intense, and destructive crown-of-thorn starfish outbreaks along the Great Barrier Reef and other Indo-Pacific Reefs, which several researchers have linked to fishing-induced reductions in fish and invertebrate predators of starfish. Crown-of-thorns are voracious corallivores normally present in small numbers on reefs where they do no harm, but are increasingly prone to explosive outbreaks in which they rapidly devour living coral reefs (Jackson et al. 2001; Sapp 1999).

While the catastrophic sea urchin die-off was the last straw in the collapse of Jamaican and other Caribbean reefs, the situation endangering the reefs in far off Kenya was nearly the opposite. Careful research there demonstrated fishing-induced loss of a few key urchin predators caused explosive growth of a specific sea urchin, *Echinometra mathaei,* which in high numbers reduced coral recruitment, increased bio-erosion of coral reefs, and badly damaged reefs through its grazing activities (see also chapter 11; McClanahan and Arthur 2001; McClanahan and Obura 1995). In the Caribbean case, the urchin (*D. antillarum*) die-off coming on the heels of decades or centuries of overfishing that had removed other grazers, threatened reefs by removing one of the key remaining effective grazers and further shifting the competitive balance between corals and algae toward fleshy algae. In the Kenyan case, fishing targeted on a key urchin predator, a triggerfish, allowed an unnatural explosion of urchins (*E. mathaei*) that directly damaged coral reefs through their grazing activities on both corals and their associated algae. Interestingly, recent studies in Jamaica suggest that *D. antillarum* urchins are recovering in some areas and shifting the competitive balance between corals and algae back toward corals (Edmunds and Carpenter 2001), but that the recovery of *D. antillarum* and the associated transition back to a coral dominated system may be mediated by the grazing of another urchin (*E. viridis*) and that the recovery of Caribbean reefs may require the restoration of a diverse herbivore (and reef) community (Bechtel, 2002).

Rampant overfishing of reef fish and invertebrates remains the rule, rather than the exception, in both developed and developing countries, and impacts on reef fish communities can be felt in remarkably short time frames. For example, a remarkable and well-documented example of rapid reef fish depletion provides an interesting contrast and complement to the long-term historical studies previously cited here, and sheds some light on what unfished reef fish populations might look like. In June 1985, within the U.S. South Atlantic Fisheries Management Council (SAFMC) jurisdiction, fishermen discovered and commenced fishing on a virgin (previously unexploited) population of snowy grouper (*Epinephelus niveatus*) on a reef offshore of North Carolina. At the time, snowy grouper were the dominant grouper and top-level predator on this area's deep water reefs and constituted over 90 percent of their fish landings. In just three months, an intense, efficient, relatively low-tech, hook and line fishery reduced grouper biomass on the reef by over 70 percent. The estimated density of grouper on the reef as the fishery developed was a remarkable 11 kilograms per square meter. Catch levels, grouper density, and fishing effort all declined as the resource was depleted and within a year approached levels similar to adjacent, previously fished reefs. Nonetheless, the initial surge of landings from this one previously unfished reef made up over 30 percent of the state's total landings for that year (Fig. 2.5; Epperly and Dodrill 1995). Similar rapid depletions of reef fish are more the rule than the

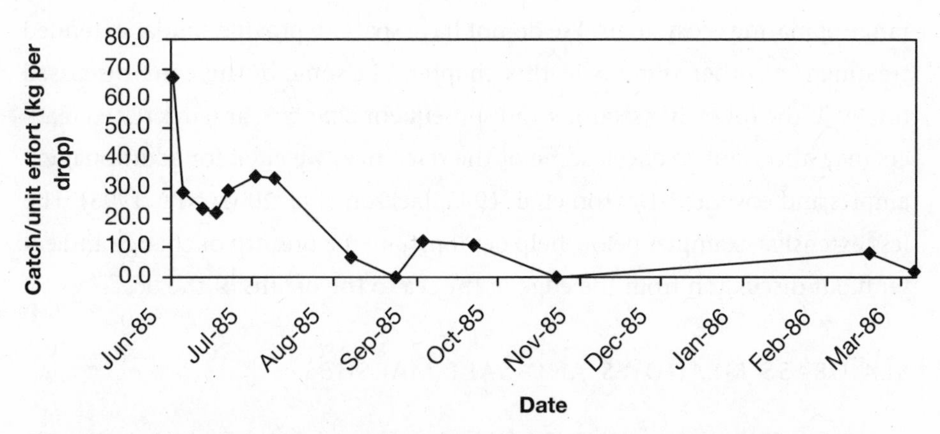


**FIG. 2.5 Rapid Decline of Snowy Grouper, *Epinephelus niveatus,* from Previously Unexploited Reef in Onslaw Bay, North Carolina, Following Start of Fishing.** Intense, efficient, hook-and-line fishery reduced grouper biomass over 70 percent in just three months. Source: Adapted from Epperly and Dodrill, 1995 and reprinted with permission from *Bulletin of Marine Science* 56:450–461.

exception, but are rarely so well documented and are often masked by shifts in targeted species or locations.

Birkeland (2001) summarized the following similar examples from the Pacific of rapid, severe, fishing-induced depletion of reef fish and invertebrate populations with no apparent signs of recovery many years later. In 1927, over a hundred tons of an oyster, *Pinctada margaritifera,* were mined from Pearl and Hermes Reefs in the Northwestern Hawaiian Islands (NWHI); but subsequent scientific surveys yielded only a few hundred survivors in 1930, only a few in 1993, and just six, despite an intense effort, in 2000. In the early 1950s, at French Frigate Shoals, also in the NWHI, a single *kupuna* or master fisher, Louis Agard, remembers discovering and netting a school of large moi, *Polydactylus sexfilis,* and then never seeing them again despite continuing to fish the area for another ten years. In the late 1930s, hundreds of tons of sea cucumbers were harvested from Truk Lagoon, but a 1988 survey of eight sites there turned up just two individuals of the valuable species, *Holothuria nobilis*. In 1967, a previously unknown pinnacle off northwestern Guam was discovered, monitored, and fished down within six months and has shown no signs of recovery in over three decades. In 1986, a single Taiwanese fishing vessel discovered and apparently extirpated a grouper spawning aggregation off of Palau that has similarly showed no signs of recovery to date.

The alteration of coral reef systems is typical, not unique, in the ocean realm. They exemplify the accelerating rate of change we are seeing across a range of marine ecosystems. We do not have space to provide similar extended treatment of other systems in this chapter, but some of these are discussed briefly in the following sections and subsequent chapters, and interested readers may also want to check some of the references we cited for additional examples and coverage (Dayton et al. 1995; Jackson et al. 2001; NRC 1995). The less extensive examples below help demonstrate the breadth of change and extend our discussion from the edge of the sea to the depths of the ocean.

## SEA GRASS MEADOWS AND SALT MARSHES

In discussing the degradation of coral reef ecosystems here, we touched upon the impacts of removing large numbers of sea turtles and manatees, even before the subsequent rampant removal of reef fish and invertebrates. The direct impact of sea turtle and manatee removal upon sea grass beds was likely more direct and even greater in scope. It was included in the preceding discussion because of the frequent linkage among tropical coral, reef, sea grass, and mangrove habitats, which are often considered as a single linked system. However, sea grass beds are widespread in tropical, subtropical, and temperate environments, often not linked to coral reefs, and are also important ecosystems in their own right. Jackson et al. (2001) and Jackson (1997) provide a good overview of the impacts of fishing and other human impacts on sea grass beds associated with Caribbean coral reefs and more generally.

More recently, Silliman and Bertness (2002) studied and described the potential cryptic linkage between the blue crab (*Callinectus sapidus*) fishery and large-scale die-offs of *Spartina alterniflora* salt marshes across the southeastern United States. Such salt marshes are among the most productive systems in the world. Based on experimental manipulation of these marshes' dominant grazer, the periwinkle (*Littoraria irrorata*) and its consumers, including blue crabs, they demonstrate that these grazers as regulated by their predators largely control plant biomass and production. They conclude that (1) periwinkle grazing can convert one of the most productive grasslands to a barren mud flat within an eight-month period; (2) marine predators, especially blue crabs, regulate the abundance of this plant-grazing snail; (3) top-down predator control of grazer density regulates marsh grass growth; (4) this trophic cascade regulates salt marsh production; and (5) overharvesting of snail predators, such as the blue crab, may be an important factor contributing to the observed large-scale decline of regional salt marshes. In so doing, they refute a long held, but largely untested, belief that nutrient supply, as opposed to grazing, is the key determinant regulating salt marsh production. Their work also contributes to a growing body of evidence regarding the critical importance of predator control and resulting trophic cascades in regulating marine macrophyte production in a variety of marine systems, as further detailed following here.

## KELP FORESTS

Kelp forest ecosystems share a number of important attributes with coral reefs. They too are critically important, broadly distributed, highly diverse, delicately balanced, and very dynamic marine ecosystems. They characterize sunlit habitats above rock and other hard substrate habitat ranging around the globe from warm temperate to colder subarctic waters in both Northern and Southern Hemispheres. Like coral reefs, kelp forests provide a structural framework and critical habitat for diverse assemblages of associated marine life with which they are interwoven and interdependent. Consequently, kelp systems are also fragile, are greatly impacted by fishing activities, and can fluctuate between very different alternative stable states as a result of such impacts. Removal of apex and other high-level predators and (some) large grazers from kelp forests worldwide has grossly altered their states, with changes including reduction of trophic levels, changed energy flow and grazing patterns, herbivore (urchin) booms, and shifts from large, complex, three-dimensional kelp forests to tightly cropped, nearly flat, "sea urchin barrens."

The classic marine ecology paradigm about the tight coupling among North Pacific sea otters, urchins, and kelp provides a good simplified model of how fishing can impact, unravel, and destabilize kelp systems, causing them to flip between different states, but real systems have proven more complex. In the simple scenario, sea otters functioned as the keystone predator in naturally balanced, intact, kelp forests. They fed on sea urchins, the principal herbivores, which in turn, fed on the primary producers, kelp, and the system stayed in balance. Then humans entered the system, decimated the otters, hunting (fishing) them close to extinction. Released from their predator, urchin populations exploded, leading to vast urchin herds that devoured kelp and left behind what became known as urchin barrens, devoid of kelp and its associated biota. Humans did not initially notice these dramatic underwater changes and may have remained ignorant about their impact were it not for the subsequent cessation of hunting and recovery of otters, just prior to their extinction. As otters returned to feed on urchins, so did the kelp forests and the myriad of life they support. The subsequent fishery targeting sea urchin enhanced and extended this trend.

The kelp–otter–urchin ecological paradigm remains largely accurate, but, as with coral reefs, the kelp forest story has proven more complex. First, the major alterations of some California kelp beds does not track with the timing of the sea otter disappearance and occurred only after fishing had also greatly reduced other urchin predators, including large spiny lobsters and California sheepshead. Second, another important North Pacific grazer, Steller's sea cow, had already been fished to extinction in a remarkably short time after being discovered by European fishers. Third, in Alaska's Aleutian Islands, recent fishing impacts on pinnipeds (seals and sea lions) and their fish prey appear to have altered the feeding behavior of killer whales to consume more otters, reversed the otters' increasing population trend, and caused the subsequent collapse of associated kelp ecosystems. Fourth, fishing has greatly altered other kelp systems where otters were never present. In New England, massive overfishing resulting in the ecological extinction of large ground fish (such as cod, haddock, and wolf fish, which were important urchin predators) released urchins and other invertebrates to greatly expand, leading to disappearance of kelp forests, which have recently reappeared following the development of a major urchin fishery. Similarly, New Zealand, which also never had sea otters, lost much of its extensive kelp forests since the 1950s, likely linked to the development of major fin and shellfisheries (see also chapters 9 and 11; Ballantine 1991; Dayton et al. 1998; Estes et al. 1998; Jackson et al. 2001; Steneck 1998).

## ROCKY COASTLINES

The thin living ribbon of life set at the oceans' margin surrounds our continents, contains some of the Earth's most special places, provides special values to humans as terrestrial organisms, and is set within the context of the ocean's larger web of life. Coral reefs and three-dimensional kelp forests are among our most charismatic, valuable, aesthetic, appreciated, interesting, utilized, diverse, productive, and dynamic marine ecosystems, but they share these traits with many nearshore systems. Here we briefly focus on the changing face and altered architecture of rocky coasts, perched at the edge of the land–sea interface where much life clings. Chile's long (6435 km), well studied, and rocky coastline provides an excellent lens for viewing such human driven modifications, but recognizing and understanding these changes would likely have been impossible were it not for Chilean marine reserves.

Work on prehistoric shell middens and other studies trace human exploitation of predation on marine life along Chile's extensive rocky coasts back close to 10,000 years. Humans continue to harvest many of the same species harvested historically, but fishing intensified dramatically in the past few decades. During the 1980s, focused ecological research on a few newly created reserves showed marked ecological changes and began to reveal the extent of human alteration to the Chilean coast. Particularly noteworthy was the recovery of the heavily fished predatory gastropod *Concholepus concholepus* within the marine reserves and the resulting cascading effects it produced that suddenly and dramatically transformed the area's entire intertidal landscape in unpredictable ways rendering it almost unrecognizable and more reminiscent of similar remote isolated areas with limited fishing than adjacent ones.

The results of nearly two decades of research since reveal that human fishing has similarly transformed much of the Chilean rocky coastline in fundamental ways. A recent review characterizes three distinct patterns of intertidal landscape alteration depending on the trophic level at which human fishing is operating: (1) humans act as herbivores collecting seaweeds, defoliating accessible sites of vegetative cover, and leaving sites bare or nearly so; (2) humans act as mesocarnivores targeting a specific ecologically important herbivorous gastropod, resulting in proliferation of a specific red alga dominating its habitat with nearly 100 percent cover; and (3) humans act as top level predators targeting the previously mentioned carnivorous gastropod leading to domination by bivalves, which completely cover its habitats with multiple layers. Much more remains uncertain or unknown regarding patterns of human impact, interactions among them, and the full extent of human alteration. Yet such knowledge remains critically important, is urgently needed, and is becoming increasingly valuable both to locate reserves and to mitigate impacts of a new coastal highway (see also chapter 11; Castilla et al. 1994; Durán and Castilla 1989; Moreno 2001).

Chilean rocky coasts are not unique; similar human alterations to these habitats have also been documented in North America, South Africa, and elsewhere. One noteworthy recent study from southern California's rocky intertidal with an interesting tie back to the preceding Chilean discussion and prior work in South Africa and Chile involves complex interactions among humans, a bird, limpets, and erect, fleshy, algae. The American black oystercatcher (*Haematopus bachmani*) preys on a variety of limpets, including the large territorial limpet (*Lottia gigantea*), impacting their density, which in turn grazes on the algae impacting both its density and the density of smaller limpets (*Lottia* spp.). However, oystercatchers cannot reach limpets on most vertical surfaces. Some humans also like to eat the larger limpet, but usually not the smaller ones, and may compete with oystercatchers for the larger limpet in some areas. Other human activity can also disturb bird-feeding behavior, especially at high levels. Humans appear capable of precluding oystercatchers from areas through either substantial removal of large limpets or high levels of other activity or both. Oystercatchers are rare where human activity is high or when it involves substantial removal of large limpets. Even in this relatively simple system, the resulting community dynamics among humans, oystercatchers, and the limpet–algae community below them are complex and often not easily modeled. Small limpets and closely grazed surfaces dominate both vertical and horizontal surfaces along many southern California rocky intertidal shores today due to human exploitation of large limpets and disturbance of oystercatchers' predatory behavior, with large limpets and oystercatchers largely absent. In areas largely protected from human disturbance, oystercatchers are present and feed mostly on horizontal surfaces leaving these surfaces algae-covered with few limpets and vertical surfaces well grazed by large limpets. In areas where human exploitation is limited, but other human activity disrupts oystercatcher feeding behavior and limits their presence, both horizontal and vertical surfaces are dominated by the larger limpets and are well grazed. Similar human–limpet–algae interactions and cascading human impacts have been documented in both South Africa and Chile and are likely present elsewhere. Human exploitation of intertidal limpets and other marine life is much more intense in Chile, and research there suggests that competition between humans and the Canarian black oystercatcher drove it to extinction (Lindberg et al. 1998).

## THE DEEP SEA AND SEAMOUNTS

> Deep environmental work is vitally important yet very difficult to do, as suggested by a comparison to tropical rain forests. It is relatively easy to show that the Earth's jungles are under siege and that their residents are going extinct at an alarming rate. But what of the darkness below? We currently have only the sketchiest understandings of its structure and inhabitants and food chains, much less the alterations being made by deep fishing and planetary heating and the seventy thousand or so synthetic chemicals that we have managed to inject in the global environment. By any measure, our ignorance is almost as boundless as the deep.
>
> —Broad 1998

What is it about the deep sea that inspired and intrigued William Broad, *New York Times* science writer, editor, and two-time Pulitzer Prize winner, to pen the preceding quote and write a whole book about it? The deep sea is indeed a charmed place worthy of his book, which captures its magic well. A still little known, but increasingly well circulated fact about the ocean is that it is home to 99.5 percent of Earth's habitable space; less well known is that the deep ocean comprises 78.5 percent of it. The deep sea remains, quite literally, a huge black box, largely out of sight and out of mind, yet its role in global climate and numerous other processes makes it vital to all life on Earth. Beyond the oceans' immense size and global importance, the ocean floor is home to the world's largest mountains and canyons, other special features, some of Earth's most diverse and productive places, and some of its richest concentrations of life. Yet it is neither untouched nor unscarred by humans.

Deep ocean ecosystems are neither immune from nor unscathed by human impact. Exploration, discovery, and knowledge of the deep sea, its biodiversity, and its resources is advancing slowly and gathering momentum but has still just scratched the surface. We've learned enough to recognize that the deep sea is not the uniform desert we once believed, but rather a varied landscape with high levels of diversity and productivity, and rich concentrations of life in at least some locations. We know many deep sea organisms grow very slowly and successfully reproduce infrequently, and that many deep sea environments are fragile, easily disturbed, and slow to recover. We lack the information to determine whether any level of exploitation is truly sustainable for these resources, let alone enough to actually manage them sustainably. Yet even as logic says to wait or at least go slow, new deep sea treasures are discovered and immediately and inexorably drawn into the race for fish as depletion of more accessible inshore resources worsens, and fisheries effort is refocused further and further offshore, targeting these new resources as quickly as they are discovered.

At about the same time deep sea researchers began discovering, documenting, and quantifying the existence of large aggregations of fish in deep ocean waters in the late 1960s and early 1970s, commercial-scale fisheries started routinely targeting them. New scientific information, improved technology, expansion of exclusive economic zones (EEZs) to 200 miles around maritime countries, limited regulation, and innovative marketing have stoked these growing deepwater fisheries since that time. The creation of New Zealand's large EEZ in 1978, new information about fish aggregations on the Chatham Rise, a prominent submerged feature contained within it, use of specialized deep water trawling gear, and a smart decision to market a species of slimehead (family of deepwater fishes previously sharing this common name) as orange roughy, illustrates how these factors led to the boom and subsequent bust of this fishery. The collapse came about after an intense fishery rapidly developed and quickly depleted stocks of this slow-growing, late-maturing fish species that included fish more than one hundred years old (Broad 1998).

How do such deepwater trawl fisheries alter deep sea systems? A recent study surveyed and analyzed such impacts from a similar orange roughy fishery to a group of small seamounts south of Tasmania, Australia. Their fauna appeared highly diverse and endemic. This lone survey identified 262 invertebrate species and 37 fish species compared to a total of only 598 invertebrate species previously reported from all seamounts worldwide. Depending on taxa, 24 to 43 percent of invertebrate species collected were new to science and 16 to 33 percent appeared to be restricted to the seamount environment. Trawl operations produced gross changes to the seamount benthic environment and associated benthic communities. Trawl operations in heavily fished areas effectively eliminated three-dimensional "reef aggregate" structure and habitat including both living and dead aggregates of the dominant coral (*Solenosmilia variabilis*), leaving in its place bare rock and pulverized rubble. Benthic biomass was more than twice as high and diversity 1 1/2 times greater on unfished seamounts than on heavily fished seamounts. The dramatic and devastating trawling impacts documented to these complex seamount reefs, including the near complete removal of the coral substrate and its associated community, are consistent with documented trawling damage to other similar habitats. Such devastation inflicted by a short-lived and possibly ephemeral fishery on such complex, critical habitats and associated species assemblages, about which very limited information has been gathered, is highly questionable and worth reconsidering (Koslow et al. 2001).

## CONCLUSION

Half a century after Aldo Leopold preached about land conservation and the need for a new land ethic, we are also making him a prophet for the oceans conservation and the need for a new ocean ethic as well. As we set the course for a new millennium, multiple stressors are mounting cumulative challenges to our most precious ocean ecosystems, not unlike those he observed facing North American terrestrial landscapes in the middle of the twentieth century. Marine scientists, among those closest to the oceans, are warning us of what we have lost, what we are losing, and what we will lose. Already, we have begun losing some of the key cogs and wheels that once kept our marine ecosystem engines running smoothly in all kinds of weather, and these systems are sputtering and collapsing with greater frequency. We continue to tinker, but not intelligently. The complex fabric of marine life that once knitted together the patchwork of our oceans is fraying and we are losing bigger and bigger pieces, perhaps more and more permanently.

Human activities have long impacted our oceans and their complex living systems, both in ways that we understand and in ways that we do not yet fully comprehend. The rate and scale of such change continue to accelerate. Our capacity to fundamentally alter the marine environment grows with our technological prowess. Our acumen for controlling and applying these wisely does not seem to do likewise. Intact, natural, marine ecosystems are diverse and dynamic, but also resistant to and resilient from both human and natural perturbations, depending on scale, intensity, and frequency. Unprecedented rates, scales, intensity, and frequency of anthropogenic disturbance to marine ecosystems are reducing their diversity, affecting their dynamics, overcoming their resistance, and weakening their resilience. We are profoundly altering the marine landscape and adversely impacting its ability to respond to future stresses.

This chapter detailed some of the profound human impacts to the ocean realm and its many and diverse ecosystems. Human activities now affect the full range of marine landscapes from relatively well known, shallow, coastal waters to still barely explored abyssal depths. From priceless coral reefs, to luxuriant kelp forests, awesome rocky coasts, productive salt marshes, and the still mysterious deep ocean; such impacts span the whole spectrum of marine ecosystems. Multiple, interacting, and sometimes synergistic, anthropogenic stressors are contributing to ocean ecosystem change, but fishing and related extractive activities are clearly primary factors. In the next chapter, we explore more fully the role of fishing in marine ecosystem degradation, how it impacts multiple levels of biodiversity, why it sometimes goes unnoticed or underestimated, and why there are reasons for optimism and the need to link more closely the fields of ecosystem protection and fisheries management.

## REFERENCES

Anderson, R. A. 1992. Diversity of eukaryotic algae. *Biodiversity and Conservation* 1:267–292.

Ballantine, W. J. 1991. *Marine Reserves for New Zealand.* Auckland: University of New Zealand.

Bechtel, J. D. 2002. *The Recovery of* Diadema antillarum *in Discovery Bay, Jamaica: Impacts and Implications for Reef Management.* Ph.D. Dissertation, Boston University, Boston, MA.

Birkeland, C. 2001. Can ecosystem management of coral reef fisheries be achieved? In

Best, B., and A. Bornbusch, eds. *Global Trade and Consumer Choices: Coral Reefs in Crisis,* 15–18. Papers Presented at the Symposium on Global Trade and Consumer Choices: Coral Reefs in Crisis, held at the 2001 Annual Meeting of the American Association for the Advancement of Science. Washington, DC: American Association for the Advancement of Science.

Broad, W. J. 1998. *The Universe Below: Discovering the Secrets of the Deep Sea.* New York: Touchstone.

Brooks, W. K. 1996. *The Oyster.* Baltimore, MD: Johns Hopkins University Press.

Carpenter, R. C. 1988. Mass mortality of a Caribbean sea urchin: Immediate effects on community metabolism and other herbivores. *Proceedings of the National Academy of Sciences, USA* 85(2):511–514.

Castilla, J. C., G. M. Branch, and A. Barkai. 1994. *Exploitation of Two Critical Predators: The Gastropod,* Concholepus concholepus, *and the Rock Lobster,* Jasus lalandii. Heidelberg, Germany: Springer-Verlag.

Choat, J. H. 1991. The biology of herbivorous fishes on coral reefs. In Sale, P. F. *The Ecology of Fishes on Coral Reefs,* 120–155. San Diego, CA: Academic.

Davis, G. E. 1977. Fishery harvest in an underwater park. *Proceedings of the 3rd International Coral Reef Symposium* 2:605–608.

Dayton, P. K., S. F. Thrush, M. T. Agardy, and R. J. Hofman. 1995. Environmental effects of marine fishing. *Aquatic Conservation: Marine and Freshwater Ecosystems* 5:205–232.

Dayton, P. K., M. J. Tegner, P. B. Edwards, and K. L. Riser. 1998. Sliding baselines, ghosts, and reduced expectations in kelp forest communities. *Ecological Applications* 8(2):309–322.

Durán, L. R., and J. C. Castilla. 1989. Variation and persistence of the middle rocky intertidal community of central Chile, with and without human harvesting. *Marine Biology* 103:555–562.

Edmunds, P. J., and R. C. Carpenter. 2001. Recovery of *Diadema antillarum* reduces macroalgal cover and increases abundance of juvenile corals on a Caribbean reef. *Proceedings of the National Academy of Sciences, USA* 98(9):5067–5071.

Epperly, S. P., and J. W. Dodrill. 1995. Catch rates of snowy grouper, *Epinephelus niveatus,* on the deep reefs of Onslow Bay, southeastern U.S.A. *Bulletin of Marine Science* 56(2):450–461.

Estes, J. A., and P. D. Steinberg. 1988. Predation, herbivory, and kelp evolution. *Paleobiology* 14(1):19–36.

Estes, J. A., M. T. Tinker, T. M. Williams, and D. F. Doak. 1998. Killer whale predation on sea otters linking oceanic and nearshore ecosystems. *Science* 282:473–476.

Gardner, T. A., I. M. Cote, J. A. Gill, A. Grant, and A. R. Watkinson. 2003. Long-Term Region-Wide Declines in Caribbean Corals. *Science* (Washington) Vol. 301, no. 5635, pp. 958–960. 15 Aug 2003.

Goeden, G. B. 1982. Intensive fishing and a "keystone" predator species: Ingredients for community instability. *Biological Conservation* 22:273–281.

Hay, M. E. 1991. Fish-Seaweed Interactions on Coral Reefs: Effects of Herbivorous Fish and Adaptations of Their Prey. In P. E. Sale, ed., *The Ecology of Fishes on Coral Reefs.* Academic Press, San Diego, California, 96–119.

Horton, T., and W. M. Eichbaum. 1991. *Turning the Tide: Saving the Chesapeake Bay.* Washington, DC: Island Press.

Hughes, T. P. 1994. Catastrophes, phase shifts, and large-scale degradation of a Caribbean coral reef. *Science* 265:1547–1551.

International Coral Reef Initiative (ICRI). 1996 (June). *International Coral Reef Initative Call to Action.* Dumaguate City, Philippines: International Coral Reef Initiative. Available on the Web at: http://icriforum.org/secretariat/pdf/ICRI-call%20to%action.pdf

Jackson, J. B. C. 1997. Reefs since Columbus. *Coral Reefs* 16:23–32.

———. 2001. What was natural in the coastal oceans? *Proceedings of the National Academy of Sciences, USA* 98(10):5411–5418.

Jackson, J. B. C., et al. 2001. Historical overfishing and the recent collapse of coastal ecosystems. *Science* 293:629–637.

Knowlton, N. 2001. The future of coral reefs. *Proceedings of the National Academy of Sciences, USA* 98(10):5419–5425.

Koslow, J. A., K. Gowlett-Holmes, J. K. Lowry, T. O'Hara, G. C. B. Poore, and A. Williams. 2001. Seamount benthic macrofauna off southern Tasmania: Community structure and impacts of trawling. *Marine Ecology Progress Series* 213:111–125.

Kurlansky, Mark. 1997. *Cod: A Biography of the Fish that Changed the World.* New York: Penguin Books, 294.

Leopold, A. Round River. 1953. In *The Journals of Aldo Leopold,* 173. New York: Oxford University Press.

Lessios, H. A., Glynn, P. W., and Robertson, D. R. 1983. Mass Mortalities of Coral Reef Organisms. *Science* 222 [4625]:715. Nov. 18.

Lessios, H. A., Robertson, D. R., and Cubit, J. D. 1984. Spread of Diadema Mass Mortality through the Caribbean. *Science* 226 [4672]:335–337. Oct. 19.

Lessios, H. A., 1988a. Mass mortality of *Diadema antillarum* in the Caribbean: What have we learned? *Annual Review of Ecology and Systematics* 19:371–393.

———. 1988b. Population dynamics of *Diadema antillarum* (Echinodermata: Echinoidea) following mass mortality in Panama. *Marine Biology* Berlin, Heidelberg 99(4):515–526.

Levitan, D. R. 1988a. Algal–urchin biomass responses following mass mortality of *Diadema antillarum Philippi* at Saint John, U.S. Virgin Islands. *Journal of Experimental Marine Biology and Ecology* 119(2):167–178.

———. 1988b. Density-dependent size regulation and negative growth in the sea urchin *Diadema antillarum Philippi. Oecologia* 76(4):627–629.

Lindberg, D. R., J. A. Estes, and K. I. Warheit. 1998. Human influences on trophic cascades along rocky shores. *Ecological Applications* 8(3):880–890.

Littler, D. S., and M. M. Littler. 2000. *Caribbean Reef Plants: An Identification Guide to the Reef Plants of the Caribbean, Bahamas, Florida, and Gulf of Mexico.* Washington, DC: Offshore Graphics.

McAllister, D. E. 1991. *Sea Wind* 5(14) Ottawa, Canada.

McAllister, D. E., F. W. Schueler, C. M. Roberts, J. P. Hawkins. 1994. In R. I. Miller, ed. *Mapping the Diversity of Nature,* pp. 155–175. London: Chapman and Hall, 1994.

McClanahan, T. R. and R. Arthur. 2001. The effect of marine reserves and habitat on populations of East African coral reef fishes. *Ecological Applications* Vol. 11, no. 2, pp. 559–569.

McClanahan, T. R. and D. Obura. 1995. Status of Kenyan coral reefs. *Coastal Management* 23: 57–76.

Moore, J. A. 1999. Deep-sea finfish fisheries: Lessons from history. *Fisheries* 24(7):16–21.

Moore, J. A. and Mace, P. M. 1999. Challenges and prospects for deep-sea finfish fisheries. *Fisheries* 24(7):22–23.

Moreno, C. A. 2001. Community patterns generated by human harvesting on Chilean shores: A review. *Aquatic Conservation: Marine and Freshwater Ecosystems* 11(1):19–30.

Myers, R. A. and Worm B. 2003. Rapid worldwide depltion of predatory fish communities. *Nature* 423(15 May 2003):280–283.

National Marine Fisheries Service (NMFS). 1990. *Historic Fishery Statistics Atlantic and Gulf States 1879–1989.* Silver Spring, MD: National Marine Fisheries Service.

———. 2002. Commercial Fisheries and Marine Recreational Fisheries Survey. Available online at: http://www.st.nmfs.gov/st1/ (last accessed Dec. 2003). (NMFS Fisheries Statistics and Economics website).

National Research Council (NRC). 1995. *Understanding Marine Biodiversity: A Research Agenda for the Nation.* Washington, DC: National Academy.

Newell, Roger I. E. 1988. Ecological changes in Chesapeake Bay: Are they the result of overharvesting the American oyster, *Crassostrea virginica?* In Lynch, M. P., and E. C. Krome, eds. *Understanding the Estuary: Advances in Chesapeake Bay Research,* 536–546. Proceedings of a Conference, 29–31 March 1988, Baltimore Maryland. Solomons, MD: Chesapeake Research Consortium.

Ogden, J. C., and P. S. Lobel. 1978. The role of herbivorous fishes and urchins in coral reef communities. *Env. Biol. Fish* 3(1):49–63.

Reaka-Kudla, M. L. 1997. The Global Biodiversity of Coral Reefs: A Comparison with Rain Forests. In *Biodiversity II: Understanding and Protecting Our Biological Resources*, M. L. Reaka-Kudla, D. E. Wilson, and E. O. Wilson, eds. John Henry/National Academy Press, Washington, DC., pp. 83–108.

Sale, P. F. 1991. *The Ecology of Fishes on Coral Reefs.* San Diego, CA: Academic Press.

Sammarco, P. W., J. S. Levington, and J. C. Ogden. 1974. Grazing and control of coral reef community structure by *Diadema antillarum Philippi* (Echinodermata: Echinoidea): A preliminary study. *Journal of Marine Research* 32(1):47–53.

Santayana, George. 1905. *The Life of Reason*, Amherst, New York: Prometheus Books (1998 edition).

Sapp, J. 1999. *What Is Natural? Coral Reef Crisis.* Oxford University Press, New York, 1999.

Sepkoski, J. J. 1995. Large scale history of biodiversity. In *Global Biodiversity Assessment,* 202–212. Cambridge, UK: Cambridge University Press.

Silliman, B. R., and M. D. Bertness. 2002. A trophic cascade regulates salt marsh primary production. *Proceedings of the National Academy of Sciences* 99(16):10500–10502.

Steneck, R. S. 1988. Herbivory on coral reefs: A synthesis. Proceedings of the Sixth International Coral Reef Symposium, Townsville, Australia, 8th–12th AUGUST 1988. Volume 1: Plenary Addresses and Status Reviews, pp. 37–49.

———. 1998. Human influences on coastal ecosystems: Does overfishing create trophic cascades? *Trends in Ecology and Evolution* 13:429–430.

United States Coral Reef Task Force (USCRTF). 2000. *The National Action Plan to Conserve Coral Reefs.* Washington, DC: U.S. Environmental Protection Agency (EPA).

Wilkinson, C. R. 1998. *Status of Coral Reefs of the World: 1998.* Townsville, Australia: Global Coral Reef Monitoring Network and Austalian Institute of Marine Science.

———. 2000. *Status of Coral Reefs of the World: 2000.* Townsville, Australia: Global Coral Reef Monitoring Network and Austalian Institute of Marine Science.

———. 2002. *Status of Coral Reefs of the World: 2002.* Townsville, Australia: Global Coral Reef Monitoring Network and Austalian Institute of Marine Science.

THREE

# Fishing and Its Impacts

> Ecological extinction caused by overfishing precedes all other pervasive human disturbance to coastal ecosystems, including pollution, degradation of water quality, and anthropogenic climate change.
>
> —Jackson et al. 2001

Fishermen are among the most ingenuous, resourceful, and hardworking natural resource gatherers ever to have walked the earth. But their proficiency, and the demand for their products, has caused a series of collapses in the past decade of the very resources (e.g., New England groundfish) on which they depend. This has ignited a resurgence of interest within the marine science community to study and better understand the adverse impact of fishing on marine species and ecosystems and its role and significance in such collapses. Many recent publications provide a good overview of the progress made (Dayton et al. 2002; Jackson et al. 2001; NRC 2001; Pauly et al. 2002). As a result, a much better recognition of the significance of such impacts, their special role in such collapses, and their paramount importance relative to other human impacts on marine systems is beginning to emerge. This chapter summarizes key issues and findings from recent studies and provides more detail on some of those most interesting and relevant to marine reserves.

In the following discussion, *fishing* refers generally to the removal or extraction of any or all marine biota, including plants, invertebrates, and vertebrates from the oceans. Fishing and the related removal of living marine resources at any level affects the targeted species and their associated habitats, species assemblages, community, and ecosystem to some extent. The point at which such impacts are noticeable, discernible, or measurable can vary and the point at which they move from acceptable to harmful or cross the line to overfishing is often blurry, subjective, or arbitrary. There can also be a lag time between the fishing activity and the full extent of its discernible impact (See section in chapter 2 on coral reefs and Hughes 1994; Jackson 1997; and Lessios et al. 1984).

The term *overfishing* refers to fishing at a level that is unsustainable, causes harm, or results in irreversible change. It involves a suite of complex biological, social, and economic factors, and its usage has been similarly complicated (Bohnsack and Ault 1996). We consider and refer to the broader impacts of fishing rather than the impacts of overfishing because *overfishing* has been defined and used variably and because significant fishing impacts often occur at levels well below some classical or widely used definitions. Traditional fisheries management has largely focused on production and primarily considered fishing impacts only as they relate to production of targeted species. The irony of this heavy focus on targeted species is its frequent failure to protect even these species or maintain their production, but the resulting adverse impacts have been much broader.

Jackson et al. (2001) provide one of the most comprehensive, integrative, and innovative reviews highlighting the critical importance and severity of fishing to marine and coastal ecosystems. Their study covered a range of temporal scales, including paleoecological records from the past 125,000 years, archaeological records from the past 10,000 years, historical documents from the past 500 years, and ecological records from the scientific literature from the past 100 years covering the global expansion of fisheries and related exploitation of marine resources. The geographic scope of their study was global and it included diverse ecosystems ranging from kelp forests and coral reefs to sea grass beds and estuaries.

The review demonstrated that fishing impacts had primacy over other human-induced stresses to marine systems. Furthermore, it also concluded that (1) ecological changes to marine systems as a result of fishing were devastating, remarkably similar in general, but different in detail across ecosystems; (2) these changes included enormous loss of biomass and abundance of large animals, now effectively extinct from most coastal ecosystems; (3) overfishing can act synergistically with other human disturbances such as eutrophication, disease outbreaks, and exotic species introductions and may be a prerequisite for them to occur, and (4) climate change is unlikely to be the primary cause for observed disease outbreaks but may be an important secondary factor.

## HOW FISHING IMPACTS MARINE SYSTEMS AND FISH

Fishing impacts the living and nonliving components and biodiversity of marine systems directly and indirectly in a variety of ways and at several levels, including ecosystems, species, and genetics. Fishing alters the genetic structure of some marine species, threatens or endangers the continued viability of others, and modifies complex marine ecosystems, including their associated species assemblages and physical environment. Equally significant, the increasingly ubiquitous geographic scope of fishing impact is no longer confined to nearshore, shallow water, or developed areas (Dayton et al. 2002, 1995; Jackson et al. 2001; Murray et al. 1999; NRC 2001, 1995; Pauly et al. 2002; Sumalia et al. 1999; Thrush et al. 1998). These hierarchical levels of biodiversity are clearly overlapping and interdependent. An altered ecosystem often impacts biodiversity at the species and genetics level, and the converse is also true. The collapse of an ecosystem threatens all of its unique contained species and their genetics, and the loss of a species clearly removes any unique genetic material. The loss of genetic and species diversity destabilizes and diminishes an ecosystem's resilience, and the loss of genetic diversity similarly impacts species.

How does fishing impact each level of biodiversity? At the species level, fishing often reduces, alters, or destabilizes the abundance, size, and biomass of both targeted and nontargeted species both directly through fishing mortality and indirectly through habitat, food web, and other ecological alterations. At the genome level, fishing is most likely to alter targeted species through artificial selection, reduction of population size to create bottlenecks (i.e., population sizes small enough to lose genetic diversity through random or chance events), or extirpation of genetically unique populations. At the ecosystem level, fishing impacts often include changes to both ecosystem structure and function as a result of the species-level impacts already described due to fishing mortality and compounded by fishing gear–induced habitat alteration and a complex set of biological interactions, only a few of which are well understood. Bycatch, incidental take, habitat alteration, and biological interactions complicate and exacerbate the impacts at each level as well as our understanding of those impacts.

At the species level, fishing impacts both targeted and nontargeted species directly and indirectly in several similar ways. The key difference is that "normal" fishing activities, standard fisheries management, and resulting "directed takes" also intentionally impact targeted species directly. The stated or intended goal of traditional fisheries management, "to achieve maximum sustainable yield (MSY)," aims to fish down, remove, or reduce targeted species' adult biomass significantly, typically between 50 and 80 percent of unfished levels. Even when such plans are effectively implemented and "successful" (often they aren't), the resulting fish population(s) can be at a small fraction of their natural levels. Such reductions also reduce spawning capacity or reproductive output. Fishing also impacts targeted and nontargeted species directly via incidental take, bycatch, and habitat alteration via certain gear and techniques, and indirectly through the secondary, higher order, and cascading impacts due to both these and directed takes. By standard definitions, all of these impacts are considered the result of fishing, not overfishing. Management measures often fall short of their goals, resulting in even greater reductions to these targeted species from directed take. Traditionally, only these overages or "excess" removals are considered the result of overfishing, a complex term with several meanings (Bohnsack and Ault 1996).

By design, fishing directly reduces the abundance, size, and biomass of targeted species, often beginning with large predators and consumers and affecting a broader suite of species later. These direct impacts often ripple outward, with secondary impacts to prey, predators, symbionts, and competitors of targeted species, and cascading impacts affecting more complex food web interactions. Targeted species often include highly vulnerable, large, high-level predators that help structure communities; easily harvested, slow moving, benthic grazers that help control aquatic plant and algal growth; and stationary, structure-producing organisms that provide important habitats and can protect water quality via filtration. Clearly, the removal of such organisms should be expected to have wide-ranging impacts on ecosystems, even before they are overfished or endangered. Unfortunately, fishing frequently occurs at unsustainable levels and results in overfishing, fishery collapses, and even endangerment. At high levels of intensity, fishing can lead to ecological or economic extinction of targeted species or suites of species and the role(s) they play in ecosystems (Jackson et al. 2001). As discussed in chapter 1, the risk of endangerment and true biological extinction is a real and growing threat for at least some marine species.

Fishing usually targets or selects older, larger individuals initially and normally results in decreased fish size and abundance over time. If fishing is intense enough, fish are removed from the population before having the chance to grow enough to maximize their value to the fishery, a condition referred to as *growth overfishing*. Because an individual's reproductive success is size dependent and a population's reproductive output is related to its adult biomass, fishing's impact on fish size and abundance can also dramatically reduce a population's reproductive output as fewer and smaller individuals survive to spawn. *Recruitment overfishing* occurs when a population produces too few offspring to sustain itself and is generally considered more serious because it threatens the population's future viability. Historically, the term *overfishing* generally referred to growth and/or recruitment overfishing. *Serial overfishing* (*depletion*) describes the frequent situation in which a fishery sequentially targets and overfishes one species or species complex after another repeating the overfishing cycle (Bohnsack and Ault 1996).

For several decades, there has been increasing scientific concern, recognition, and evidence that fishing can and already is altering genetic characteristics of some targeted species and populations and that this constitutes a growing problem. *Genetic overfishing* describes fishing-induced genetic changes that can result from targeting older, larger individuals or from other means (Bohnsack and Ault 1996). Bjorkland (1974) identified and discussed threats to marine genetic resources and the need to protect them. Polunin (1983) reviewed progress on these topics and reached similar conclusions. Scientists developing a regional plan addressed these issues in more detail for reef fish, identified loss of genetic diversity as one of the primary problems facing them, and discussed the evolutionary basis for this (NMFS 1990). An international team of scientists confirmed their findings, concluding that the risk to genetic diversity was real and that alteration was likely occurring (Roberts et al. 1995). The NRC (1995) report on marine biodiversity reached similar conclusions, extending them more broadly to other species and ecosystems. Bruton (1995) similarly stressed the threat to and importance of conserving genetic resources in marine fish. Boehlert (1996) stressed the importance of maintaining genetic diversity in marine fisheries to ensure their sustainability, and supported this with analyses of fishing impact on a variety of species. These analyses showed strong evidence of fishing impacts on the population characteristics of a number of species, including Pacific rockfish (*Sebastes* spp.), Pacific salmon, and orange roughy.

In addition to targeting large individuals, fishing may selectively target individuals with specific characteristics, such as rapid growth, aggressive feeding behavior, or specific migration patterns. If these characteristics are genetically based, such fishing can result in genetic overfishing, or the loss of specific genetic traits from the population. Loss of genetic diversity may reduce a species' resistance or resilience in the face of natural environmental or anthropogenic stresses. Thus genetic overfishing can contribute to stock collapses or prolong the recovery period of stocks rebounding from past overfishing. We explore in more depth some specific examples of fishing-induced genetic change and the role of marine reserves in mitigating this later.

Although overfishing and its derivatives, as traditionally used, are applied only to targeted species, they clearly affect all species, habitats, and ecosystems, as does fishing generally. Traditional fishery terminology and management practices have largely failed to address ecological interactions among overfishing, other fishing impacts, and the resources affected by both. Consequently, a rising chorus of voices from diverse sources are attempting to define new terms such as *ecosystem overfishing* to describe levels of fishing that affect ecosystem structure or function (Bohnsack and Ault 1996; Murawski 2000) and calling for an ecosystem-based approach to management. The examples provided earlier in chapter 2 and the Jackson et al. (2001) review emphasized these types of fishing impact. They are among the most critical, widespread, and severe impacts of fishing but have received little attention from traditional fishery management and until recently were largely neglected. A full review of ecosystem-level fishing impacts is beyond the scope of this book, but interested readers are directed to the references cited earlier (in particular, Dayton et al. 2002; Jackson et al. 2001; Pauly et al. 2002). A summary of ecosystem impacts is provided following here.

Exploitation may impact marine ecosystems in a number of ways beyond simple, direct reductions of or impacts on target species (Fig. 3.1). By reducing the abundance or changing the population structure (e.g., size frequency, ratio of males to females) of target species in an ecosystem, fishing may have a number of indirect effects on nontarget species and the entire ecosystem. These indirect effects result from the fact that many target fishery species play important roles in marine ecosystems due to their abilities to outcompete other species for limited resources or to alter the distribution and abundance of their prey. If the abundance of these ecologically important target species decreases, the abundance of competitor or prey species may also change. Thus, even if fishing levels are sustainable for a particular species, reduced abundance of that species may have profound impacts on marine communities and ecosystems.

Functional changes accompany structural changes at the ecosystem level. Grazing, predation, competition, energy flow, and other species interactions and rate processes may be severely affected. Such changes may cause secondary or cascading impacts to other species, habitats, and nonliving ecosystem

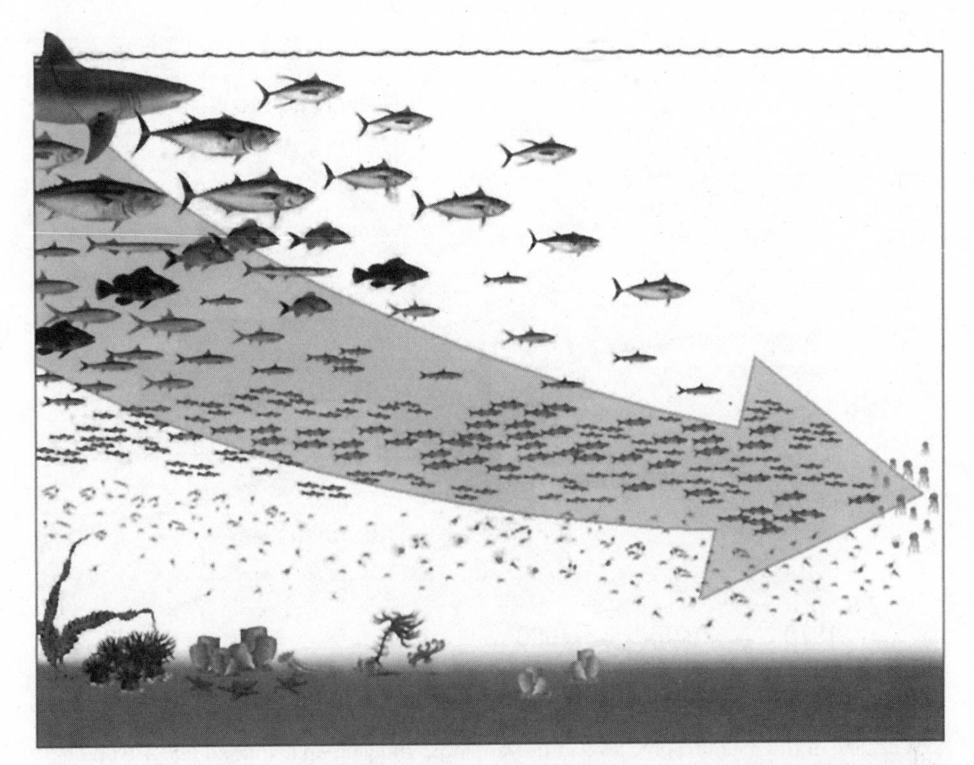

**FIG. 3.1 Ecosystem Impacts of Fishing.** Fishing directly affects the abundance of marine fish populations, as well as the age of maturity, size structure, sex ratio, and genetic makeup of those populations. Fishing affects marine biodiversity and ecosystems indirectly through bycatch, habitat degradation, and through biological interactions. Through these unintended ecological consequences, fishing can contribute to altered ecosystem structure and function. As commercially valuable populations decline, people begin fishing down the food web, which results in a decline in the mean trophic level of the world catch. Source: Adapted from Pauly et al. 1998; Goni 2000; and Dayton et al. 2002.

components. Structural and functional modifications to ecosystems that result from fishing can precondition their collapse, reduce their resilience, and synergize with other anthropogenic or natural disturbances to cause irreversible or long-term changes, degrade, or destroy them. Such severe impacts may be sudden, surprising, and catastrophic; may involve phase shifts between alternative stable states and may or may not be predictable; and may occur following a considerable lag time as in the case of coral reefs (chapter 2; Hughes 1994; Jackson 1997). Recent studies exploring fishing impacts on food webs at the large marine ecosystem or entire ocean basin scale suggest that expanded fishing effort is now altering entire food webs even at this geographic scale. This frequently results in a phenomenon termed "fishing down the food web" in which the food web is truncated, the mean trophic level is lowered, and fish-

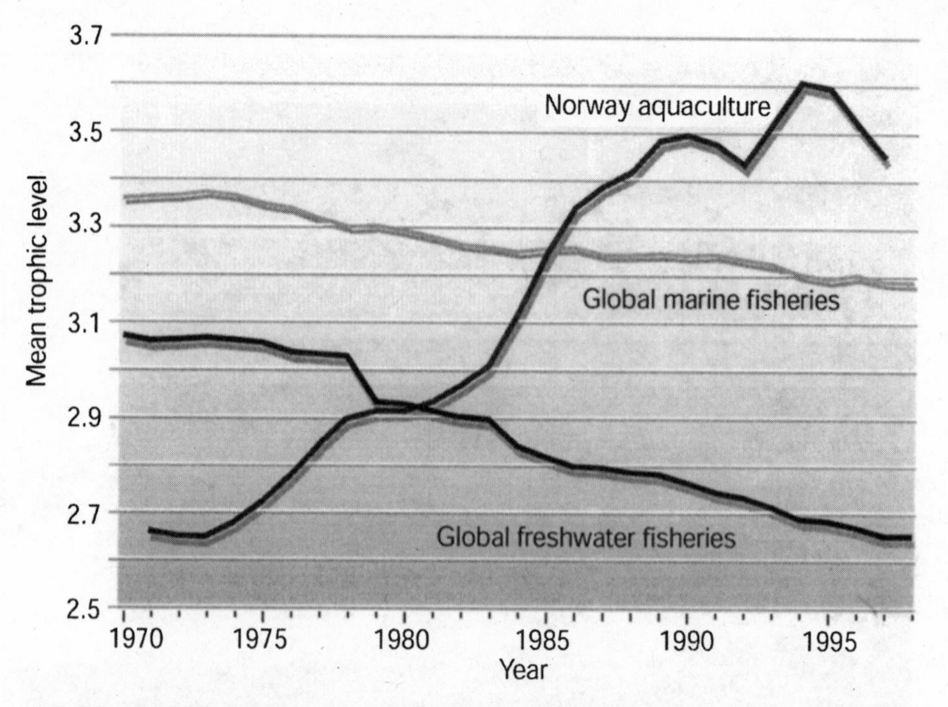


**FIG. 3.2 Altered Trophic Structure and Food Webs.** Marine fisheries are characterized by a decline of the mean trophic level in the landings, implying an increased reliance on organisms low in food webs over time. Source: Adapted from Pauly et al. 2002 and reprinted with permission from *Nature*.

ers successively fish lower and lower down the food web, usually for less and less desirable species (Dayton et al. 2002; Pauly et al. 2002, 1998; Fig. 3.2.).

Modern fisheries also impact targeted species, nontargeted species, habitats, and ecosystems via bycatch and incidental take, habitat disturbance and alteration, and destructive fishing gear and practices (Fig. 3.1; Dayton et al. 2002, 1995; Thrush et al. 1998). These related and overlapping impacts can be as or even more important and severe than the intended removal of targeted species. Morgan and Chuenpagdee (2003) employ "collateral (fishing) impacts" to embrace the whole spectrum of unintentional or incidental damage to marine life and habitat resulting from fishing activities targeting other types of sea life. The references cited above provide more detailed reviews and analyses of this broad class of impacts, which we define, describe, and discuss briefly in the following text.

Although definitions vary, *bycatch* or *incidental take* generally refers to the unintentional taking, discarding, or damaging of marine resources when fishing for targeted species. Certain fishing gear and techniques also directly impact the habitat and physical structure of marine ecosystems and can operate at the scale of the fishery. *Habitat disturbance and alteration* generally refer to habitat changes, often destructive, that result from fishing operations and can include both ecological changes resulting from altered fish communities and physical damage caused by gear or destructive fishing practices. *Destructive fishing gear and practices* often refers only to those gear and practices that have direct physical impacts on habitat or nontargeted resources, but clearly can include much more. Impacts on nontarget species and habitat can also have secondary and cascading impacts on species, communities, and ecosystems. Consequently, fishing usually alters population and community structure, and frequently trophic and physical structure within ecosystems. The selective nature of all forms of fishing and their direct impacts enhance their secondary or indirect impacts. Trophic levels may be reduced or modified and food webs simplified or otherwise altered (see Fig. 3.2; Pauly et al. 1998, 2002).

## CHANGING FISH, FISHERS, AND FISHERIES

The status of fish, fishers, and fisheries dependent on our rapidly changing marine ecosystems reflects their alteration. Traditional fisheries management tools were neither created nor designed to protect marine ecosystems or prevent species from endangerment. Consequently, it should not be overly surprising that such tools are failing to meet these goals. The fact that they are also frequently and increasingly not meeting their oft-stated goals of maximizing public benefit, optimizing yield, and sustaining fisheries is more surprising.

The quality of information available for fisheries status on a global scale is limited, questionable, and variable. The primary source for such information is the United Nations Food and Agriculture Organization (FAO), which tends to be fishing country and industry dependent, production-oriented, and conservative with respect to determination of overexploitation (overfishing). Nonetheless, according to FAO, world marine fisheries production peaked in or around 1995, has declined or hovered around that level since (depending on whether questionable Chinese data are included), and will likely stagnate or even decline over the next several decades. At a regional scale, FAO data similarly indicate that production had likely already peaked for fourteen of sixteen marine reporting areas by 2000 and was either declining or fluctuating around that peak in each. Regions with longer fishing histories tended to be in worse shape, some showing substantial long-term declines (e.g., Northwest Atlantic). Furthermore, 75 percent of the world's marine fish stocks, for which information is available, are either fully exploited (47 percent), overexploited (18 percent), or severely depleted (10 percent), and only 1 percent of the total fish stocks are showing signs of slow recovery from depletion. Since 1974, the percentage of stocks overexploited or depleted has been steadily rising, while the number of stocks less than fully utilized has been steadily declining (Fig. 3.3; FAO 2000, 2002).

During 1998, in the United States, approximately 46 percent of the 158 stocks for which there was adequate data were overutilized, and an additional 39 percent were utilized at their maximum capacity (NMFS 1999a). More recent data (NMFS 2001) for 2000 shows that for U.S. fisheries that had sufficient data for assessment the percentage of overutilized species had grown to nearly half, and, perhaps equally disturbing, nearly two-thirds could not be fully assessed due to insufficient information (Figure 3.4; Nowlis and Bollermann 2002). The most recent assessment for 2002 showed little, if any, overall im-

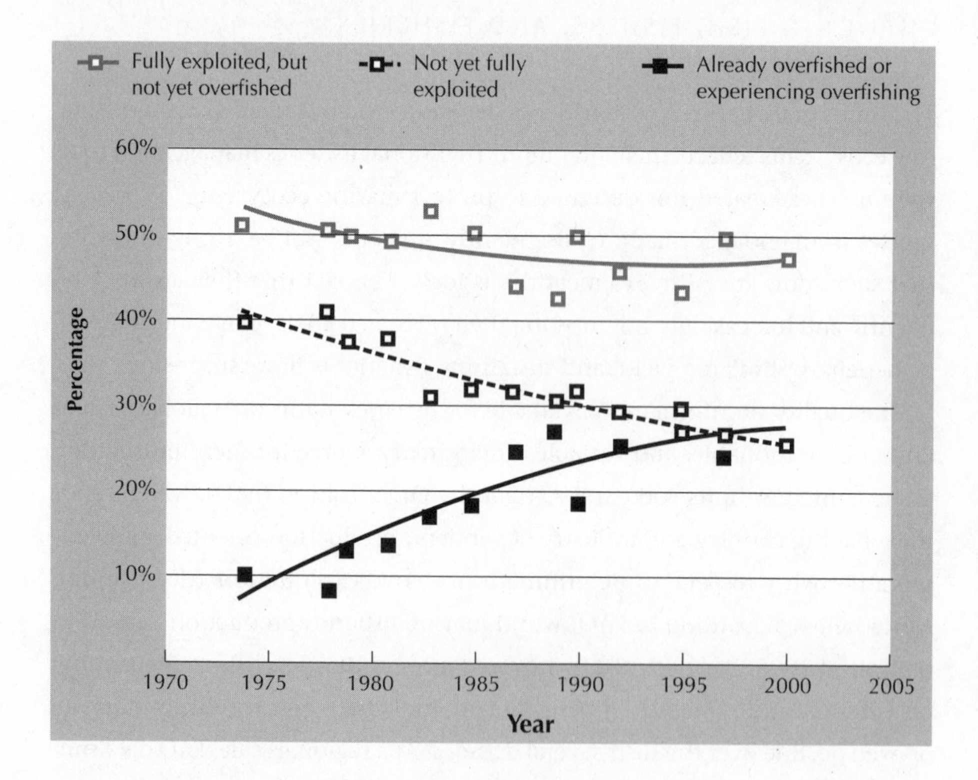


**FIG. 3.3 Global Trends in the State of Fish Stocks since 1974.** Over the past three decades, the proportion of the world's fish stocks fully exploited, but not yet overfished, has declined somewhat or remained unchanged. Over the same period, the proportion not yet fully exploited has declined steadily and the proportion already overfished or experiencing overfishing has increased steadily. These trends indicate that fisheries are targeting an increasing proportion of the world's biodiversity, an increasing proportion of which is becoming overexploited, and the proportion of fully exploited, but not yet overfished species, is either already declining or will soon be declining. Source: Adapted from FAO, 2002.

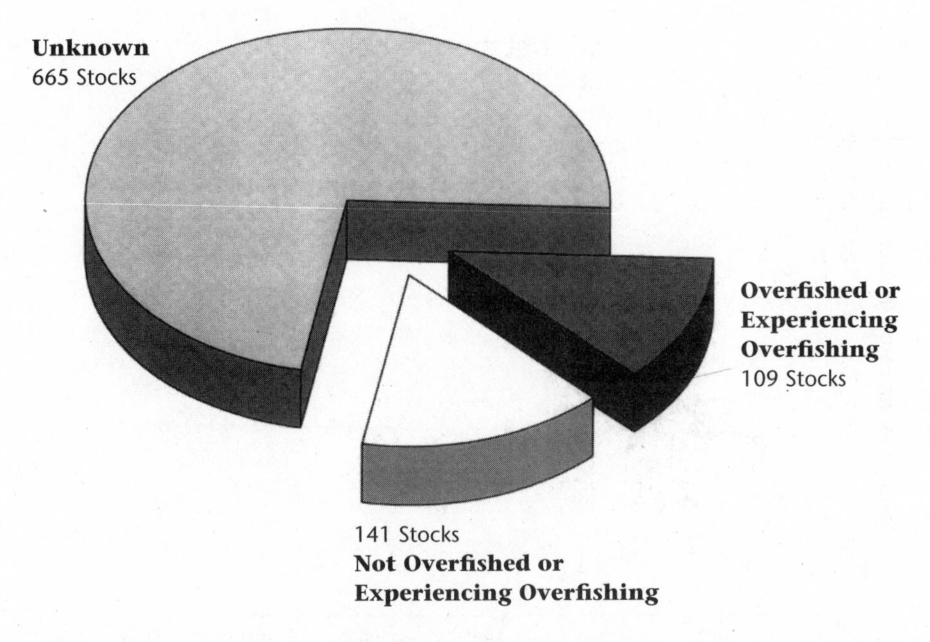


**FIG. 3.4 Status of U.S. Fisheries in 2000.** The pie chart shows that nearly half of all fish stocks for which there was adequate data available were either overfished, being overfished, or both. Equally disturbing was the fact that for nearly two thirds of all stocks adequate information was lacking to even do such an assessment. Thus, roughly 85 percent of all stocks were either known to be overfished, experiencing overfishing, or both, or lacked sufficient information to be assessed. Source: Data from NMFS 2001.

provement and continued to highlight a number of disturbing issues, including the following: (1) roughly 50 percent of the species with adequate data for assessment remain overexploited, (2) overfishing continues to occur on 65 percent of U.S. fish populations known to be seriously depleted, (3) 30 percent of seriously depleted fish populations still lack rebuilding plans, and (4) 75 percent of the fish populations in the U.S. remain without the basic biological information necessary for assessing their status or providing sound management (NMFS 2003a). Similar situations exist in Europe, Asia, and around the world. Such overfishing can lead to both biodiversity and economic losses.

As disturbing as the FAO and NMFS statistics seem, they may actually underestimate the severity of the problem and its prognosis. For example, recent studies suggest that FAO reporting minimizes or masks problems in global fisheries, and NMFS may do likewise for U.S. fisheries. Watson and Pauly (2001) analyzed FAO global fisheries statistics, found systematic distortions in global catch reporting, showed that global fisheries had been declining since the 1980s (Fig. 3.5), and concluded that poor reporting masked the severity of the

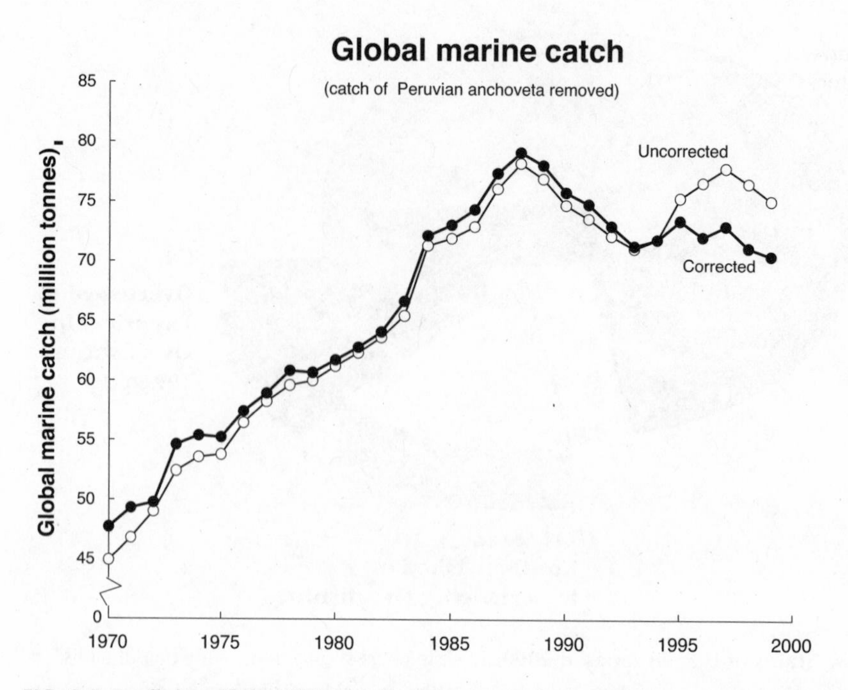


**FIG. 3.5 Declining Global Fish Production.** Graph shows decline in world capture fisheries production since 1989, if one excludes highly suspect estimates of Chinese fisheries production. Source: Adapted from FAO 2002 and Watson and Pauly 2001.

global fisheries crisis, prevented proper management, and threatened global food security. Pauly et al.'s (2002) recent global fisheries review went further, arguing that fisheries have rarely been sustainable, re-iterating that global fisheries have been in decline since the 1980s, identifying many recurring reasons for this, and highlighting changes in fisheries over time as both contributing to and masking these problems. Such changes included increasingly efficient gear, shifts in and expansion of areas targeted, and major modification of species targeted and caught. Symptoms of these trends include serial overfishing and fishing down the food web, the now decades old phenomenon in which the mean trophic level of global fisheries landings has been declining at a rate of 0.05 to 0.1 trophic levels per decade (Fig. 3.2; Pauly et al. 1998, 2002).

Several additional recent studies highlight and reinforce the fact that large predatory fish are especially prone to overfishing, depletion, and endangerment. Myers and Worm (2003) constructed trajectories of community biomass and composition of large predatory fishes for both continental shelf and open ocean systems from the start of industrialized fishing. They found that such fisheries typically reduce community biomass of large predators by 80 percent within fifteen years and that current large predatory fish biomass is less than

10 percent of what it was prior to their introduction. They concluded that sharp reductions in large predatory fish were a global phenomenon with potentially serious consequences for marine ecosystems. Baum et al. (2003) showed similar large rapid declines, up to 75 percent over the last fifteen years, for both coastal and oceanic shark populations in the Northwest Atlantic with recommendations for marine reserves to benefit multiple threatened species. Finally, Friedlander and DeMartini (2002) also demonstrated the effect of fishing down apex predators in the Hawaiian Islands by contrasting the fish communities in the heavily fished main Hawaiian Islands (MHI) with those of the lightly fished and now partially protected Northwestern Hawaiian Islands (NWHI). Total average fish biomass was more than two and a half times higher in the NWHI than in the MHI, but differences in large predators and community structure were even more striking. Biomass of top predators was orders of magnitude greater in the NWHI, where they made up over 50 percent of the fish biomass compared to less than 3 percent in the MHI. The NWHI provide one of the few remaining large, intact, predator-dominated reef systems in the world and a tremendous opportunity for their protection and study.

Fisheries crises and management failures span the globe, include examples from temperate and tropical waters and developed and developing countries, and increasingly are more the rule than the exception. Temperate failures have ranged from the well-publicized collapse of large fisheries including North Atlantic groundfish (cod and other species) and oysters, to California rockfish, Peruvian anchovetta, New Zealand snapper, and orange roughy. Overfishing and ecosystem alteration, however, are not limited to large-scale commercial fisheries in temperate regions. Coral reef fisheries are also clearly vulnerable to these problems despite fishing effort being spread across a large number of species. In the Caribbean, Nassau grouper (*Epinephelus striatus*) (IUCN 1998), spiny lobster (*Panularis argus*), and queen conch (*Strombis gigas*), traditional fisheries mainstays and cultural icons across the region, are all now greatly depleted or endangered throughout it. Other large groupers (*Serranids*), wrasse (*Labrids*), and porgy or bream (*Sparids*) have met similar fates around the globe. In Indonesia, a species of cardinalfish, *Pterapogon kauderni,* may be extinct on reefs in the wild due to collection by aquarists (see also chapter 1; Roberts and Hawkins 1999).

Overfishing of individual stocks and of multiple stocks may occur in a variety of ways. In some cases, a stock may be harvested for years before it is gradually depleted to the point where it is no longer productive. In such cases, stock declines may go virtually unnoticed for years before marine resource managers realize that exploitation rates are exceeding sustainable levels. In other cases, the depletion of a stock may occur on extremely short time scales. For example, as shown in chapter 2, a previously unexploited population of snowy grouper (*Epinephelus niveatus*) from a reef offshore from North Carolina was reduced by 70 percent within three months (Epperley and Dodrill 1995). Such intense fishing, particularly for long-lived, late-maturing fish, cannot continue for long periods of time without the population being overfished. Nevertheless, such intense overfishing is common in fisheries around the world.

*Serial overfishing* (*depletion*) is an increasingly common problem as well. As stocks collapse, fishing pressure on previously untargeted stocks or lightly fished stocks increases and the overfishing cycle repeats itself. Off the California coast, for example, a pattern of serial overfishing is evident in the fishery for abalone and other benthic invertebrates (Dugan and Davis 1994). Landings of red and pink abalone increased in the 1940s and remained at high levels until the mid-1970s, at which time their landings decreased due to overfishing, and landings of black, green, and white abalone increased. By the late 1970s, however, landings of these species dropped precipitously to be replaced by increased landings of sea urchins. In the 1990s urchin landings, in turn, have decreased and recruitment has declined in some places. Another benthic invertebrate fishery, sea cucumbers, has been increasing since the 1980s however, and may replace urchins if their stocks collapse (Estes and Peterson 2002). For finfish, Pacific rockfish (*Sebastes spp.*) demonstrate a similar pattern for finfish off the North American West Coast, both with respect to species and location (see Fig. 3.6 and chapter 9), New England groundfish, and southeastern U.S. reef fish provide other good examples of serial overfishing.

The impacts of commercial fishing have traditionally garnered much more attention than those of recreational fishing. Historically, this was probably justified because commercial landings and impacts greatly exceeded recreational ones in most areas for most fisheries. As coastal populations have grown, so has the number of recreational and subsistence fishers and their catch, often bringing them in conflict with commercial fishers. In the Florida Keys for instance, recreational fishing effort, measured in terms of the number of registered recreational fishing boats, increased more than threefold between 1965 and 1993 (Ault et al. 1997). This situation is not unique to the Florida Keys, however. In the United States alone, there were over 17 million recreational anglers who made over 68 million fishing trips in 1997 (NMFS 1999b). Recent polling in the United States indicates that the public continues to believe that recreational fishing poses little threat to marine fish and ecosystems (Sea Web

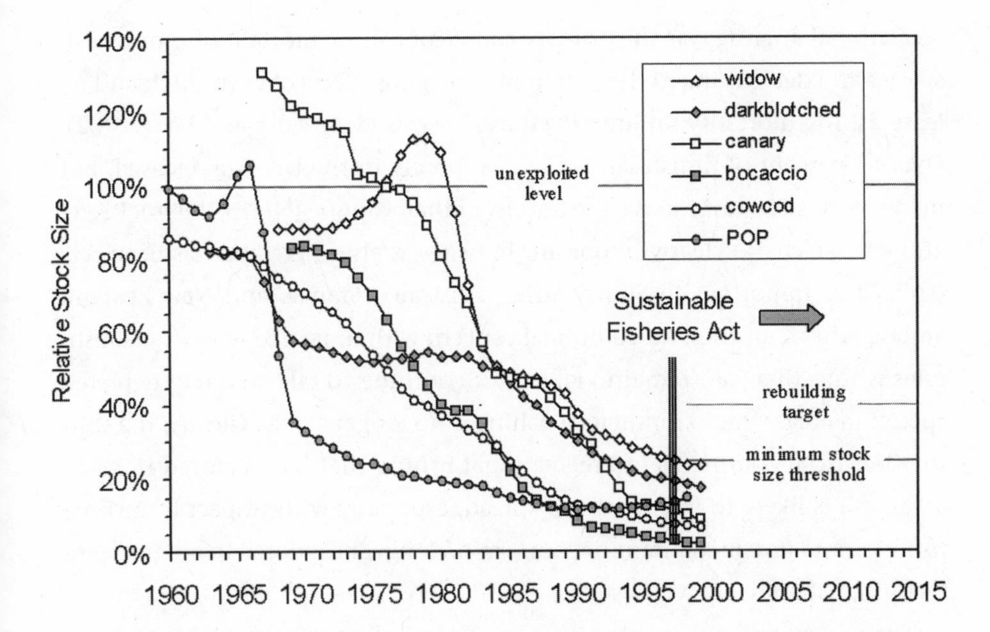


**FIG. 3.6 Serial Depletion of Pacific Rockfish off western North America.** Graph shows long-term depletion of Pacific rockfish (*Sebastes* spp.) spanning at least four decades. Some heavily targeted species declined earlier (e.g., Pacific ocean perch, *Sebastes alutus*) and others more recently (e.g., widow rock fish, *Sebastes entomelas*). None were considered overfished in 1970, but all species shown here were overfished by 2000. Source: Steve Ralston, NMFS, 2002.

2002). Nonetheless, the relative landings and impact of recreational versus commercial ones has shifted dramatically for many species in many locations, as commercial landings have stagnated or declined and recreational ones have increased. Recreational and subsistence fishing can have a substantial impact on fish stocks by significantly reducing abundance, size, and biomass of recreational species (Ault et al. 2001; Eggleston and Dahlgren 2001; Johnson et al. 1999; Schroeder and Love 2002).

The impact of high concentrations of recreational fishers can be great. In the United States, within several states (e.g., California, Florida, New Jersey) and regions (e.g., the Southeast), many, if not most, inshore fisheries are now dominated by recreational, not commercial, take (Fig. 3.7; NMFS 2002; Schroeder and Love 2002; TOC 2002). Chapter 1 touched briefly on the importance and impact of the Florida Keys' two-day, recreational-only, spiny lobster (*Panulirus argus*) "mini-season" that can reduce legal-sized lobster populations by more than 80 percent in inshore areas (Eggleston and Dahlgren 2001). A recent review of impacts to California nearshore fish concluded that recreational angling was the primary source of fishing mortality for most fish species and demonstrated the potential damaging effects of even catch and release fishing mortality on long-lived fish species (Schroeder and Love 2002). Over 55 percent of finfish landed recreationally in the U.S. are released, but many of those do not survive the trauma of their capture (NMFS 1999b). Recreational take is also clearly important in fisheries elsewhere and has been recognized as important in South Africa, Australia, France, and New Zealand among other countries. An additional concern with respect to recreational fisheries is that they can remain viable and continue to take severely depleted species in areas where commercial fishing is no longer viable. Clearly, the shift in effort from commercial to recreational fishing that has occurred in some areas and is likely to continue and spread, especially with respect to inshore fisheries, and its concomitant impact, can no longer be ignored in order to protect fish, fisheries, and ecosystems.

The related and overlapping impacts of bycatch and incidental take, habitat disturbance and alteration, and destructive fishing gear and practices are also all of increasing concern (Auster 1998; Dayton et al. 1995, 2002; Thrush et al. 1998; Watling and Norse 1998). As the geographic scope of fisheries spreads, natural refuges disappear and areas free from these and other fishing impacts are rapidly vanishing. As fishing intensity increases, fish size and abundance decrease, and more powerful, but less selective, gear is frequently employed to make up for the scarcity of targeted species and commonly results in increased bycatch mortality, incidental take, habitat damage, and use of destructive gear and practices. In addition to altering marine ecosystems directly and indirectly by removing target species, fishing can also impact marine ecosystems through its direct impact on nontarget stocks and habitats. In most fisheries, nontarget species or undersized individuals of target species are incidentally captured and often killed because the fishing gear or techniques used do not efficiently select for the target stock. Between 18 and 40 million tons of fish, equivalent to roughly 20 percent of the world's marine harvest, is discarded as bycatch each year (FAO 1997).

These impacts are particularly problematic for certain gear types. Trawls, dredges, and other mobile gears that are dragged across the seafloor are particularly nonselective and damaging to habitat. The Gulf of Mexico and other shrimp trawl fisheries may employ the most wasteful fishing gear in the world. Conservative gulf-wide estimates (Branstetter 1997, 2002) suggest that, on average, for every pound of targeted shrimp produced in this fishery, four pounds of unwanted "trash" species or bycatch is killed, but the ratio may be as high

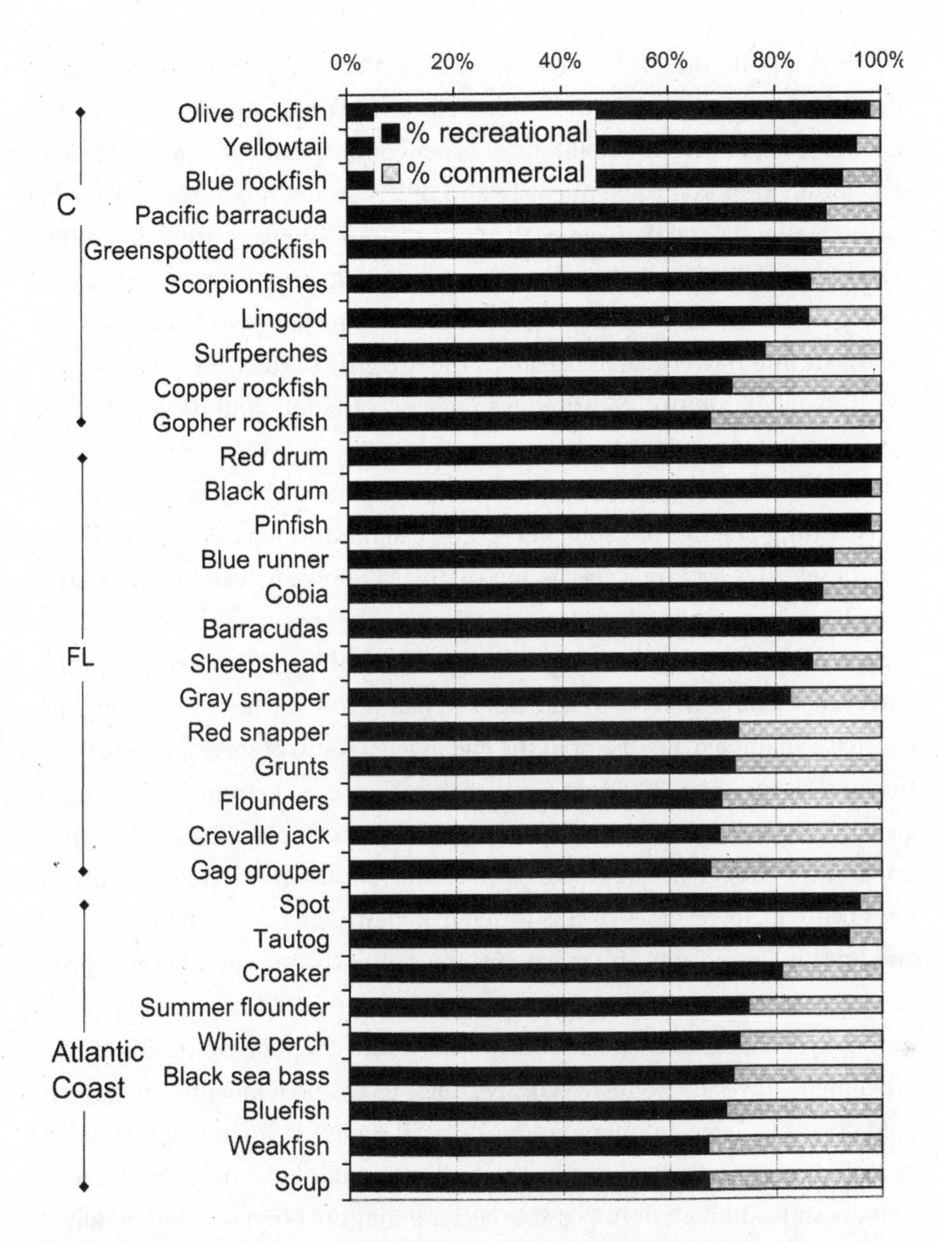


**FIG. 3.7 Significance of Recreational Take in Some U.S. Fisheries.** Graph shows growing significance and dominance of recreational take for many inshore fish species in California (C), Florida (FL), and along the Atlantic coast where recreational effort is focused. Source: Data from NMFS at (http://www.st.nmfs.gov/st1/).

as 10:1. Because shrimp grounds and critical nursery habitats for juvenile fish often overlap, this fishery strongly impacts many fish species. Some species killed as bycatch play key roles in ecosystem health, support valuable fisheries, and/or are endangered and therefore of special concern. The Gulf of Mexico red snapper (*Lutjanus campechanus*) plays key roles in the Gulf's ecological health, supports one of its most valuable fisheries, and is among its most threatened species due to both directed take and bycatch. An estimated 42 million red snapper juveniles; billions of other commercially, recreationally, and ecologically important juvenile fish; and nearly a billion pounds of fish total perished as collateral damage of the Gulf shrimp fishery in 1997 alone (Dobrzynski et al. 2002). Shrimp bycatch also remains a major threat to the recovery of several threatened or endangered sea turtle species (Crouse et al. 1999). Shrimp trawl bycatch threatens the integrity of the Gulf ecosystem, alters its fish community, contributes to overexploitation, and prevents the recovery of endangered species.

Trawl fisheries are not the only gear type where bycatch is a problem; nearly every fishing practice has some level of bycatch. Most nets, trawls, and traps are not selective for target species but catch a variety of fish and invertebrates. Long lines commonly catch a diverse assortment of nontarget species, including marine mammals, sea turtles, billfish, and other highly valued fish species, and even sea birds that try to take hooked bait before the lines sink. Long-line bycatch remains a major factor in the endangered status of several sea birds and turtles (Dayton et al. 2002). In some tropical systems, extremely destructive gear and techniques including dynamite, bleach, and poison have widespread impacts affecting many nontarget species and habitats (further discussion follows below). However, even less inherently destructive gear including hook and line and even catch and release fishing can have important bycatch problems. Bycatch is often problematic in multispecies fisheries, where a particular gear type takes many species. Such fisheries are common on coral reefs and other highly diverse systems. Aggressive, high-level, and frequently endangered predators (e.g., large groupers and rockfish) in such systems are especially vulnerable. Protecting such species using conventional fishery management techniques can be difficult or impossible because they are often caught and killed, even when legally protected, if any fishing is permitted in areas where they are present. Even if the take or possession of a particular species is illegal because it is overfished or endangered, it may be captured and inadvertently killed during fishing for other species. Even catch and release fishing can be harmful, especially in deeper waters, where the rapid change in pressure can cause severe trauma and damage internal organs or greatly increase mortality after release. Thus, even catch and release recreational fishing may be harmful to nontargeted marine life.

Incidental catch or bycatch of nontarget species may be so severe that it threatens such species with extinction. For example, the barndoor skate (*Raja laevis*), a large skate found in the North Atlantic, is incidentally taken in the trawl fishery for cod and other "groundfish" of New England and Eastern Canada. Since the 1950s the abundance of this species has seen a steady decline to the point where it teetered on the verge of extinction (Casey and Myers 1998). More recently, the U.S. government listed the smalltooth sawfish (*Pristis pectinata*) as endangered under the Endangered Species Act (ESA), after a government review implicated bycatch in net fisheries as the primary cause of its dramatic, century-long decline (see also chapter 1; Dayton et al. 2002; NMFS 2003b).

Many of the same fishing practices that result in high rates of bycatch are also destructive to marine habitats that are essential to the growth and survival of both target and nontarget species. Trawling and dredging, for example can alter characteristics of the substrate and destroy a variety of habitats, including biogenic reef structures built by corals, polychaetes, and oysters; sea grass beds, and bryozoan colonies, as well as physical structures ranging from sand waves to piled boulders (Fig. 3.8; Auster 1998; reviews by Dayton et al. 1995; Watling and Norse 1998). Many of these habitats are important, particularly as nursery areas for juvenile stages of both exploited and nonexploited species, or as adult habitat. It may take many years before reef structures are rebuilt, if they ever are. Problems are compounded because many areas are trawled several times each year, preventing the ecological community on the seafloor from recovering (Table 3.1; Watling and Norse 1998). Such intense trawling is expected to have a severe impact on benthic habitats and communities and prohibit them from recovering. By altering habitats and benthic communities, trawling and other fishing practices may make an area uninhabitable for many species and may have long-term effects on the health of the marine ecosystem.

Equally or even more destructive fishing practices are often employed in coral reef ecosystems. Dynamite and other explosives used to kill or stun fish not only kill both exploited and nonexploited species alike but also destroy the reef itself. Other common fishing techniques used in coral reef systems also destroy the reef structure. Muro-ami is a destructive fishing technique used in the Philippines and elsewhere in the Indo-Pacific that involves the use of weighted scarelines to drive fish into nets. Swimmers driving the fish with scarelines continuously lift the weighted lines and drop them on the reef to scare fish into nets, killing living corals and reducing fragile coral species to rubble (Russ 1991). The use of cyanide, bleach, and other poisons to stun or kill fish and invertebrates on coral reefs can also kill and damage corals and other habitats, as well as other reef residents. Trap fishing can also destroy the reef habitat when

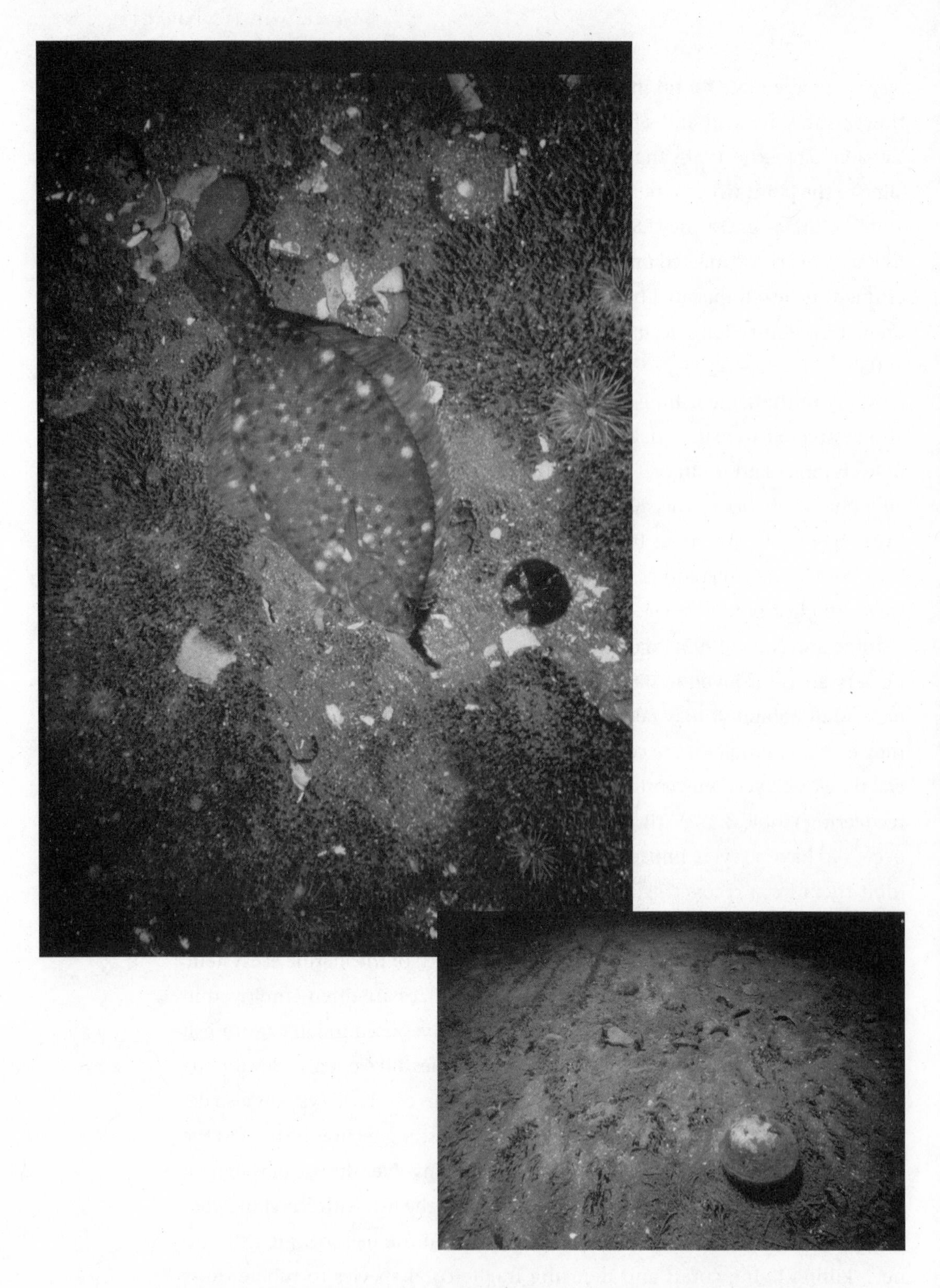

FIG. 3.8 Contrasting images of seafloor and benthic community off New England from (A) before bottom trawling and (B) after bottom trawling demonstrate the capacity of this activity to alter physical structure, biodiversity and habitat. Photos courtesy of Peter Auster, NURC, University of Connecticut.

Table 3.1 Frequency of Bottom Trawling in Different Locations

| Location (area) | Percent Trawled Annually (years) | Reference (original source) |
|---|---|---|
| Limfjord, Denmark | 200 | Riemann and Hoffman 1991 |
| Irish Sea (3 ICES rectangles) | 4, 12, 50 | Kaiser et al. 1996 |
| Southern North Sea | 150–200 | Lindeboom and de Groot 1998 |
| Georges Bank (37,000 km²) | 21 (1970) | Caddy 1973 |
| Georges Bank (40,806 km²) | 200–400 (1976–1991) | Auster et al. 1996 |
| Gulf of Maine (65,013 km²) | 100 (1976–1991) | Auster et al. 1996 |
| Gulf of Maine and Georges Bank (U.S. vessels only) | 0–450 (1993) | Pilskaln et al. 1998 |
| U.S. Continental shelf south of Nantucket and Nantucket Shoals (Individual 30° × 30° grid areas) | Maximum > 400 (1985) | Churchill 1989 |

Adapted from Watling and Norse (1998).

carelessly dropped traps land on coral heads, or when corals are used to weight or camouflage traps (Russ 1991).

The degradation of coral reefs by fishing gear and other threats can result in long-term detrimental effects similar to those discussed previously for overfishing. Loss of living coral and reef structure can reduce reef fish abundance and change coral-dominated communities to algal-dominated ones, particularly when the reef is subject to other stresses. The structure provided by the reef also provides essential habitat for both target and nontarget reef fish species. If fishing destroys the reef, the availability of food and shelter for reef-associated species may change to the point where some species can no longer survive there, and others may suffer dramatic declines in abundance. Because coral recruitment and growth rates are quite low, the ecosystem may not recover from any of these destructive fishing practices for decades or even centuries. The loss of habitat and important species can alter the ecosystem to such an extent that it may not return to its previous state and may be incapable of supporting reef fish assemblages in the future (Russ 1991).

## CHANGING GENETICS

The prior section included a summary of the strong theoretical underpinnings and touched on some of the empirical evidence regarding predicted and observed impacts of fishing to genetic diversity in targeted fish populations and the potential for marine reserves to address these and conserve their genetic diversity. Several examples of documented genetic alteration to marine fish populations from targeted fishing effort follow here to illustrate how important these can be.

Pacific salmon are remarkable and fascinating creatures and provide an ideal model for how intense fishing pressure can interact with specific life history characteristics to alter a species genetic diversity and makeup and affect its evolution. Due to their economic importance and other factors, an unusually strong long-term time series and database exist for their fisheries. Based on this long-term record, Ricker (1981) found that (1) all five species showed decreases in size and age over the time span of the fishery, (2) coho and pink salmon exhibited the greatest changes in size and age, (3) these changes were almost certainly due to the cumulative genetic impacts caused by commercial trolls and gill nets removing fish of larger than average size, and (4) for sockeye and possibly other species, selection increased both the percentage of young, lightly taken "jack" salmon (precocious male salmon that return a year or more earlier to spawn at a smaller size), and the size difference between jacks and "hooknose" (larger males that spend additional time in the ocean and return to spawn at a larger size). Why is this significant? It demonstrates empirically that fishing pressure can alter the genetics of wild fish populations. This becomes more significant when coupled with findings from a separate set of studies focusing on coho salmon reproductive behavior, alternative life history strategies, and intraspecific competition that demonstrate how such changes can impact the long- and short-term evolutionary traits and success of a species.

Anadromous Pacific salmon lay their eggs in freshwater where there young develop, migrate downstream as juveniles for an extended ocean phase, return upstream as adults to spawn, complete their life cycle, and die. Males of Pacific species are dimorphic and also employ complex conditional reproductive strategies. In the specific case of coho salmon, faster growing males tend to mature and return to spawn as smaller jacks after roughly six months at sea, and slower growing males tend to return as larger hooknose males after about eighteen months at sea at a larger size. Coho males must also choose between two alternative modes of reproductive behavior on their freshwater spawning grounds. Most of the larger hooknose males employ a territorial "fighting" strategy battling other males for the premier spawning sites to attract, court, and mate with females. Most of the smaller jacks employ a "sneaking" strategy, where they lie in wait, hiding near the premier beds, time their move, then rush out at the last moment to fertilize a female's eggs, just before a courting hooknose can. However, both jack and hooknose males can choose either strategy depending on their spawning environment. The jack and hooknose traits are strongly heritable, coevolved in delicate balance based on density-dependent environmental trade-offs, and are sensitive to human impact. Pacific salmon also possess complex, diverse, and highly differentiated genetic patterns and are threatened by a range of human activities throughout their lives.

Documented fishing-induced genetic impacts to salmon include increasing the percentage of smaller jack males spawning, greatly decreasing the average size of spawning males (and females) overall, and decreasing the average size of salmon caught in most fisheries. Predicting with any certainty the long-term impact on these traits and trade-offs is impossible, but it is clear that humans already have and are continuing to profoundly alter these and likely other salmon traits through fishing and other activities. Furthermore, human activity is rapidly altering the marine, estuarine, and freshwater environments on which salmon (and other species) depend. Such alterations can also be expected to produce long-term evolutionary changes in the proportion of males maturing as jacks. Research suggests that in some cases male salmon populations once characterized by large, aggressive, fighting, hooknose males may be transformed into ones dominated by much smaller, precocious jacks. A variety of other outcomes is also likely, and it is entirely possibly that in the future, ocean caught salmon will consist solely of females (Gross 1985, 1991; Gross et al. 1988).

More recently, Conover and Munch (2002) conducted a more carefully controlled laboratory study to examine the impact of size-selective fishing on targeted fish population genetics. They raised six tanks (populations) of the exploited Atlantic silverside (*Menidia menidia*), subjecting two each to three different levels (treatments) of simulated fishing. When the fish reached adulthood, the scientists removed 90 percent of the fish in each treatment group, taking the largest fish from two tanks (large-harvested), the smallest fish from two tanks (small-harvested), and a random sample from the remaining two tanks. The remaining fish in each tank were than allowed to reproduce and the treatments were continued for a total of four generations. Within just four generations (years), the three experimental groups evolved rapidly in different directions in response to the size-selective fishing pressures and showed significant changes and differences in average length, weight, and growth rates. The weight of the fish removed (caught) in successive generations responded similarly, also showing significant changes that ran counter to the imposed size selection. The large-harvested populations became smaller and grew more slowly over time; the small-harvested populations became larger and grew more quickly, and the control group showed no significant change. With respect to catch, the large-harvested fish population started highest, but declined over time. The small-harvested fish population showed the reverse trend, and the control was intermediate and showed no significant change. The authors demonstrated that these shifts were the result of the imposed selection acting on the genotypes favoring faster or slower growth and concluded that fisheries management tools must preserve natural genetic variation to ensure long-term sustainable yield. Perhaps equally important, they demonstrated that levels of fishing similar to those imposed on many wild fish populations, acting on a species sharing many characteristics with ones typically targeted, can rapidly alter the genetic characteristics of that species.

While Conover and Munch were busy with their carefully controlled experiment in New York, another group of researchers focused modern analytical techniques on a rare opportunity to investigate the historical fishing impacts on the genetics of wild fish populations half a world away in New Zealand. Hauser et al. (2002) studied the genetic effects of fishing-induced reductions in biomass on two populations of New Zealand "snapper" (*Pagrus auratus*) (actually a Sparid or porgy), by analyzing long-term historical scale collections using microsatellite techniques. In northern New Zealand, the Hauraki Bay snapper fishery developed in the mid-1800s and shares a similar history to many other major world fisheries. This fishery and the yield from it developed and grew slowly from its inception through the 1970s, followed by a rapid acceleration in growth, the introduction of pair-trawls since then, and a total reduction in the spawning stock of nearly 90 percent. The southern New Zealand, Tasman Bay, snapper fishery was virtually unexploited prior to about 1950 but has been similarly exploited and reduced since then. The existence of a long-term time series of scale samples for both fisheries dating back to about 1950 and their differential exploitation history enabled researchers to analyze the impacts of fishing between an unexploited and exploited population, and within the same population between preexploitation and post-exploitation phases. They demonstrated that (1) a significant decline in genetic diversity in snapper populations occurred during their exploitation history; (2) effective population sizes were five orders of magnitude smaller than census population sizes from fisheries data; (3) snapper and likely many other exploited marine species are in danger of losing genetic variability that could result in reduced adaptability, population persistence, and loss in productivity; and (4) fishing may have already caused a considerable loss of overall (genetic) biodiversity.

## CONCLUSION

The limitations of human memory and resulting "shifting baseline syndrome of fisheries" (Pauly 1995) contribute to misperceptions about the state of marine resources and consequently exacerbate the threat to them. Although Pauly defined this syndrome for fisheries scientists and managers, it also applies more broadly to the public at large. The syndrome stems from the fact that each generation accepts the species composition, abundance, and sizes they first observe as a natural baseline from which to evaluate changes. However, this baseline represents an already disturbed state. The resource then continues to decline, but the next generation resets their baseline to this newly depressed state. The result in Pauly's words is "a gradual accommodation of the creeping disappearance of resource species, and inappropriate reference points for evaluating economic losses resulting from overfishing, or for identifying targets for rehabilitation measures." He concludes, "Frameworks that maximize the use of fisheries history would help us to understand and overcome . . . the syndrome . . . and to evaluate the true social and ecological costs of fishing." The syndrome may also be termed "serial depletion of memory," to link it to one of its consequences, "serial depletion of fisheries."

For more than a century, many issues and problems concerning fish, fishing, and fisheries have surfaced and resurfaced, often with remarkably little change in the issues or management effectiveness. Certainly, fisheries management and repeated efforts to reform it have remained controversial throughout that period and longer. The controversial nature of these issues and the strong and vocal opposition to change among at least a segment of the fishing industry and fisheries management community has resulted in political inertia with respect to implementing fundamental change or truly innovative approaches and resulted in a tendency to favor the status quo. As Machiavelli noted, at about the time the groundfish industry was advancing toward the East Coast of North America:

> It must be considered that there is nothing more difficult to carry out, nor more doubtful of success, nor more dangerous to handle, that to initiate a new order of things. For the reformer has enemies in all those who profit by the order, and only

> lukewarm defenders in all those who would profit by the new order, this lukewarmness arising partly form fear of their adversaries, who have the laws in their favor; and partly from the incredulity of mankind, who do not truly believe in anything new until they have had actual experience of it. Thus, it arises that on every opportunity for attacking the reformer, his opponents do so with the zeal of the partisan, the others defend him half-heartedly, so that between them he runs great danger. (Machiavelli, 1525)

Our management of fisheries and ecosystems must be reformed. That much is clear. What remains is to determine when this will happen and how best to do it.

The times appear to be changing and the time may be now. Several trends in our scientific awareness with respect to fishing impacts are worth noting briefly here. There is growing recognition and evidence that

1. the most extreme consequences of fishing and other threats to marine species, endangerment, extirpation, and extinction are no longer remote possibilities but real and growing risks that will become realities if we stay the course;
2. fishing impacts have already contributed greatly to the collapse of some coastal and marine ecosystems, continue to be among the most important threats to these and increasingly all marine ecosystems, and act synergistically with other human and natural stressors; consequently fishing must be managed in a broader ecosystem context;
3. fishing can and has already altered genetic characteristics of some targeted species and populations and that this constitutes a growing and serious problem;
4. traditional fisheries management has often failed to achieve even its original goals and is poorly suited to address new and growing concerns, including ecosystem protection;
5. incidental-take, bycatch, and destructive fishing gear are critically important impacts that need better solutions, but it is also critical to address the ecological impact of fish removal by any gear, which is often fartherreaching;
6. recreational fishing, often viewed as benign, is already a significant and growing impact in many coastal ecosystems;
7. improved ecosystem management, including the use of marine reserves and other MPAs, is needed to address existing and growing threats to both marine ecosystems and fisheries.

There is reason for cautious optimism. One reason is the potential merging of the historically disparate pathways of environmental or ecosystem protection and fisheries management as segments within both recognize the need to move toward more ecosystem-based management, including the use of marine reserves, to accomplish the goals of both. The rising chorus of diverse voices calling for both an ecosystem-based approach to management and increased use of marine reserve networks provides some hope. The potential to provide enhanced ecosystem protection, improved fisheries, and benefits to fishers and nonconsumptive interests strengthens this hope. Finally, recent successes in designing and establishing marine reserves, the increasingly strong track record of existing reserves, and the improved documentation of these reserve successes discussed in the remainder of this book bode especially well for the future.

## REFERENCES


Ault, J. S., J. A. Bohnsack, and G. A. A. Meester. 1997. A retrospective (1979–1996) multispecies assessment of coral reef fish stocks in the Florida Keys. *Fishery Bulletin* 96:395–414.

Ault, J. A., S. G. Smith, J. Luo, G. A. Meester, J. A. Bohnsack, and S. L. Miller. 2001. *Baseline Multispecies Coral Reef Fish Stock Assessments for the Dry Tortugas Final Report.* Miami, FL: University of Miami Rosenstiel School of Marine and Atmospheric Science.

Auster, P. J. 1998. A conceptual model of the impacts of fishing gear on the integrity of fish habitats. *Conservation Biology* 12(6):1198–1203.

Auster, P. J., R. J. Malatesta, R. W. Langton, L. V. P. C. Watling, C. L. S. Donaldson, E. W. Langton, A. N. Shepard, and I. G. Babb. 1996. The impact of mobile fishing gear on seafloor habitats in the Gulf of Maine (Northwest Atlantic): implications for conservation of fish populations. *Reviews in Fisheries Science* 4:185–202.

Baum, J. K., R. A. Myers, D. G. Kehler, B. Worm, S. J. Harley, and P. A. Doherty. 2003. Collapse and conservation of shark populations in the northwest Atlantic. *Science* 299(5605):389–392.

Bjorkland, M. I. 1974. Achievements in marine conservation: marine parks. *Environmental Conservation* 3:205–223.

Boehlert, G. W. 1996. Biodiversity and the sustainability of marine fisheries. *Oceanography* 9:28–35.

Bohnsack, J. A., and J. S. Ault. 1996. Management strategies to conserve marine biodiversity. *Oceanography* 9:72–82.

Branstetter, S. 1997. Status of research leading to the reduction of unwanted bycatch in the shrimp fishery of the southeastern United States in *Fisheries Bycatch: Consequences and Management.* Proceedings of the Symposium on the Consequences and Management of Fisheries Bycatch. Alaska Sea Grant College Report No. 97–02. Fairbanks, Alaska: University of Alaska, Fairbanks.

———. 2002. *Report to Gulf of Mexico Fishery Management Council on Shrimp Bycatch.* National Marine Fisheries Service, St. Petersburg, Florida.

Bruton, M. N. 1995. Have fishes had their chips? The dilemma of threatened fishes. *Environmental Biology of Fishes* (The Hague) 43(1):1–27.

Caddy, J. F. 1973. Underwater observation on tracks of dredges and trawls and some effects of dredging on a scallop ground. *Journal of the Fisheries Research Board of Canada* 30:173–180.

Casey, J. M., and R. A. Myers. 1998. Near extinction of a large, widely distributed fish. *Science* 281(5377):690–692.

Churchill, J. H. 1989. The effect of commercial trawling on sediment resuspension and transport over the Middle Atlantic Bight continental shelf. *Continental Shelf Research* 9:841–864.

Conover, D. O., and S. B. Munch. 2002. Sustaining fisheries yields over evolutionary time scales. *Science* 297(5578):94–96.

Crouse, D. T. 1999. *The Consequences of Delayed Maturity in a Human-Dominated World.* Musick, J. A. 5410 Grosvenor Ln. Ste. 110 Bethesda, MD: American Fisheries Society.

Dayton, P. K., S. F. Thrush, M. T. Agardy, and R. J. Hofman. 1995. Environmental effects of marine fishing. *Aquatic Conservation: Marine and Freshwater Ecosystems* 5:205–232.

Dayton, P. K., S. Thrush, and F. C. Coleman. 2002. *Ecological Effects of Fishing in Marine Ecosystems of the United States*. Arlington, VA: Pew Oceans Commission.

Dobrzynski, T., C. Gray, and M. Hirshfield. 2002. *OCEANS AT RISK: Wasted Catch and the Destruction of Ocean Life*. Washington, DC: OCEANA.

Dugan, J. E. and G. E. Davis. 1993. Applications of marine refugia to coastal fisheries management. *Canadian Journal of Fisheries and Aquatic Sciences* 50:2029–2042.

Eggleston, D. B., and C. P. Dahlgren. 2001. Distribution and abundance of Caribbean spiny lobsters in the Key West National Wildlife Refuge: Relationship to habitat features and impact of an intensive recreational fishery. *Marine Freshwater Research* 52:1567–1576.

Epperly, S. P., and J. W. Dodrill. 1995. Catch rates of snowy grouper, *Epinephelus niveatus,* on the deep reefs of Onslow Bay, southeastern U.S.A. *Bulletin of Marine Science* 56(2):450–461.

Estes, J. A. and C. H. Peterson. 2000. Marine ecological research in seashore and seafloor systems: Accomplishments and future directions. *Marine Ecology Progress Series*. Vol. 195, pp. 281–289.

Food and Agriculture Organization (FAO) Fisheries Department. 2000. *The State of World Fisheries and Aquaculture 2000.* Rome, Italy: FAO.

———. 2002. *The State of World Fisheries and Aquaculture 2002.* Rome, Italy: FAO.

Food and Agriculture Organization (FAO) Fisheries Department, Fishery Resources Division Marine Resources Service. 1997. *Review of* The State of World Fishery Resources. Rome, Italy: FAO. FAO Fisheries Circular No. 920 FIRM/C920(En).

Friedlander, A. M., and E. E. DeMartini. 2002. Contrasts in density, size, and biomass of reef fishes between the Northwestern and the main Hawaiian Islands: The effects of fishing down apex predators. *Marine Ecology Progress Series* 230:253–264.

Goni, R. 2000. Fisheries effects on ecosystems. In *Seas at the Millennium: An Environmental Evaluation*, C. R. C. Sheppard, ed. Pergamon, Amsterdam, The Netherlands, 118 pp.

Gross, M. R. 1985. Disruptive selection for alternative life histories in salmon. *Nature* 313(5997):47–48.

———. 1991. Salmon breeding behavior and life history evolution in changing environments. *Ecology* 72(4):1180–1186.

Gross, M. R., R. M. Coleman, and R. M. McDowall. 1988. Aquatic productivity and the evolution of diadromous fish migration. *Science* 239(4845):1291–1293.

Hauser, L., G. J. Adcock, P. J. Smith, J. H. B. Ramirez, and G. R. Carvalho. 2002. Loss of microsatellite diversity and low effective population size in an overexploited population of New Zealand snapper (*Pagrus auratus*). *Proceedings of the National Academy of Sciences, USA* 99(18):11742–11747.

Hughes, T. P. 1994. Catastrophes, phase shifts, and large-scale degradation of a Caribbean coral reef. *Science* 265:1547–1551.

IUCN Red List of Threatened Species. 2000. Gland, Switzerland: IUCN.

Jackson, J. B. C. 1997. Reefs since Columbus. *Coral Reefs* 16:23–32.

Jackson, J. B. C., et al. 2001. Historical overfishing and the recent collapse of coastal ecosystems. *Science* 293(27 July 2001):629–637.

Johnson, D. R., N. A. Funicelli, and J. A. Bohnsack. 1999. Effectiveness of an existing estuarine no-take fish sanctuary within the Kennedy Space Center, Florida. *North American Journal of Fisheries Management* 19(2):436–453.

Kaiser, M. J. and B. E. Spencer. 1996. The effects of beam-trawl disturbance on infaunal communities in different habitats. *Journal of Animal Ecology* 65:348–358.

Lessios, H. A., D. R. Robertson, and J. D. Cubit. 1984. Spread of Diadema Mass Mortality through the Caribbean. *Science* 226:335–337.

Lindeboom, H. J. and S. J. de Groot, eds. 1998. The effects of different types of fisheries on the North Sea and Irish Sea benthic eco-systems. Netherlands Institute for Sea Research. NIOZ-Rapport 198–1. RIVO–DLO Report C003/98.

Morgan, L. E., and R. Chuenpagdee. 2003. *Shifting Gears: Addressing the Collateral Impacts of Fishing Methods in U.S. Waters*. Covelo, CA: Island Press.

Murawski, S. A. 2000. Definitions of overfishing from an ecosystem perspective. *ICES Journal of Marine Science* 57(3):649–658.

Murray, S. N., R. F. Ambrose, R. F. J. A. Bohnsack, L. W. Botsford, M. H. Carr, G. E. Davis, P. K. Dayton, D. Gotshall, D. R. Gunderson, M. A. Hixon, J. Lubchenco, M. Mangel, A. MacCall, D. A. McArdle, et al. 1999. No-take reserve networks: Sustaining fishery populations and marine ecosystems. *Fisheries* 24(11):11–25.

Myers, R. A., and B. Worm. 2003. Rapid worldwide depletion of predatory fish communities. *Nature* 423(15 May 2003):280–283.

National Marine Fisheries Service (NMFS). 1990. *Historic Fishery Statistics Atlantic and Gulf States 1879–1989*. Silver Spring, MD: NMFS.

———. 1999a. *Annual Report to Congress on the Status of U.S. Fisheries—1998*. Silver Spring, MD: U.S. Dept. of Commerce, NOAA, NMFS.

———. 1999b. *1999 Accomplishment Report to Congress under the Recreational Fishery Resources Conservation Plan*. Silver Spring, MD: U.S. Dept. of Commerce, NOAA, NMFS.

———. 2001. *Annual Report to Congress on the Status of U.S. Fisheries—2000*. Silver Spring, MD: U.S. Dept. of Commerce, NOAA, NMFS.

———. 2002. Commercial Fisheries and Marine Recreational Fisheries Statistics Survey. Available online: http://www.st.nmfs.gov/st1/. (Accessed December 2003.)

———. 2003a. *Annual Report to Congress on the Status of U.S. Fisheries—2002*. Silver Spring, MD: U.S. Dept. of Commerce, NOAA, NMFS.

———. 2003b. Endangered and threatened species; final endangered status for a distinct population segment of smalltooth sawfish (*Pristis pectinata*). *Federal Register* 68(62):15674– 15680.

Nowlis, J. S., and B. Bollermann. 2002. Methods for increasing the likelihood of restoring and maintaining productive fisheries. *Bulletin of Marine Science* 70(2):715–731.

National Research Council (NRC). 1995. *Understanding Marine Biodiversity: A Research Agenda for the Nation.* Washington, DC: National Academy.

———. 2001. *Marine Protected Areas: Tools for Sustaining Ocean Ecosystem.* Washington, DC: National Academy Press.

Pauly, D. 1995. Anecdotes and the shifting baseline syndrome of fisheries. *Trends in Ecology and Evolution* 10:430.

Pauly, D., V. Christensen, J. Dalsgaard, R. Froese, and F. Torres. 1998. Fishing down marine food webs. *Science* 279:860–863.

Pauly, D., V. Christensen, S. Guanette, T. J. Pitcher, U. R. Sumalia, C. J. Walters, R. Watson, and D. Zeller. 2002. Toward sustainability in world fisheries. *Nature* 418:689–695.

Pilskaln, C. H., J. H. Churchill, and L. M. Mayer. 1998. Resuspension of Sediment by Bottom Trawling in the Gulf of Maine and Potential Geochemical Consequences. *Conservation Biology*. Vol. 12, no. 6, pp. 1223–1229.

Plan Development Team. 1990. The Potential of Marine Fishery Reserves for Reef Fish Management in the U.S. Southern Atlantic.

Polunin, N. V. C. 1983. Marine "genetic resources" and the potential role of protected areas in conserving them. *Environmental Conservation* 10(1):31–40.

Ricker, W. E. 1981. Changes in the average size and average age of Pacific salmon. *Canadian Journal of Fisheries and Aquatic Sciences* 38(12):1636–1656.

Riemann, B. and E. Hoffmann. 1991. Ecological consequences of dredging and bottom trawling in the Limfjord, Denmark. *Marine Ecology Progress Series* 69:171–178.

Roberts, C. M., and J. P. Hawkins. 1999. Extinction risk in the sea. *Trends in Ecology and Evolution* 14:241–246.

Roberts, C., W. J. Ballantine, C. D. Buxton, P. Dayton, L. B. Crowder, W. Milon, M. K. Orbach, D. Pauly, J. Trexler and C. J. Walters. 1995. *Review of the Use of Marine Fishery Reserves in the U.S. Southeastern Atlantic.* National Marine Fisheries Service, Southeast Fisheries Center. NOAA Technical Memorandum NMFS-SEFSC-376, 31 pp.

Russ, G. R. 1991. Coral reef fisheries: Effects and yields. In Sale, P. F. *The Ecology of Fishes on Coral Reefs,* 601–635. San Diego: Academic.

Schroeder, D. M., and M. S. Love. 2002. *On Recreational Fishing and Marine Fish Populations in California.* California Cooperative Oceanic Fisheries Investigations (CalCOFI) Reports.

SeaWeb. 2002. *Results from Survey on Threats to Marine Ecosystems.*

Sumalia, U. R., S. Guanette, J. Alder, D. Pollard, and R. Chuenpagdee. 1999. *Marine Protected Areas and Managing Fished Ecosystems.* Bergen, Norway: Michelsen Institute.

The Ocean Conservancy (TOC). 2002. *Health of the Oceans 2002 Report.* Washington, DC: The Ocean Conservancy.

Thrush, S. F., J. E. Hewitt, V. J. Cummings, P. K. Dayton, M. Cryer, S. J. Turner, G. A. Funnell, R. G. Budd, C. J. Milburn, and M. R. Wilkinson. 1998. Disturbance of the marine benthic habitat by commercial fishing: Impacts at the scale of the fishery. *Ecological Applications* 8:866–879.

Watling, L., and E. A. Norse. 1998. Disturbance of the Seabed by Mobile Fishing Gear: A Comparison to Forest Clearcutting. *Conservation Biology* 12(6):1180–1197.

Watson, R., and D. Pauly. 2001. Systematic distortions in world fisheries catch trends. *Nature* 414, (6863):534–536.

FOUR

# What Marine Reserves Can Accomplish

A thing is right when it tends to preserve the integrity, stability, and beauty of the biotic community. It is wrong when it does otherwise.

—Aldo Leopold, 1949

Single-species management practices are like automobile maintenance: Action is only taken when a problem occurs or a part breaks. Ecosystem-based management however is likened to airplane maintenance, where the overall goal is to maintain the total system by preventing the failure of all important systems and components (Bohnsack 1998b). Because both ecosystem management and airplane maintenance share a common goal of avoiding crashes, both require (1) understanding how various systems operate, (2) building precautionary margins of safety into all operational systems, and (3) continuous monitoring of all critical elements. Marine reserves are often acclaimed as the marine ecosystem management tool par excellence, and their potential to provide a host of benefits has been repeatedly and positively reviewed (Box 4.1). What is the factual basis for these claims?

Despite their rising popularity, the application of marine reserves to marine conservation and fisheries management remains in its infancy. As the use of marine reserves continues to grow globally, our understanding of their effects on both exploited (i.e., those species that are targeted in commercial and recreational fisheries) and unexploited marine species as well as marine communities and ecosystems has grown rapidly and will continue to do so. Although we are far from understanding the full potential for marine reserves as a management tool, there is overwhelming direct scientific evidence for many beneficial reserve effects, strong and growing evidence for others, and strong indirect evidence, theoretical, and common sense support for still more.

### 4.1 Findings from Selected Scientific Reviews of Marine Reserve Efficacy

The scientific literature on marine reserve efficacy continues to expand rapidly. Over the last decade alone, hundreds of peer-reviewed scientific articles and more than a dozen reviews of that extensive literature have been published. The following excerpts from some of those reviews provide an indication of the high degree of scientific documentation and support for reserves to accomplish stated objectives:

- Marine reserves, regardless of their size, and with few exceptions, lead to increases in density, biomass, individual size, and diversity in all functional groups. The diversity of communities and the average size of the organisms within a reserve are between 20 and 30 percent higher relative to unprotected areas. The density of organisms is roughly double in reserves, while the biomass of organisms is nearly triple. These results are robust despite the many potential sources of error in the individual studies. (Halpern 2003, p. S129)
- Based on evidence from existing marine area closures (i.e., marine no-take reserves) in both temperate and tropical regions, marine reserves and protected areas will be effective tools for addressing conservation needs as part of integrated coastal and marine area management. (NRC 2001, p. 2)
- There is compelling, irrefutable evidence that protecting areas from fishing leads to rapid increases in abundance, average body size, and biomass of exploited species . . . increased diversity of species and recovery of habitats from fishing disturbance . . . in a wide range of habitats . . . ranging from tropical to cool temperate zones. . . . Marine reserves typically lead to at least a doubling in the biomass of exploited species after three to five years . . . [and] can increase [biomass and offspring production] by orders of magnitude over levels in fishing grounds. . . . Even relatively small reserves could produce regionally significant replenishment of exploited populations. (Roberts and Hawkins 2000, p. 16–17)
- Networks of no-take marine reserves can (1) help recover fishery populations; (2) eliminate mortality of nontargeted species within protected areas due to bycatch, discards, and ghost fishing; (3) protect reserve habitats from damage by fishing gear; and (4) increase the probability that rare and vulnerable habitats, species, and communities are able to persist. (Murray et al. 1999, p. 15)
- Reserves will be essential for conservation efforts because they can provide unique protection for critical areas, they can provide a spatial escape for intensely exploited species, and they can potentially act as buffers against some management miscalculations and unforeseen or unusual conditions. Reserve design and effectiveness can be dramatically improved by better use of existing scientific understanding. (Allison et al. 1998, p. S79)
- The benefits that can reasonably be expected from an appropriate system of marine reserves are extensive and substantial. Many have been repeatedly documented and conclusively established at a number of existing reserves in a variety of environments. (Sobel 1996, p. 16)
- There is overwhelming evidence from both temperate and tropical areas that exploited populations in protected areas will recover following cessation of fishing and that spawning stock biomass will be rebuilt. (Roberts et al. 1995, p. 5)
- Marine reserves commonly support higher densities and larger sizes of heavily fished species than are found outside reserves. (Rowley 1994, p. 233)
- Evidence from existing marine reserves indicates that increased abundance, individual size, reproductive output, and species diversity occurred in a variety of marine species in refuges of various sizes, shapes, and histories in communities ranging from tropical coral reefs to temperate kelp forests. (Dugan and Davis 1993a, p. 2029)

- Although there is overwhelming evidence of increases in the abundance, size, and biomass of exploited species in protected areas, many of the outcomes of diversity and species composition are not predicted. (Jones et al. 1992, p. 29)
- It has now been well established that the abundances of and average sizes of many larger carnivorous fishes increase within protected areas. Smaller fishes and species from different trophic levels show similar patterns where they are targeted by fishermen. (Roberts and Polunin 1991, p. 82)

Despite this substantial evidence, it will be a long time, if ever, before one can conclusively prove that marine reserves will improve fish catches overall to everyone's satisfaction. Scientific study of marine reserves is complicated by many factors, including:

- "control" and replication issues (both spatial and temporal),
- natural environmental and recruitment variability,
- the complicating effects of other management measures, and
- changes in fishing patterns or effort external to the reserves.

Even under these constraints, and considering how few and how small existing reserves generally are compared to the size of fished areas, science has made much progress in demonstrating reserves' potential to provide fishing and other benefits beyond their borders. As the evidence for reserve benefits continues to mount, peoples' comfort levels with reserves continue to grow. Furthermore, as reserve size, number, and design improve, the evidence for external fisheries benefits will also continue to grow. There is much more to reserve benefits than just fishery benefits, and they may not even be the most important, but clearly demonstrable fishery benefits would likely aid stakeholder acceptance.

## THE SCIENTIFIC EVIDENCE

Among the more recent, comprehensive, and rigorous reviews of marine reserve impact, Halpern and Warner (2002) carefully analyzed published results from eighty-nine separate studies spanning the globe. They reviewed only studies that (1) involved strictly no-take marine reserves; (2) provided survey data from before and after reserve creation or from inside and outside the reserve, or both; (3) addressed one or more of four key biological measures (density, biomass, size, and diversity of organisms); and (4) focused on reserves where no known harvesting had occurred. The results provide clear, strong, and unequivocal evidence ($p < 0.001$) that marine no-take reserves generally produce positive changes or increases in overall density, size, and biomass of organisms, and increase biodiversity, as measured by species richness, within their boundaries. Despite the recognized limitations in available data and in their comparability, the qualitative and quantitative analyses performed leave little doubt regarding the positive impact of reserves on these biological parameters.

The Halpern and Warner study went further than prior work in estimating mean changes to these parameters at marine reserves and provides an indication of the magnitude of the changes one can expect within them. The study concludes that reserves appear on average to roughly double the density, triple the biomass, and increase organism size and diversity 20 to 30 percent compared to controls. A similar analysis restricted only to peer-reviewed studies demonstrated even greater increases (Fig. 4.1; PISCO 2003). Such averages are useful, but should be used with caution given variability among sites and species.

The burgeoning marine reserves literature identifies numerous proposed, potential, and documented benefits for reserves. Most potential benefits fall under one of four broad categories: (1) protect ecosystem structure, function, and integrity; (2) improve fisheries; (3) expand knowledge and understanding of marine systems; (4) enhance nonconsumptive opportunities. These categories should be considered overlapping rather than discrete. For example, protecting ecosystem structure, function, and integrity clearly contributes to and overlaps with all three of the other broad categories. Likewise, the enhanced understanding and knowledge of marine systems stemming from marine reserves and their study are essential to or contribute to each of the other broad categories. Box 4.2 provides a list of approximately fifty identified benefits, classified under these four broad headings. These benefits were identified by marine scientists and other reserve experts as ones that can be reasonably expected to result from an effective system of marine no-take reserves complemented by other management measures. In some cases, the degree of benefit may vary for individual species and be dependent on both their life histories and the reserve design (Bohnsack 1998a; Sobel 1996).

In the sections that follow, we review the experience with marine reserves and their efficacy for achieving specific benefits under these four broad goals as well as the mechanisms by which marine reserves produce these benefits.

## PROTECTING ECOSYSTEM STRUCTURE, FUNCTION, AND INTEGRITY

As an integral component of ecosystem-based management, one of the most common goals of marine reserves and marine reserve networks is to preserve

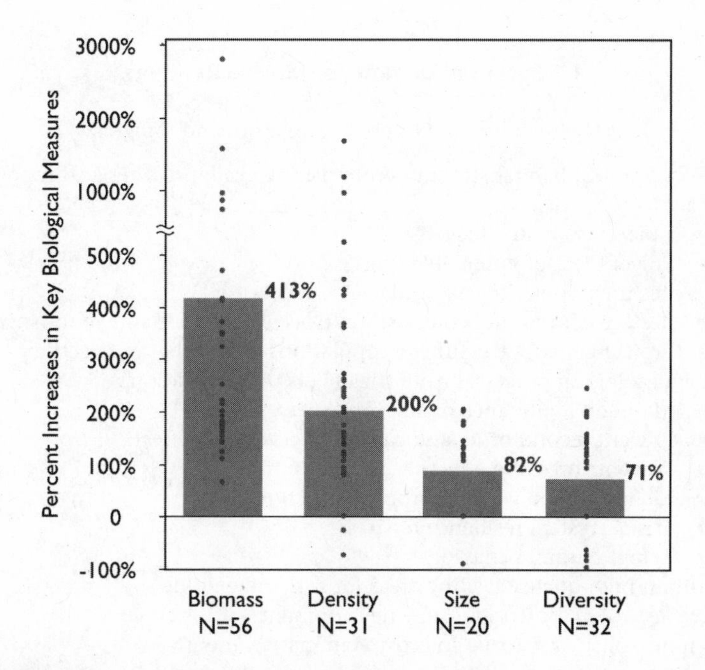


**FIG. 4.1 Quantitative Analysis of Marine Reserve Impact on Key Biological Measures.** Graph shows average percent increase in biomass (+413 percent), density (+200 percent), size (+82 percent), and diversity (+71 percent) within marine reserve borders based on analysis of results from peer-reviewed studies on marine reserves spanning the globe. Source: Adapted and reprinted with permission from PISCO 2003. Data courtesy of Halpern 2003 and Palumbi 2003.

ecosystem properties or characteristics and the processes that control or regulate them. This is true of both the ecosystems defined within the reserves and the larger ones of which they are a part. Within marine reserves, natural processes that preserve ecosystem structure have the opportunity to exercise their influence over the distribution and abundance of species, the interactions among species within an ecological community, and the pathways through which matter and energy flow.

Among the most commonly documented effects that reserves have on marine ecosystems are those in which the abundance, density, population structure, or composition of exploited species is changed within them. Because marine reserves prohibit all forms of fishing and other extractive activities within their boundaries, targeted species benefit the most from reserve protection. By eliminating fishing and other forms of exploitation from an area, a significant source of mortality is removed, and a greater number of target or exploited fish and invertebrates survive. Because extraterritorial marine reserve effects are contingent on changes occurring within the reserve it is also expected that these would be observed first.

**4.2 Potential Benefits of Marine Reserves**

Protect Ecosystem Structure, Function, and Integrity

- Protect physical habitat structure from fishing gear and other anthropogenic impacts
- Protect biodiversity at all levels
    - Prevent loss of vulnerable species
    - Restore population size and age structure
    - Restore community composition (species presence and abundance)
    - Protect genetic structure of populations from fisheries selection
- Protect ecological processes from the effects of exploitation
    - Maintain abundance of keystone species
    - Prevent second-order and cascading ecosystem effects
    - Prevent threshold effects
    - Maintain food web and trophic structure
    - Ensure system resilience to stress
    - Retain natural behaviors
- Maintain high-quality feeding areas for fish and wildlife
- Leave less room for irresponsible development
- Promote holistic approach to ecosystem management

Enhance Nonconsumptive Opportunities

- Enhance and diversify economic opportunities
- Enhance and diversify social activities
- Enhance personal satisfaction by
    - Improving peace of mind for naturalists, conservationists, and other passive users
    - Enhancing aesthetic experiences
    - Enhancing spiritual connection to natural resources
- Enhance nonconsumptive recreational activities
- Create opportunities for wilderness experiences
- Enhance educational opportunities
- Promote ecotourism
- Create public awareness about environment
- Build conservation ethic
- Increase sustainable employment opportunities
- Stabilize the economy

Improve Fisheries

- Increase abundance of overfished stocks
- Reduce fishing for vulnerable species
- Reduce bycatch and incidental fishing mortality
- Simplify enforcement and compliance
- Reduce conflicts between users
- Provide resource protection with little data and information needs
- Enhance reproduction
    - Increase spawning stock biomass
    - Increase spawner density
    - Provide undisturbed spawning sites
    - Increase spawning potential and stock fecundity
    - Provide export of eggs and larvae

- Enhance fisheries via spillover of juveniles and adults
- Provide insurance against management failure
- Accelerate recovery after stock collapse
- Support trophy fisheries
- Provide data for improved fisheries management
- Increase understanding of management
- Facilitate stakeholder involvement in management
- Protect population genetics and life history characteristics from selective fishing
- Enhance recruitment

Expand Knowledge and Understanding of Marine Systems

- Foster understanding of natural systems
- Provide long-term monitoring sites free of human impacts
- Provide continuity of knowledge in undisturbed sites
- Reduce risk to long-term experiments
- Provide experimental sites needing natural areas
- Provide focus for study
- Enhance synergies from cumulative studies at one site over time• Provide undisturbed natural references for studies of fisheries and other anthropogenic impacts
- Allow studies of natural behaviors
- Provide natural sites for education

Adapted from Sobel 1996 and Bohnsack 1998a.

Studies of marine reserves of all shapes and sizes, in a variety of temperate and tropical marine systems, have documented a greater relative abundance or density of exploited fish and invertebrate species within reserves than outside reserves in areas subject to fishing (Fig. 4.1 and 4.2). Although few individual studies of this sort can conclusively demonstrate that observed differences are due solely to the protection afforded by the reserve rather than to other factors that may differ between reserve and nonreserve areas (e.g., habitat quality), the large number of studies involving different species from different places that show the same pattern of higher densities in reserves provides strong evidence that reserve protection is responsible for observed differences (Dugan and Davis, 1993a; Halpern and Warner 2002; NRC 2001; Roberts and Hawkins 2000; Roberts et al. 1995; Ward et al. 2001).

Considerable evidence that marine reserves are capable of increasing the abundance or density of exploited species is also provided by studies that have examined the change in the abundance or density of exploited species in an area over time before and after a reserve has been created. These studies typically show an increase in the abundance of exploited species within the reserve area for several years following its designation. As with spatial comparisons of species abundance or density within versus outside reserves, comparisons of

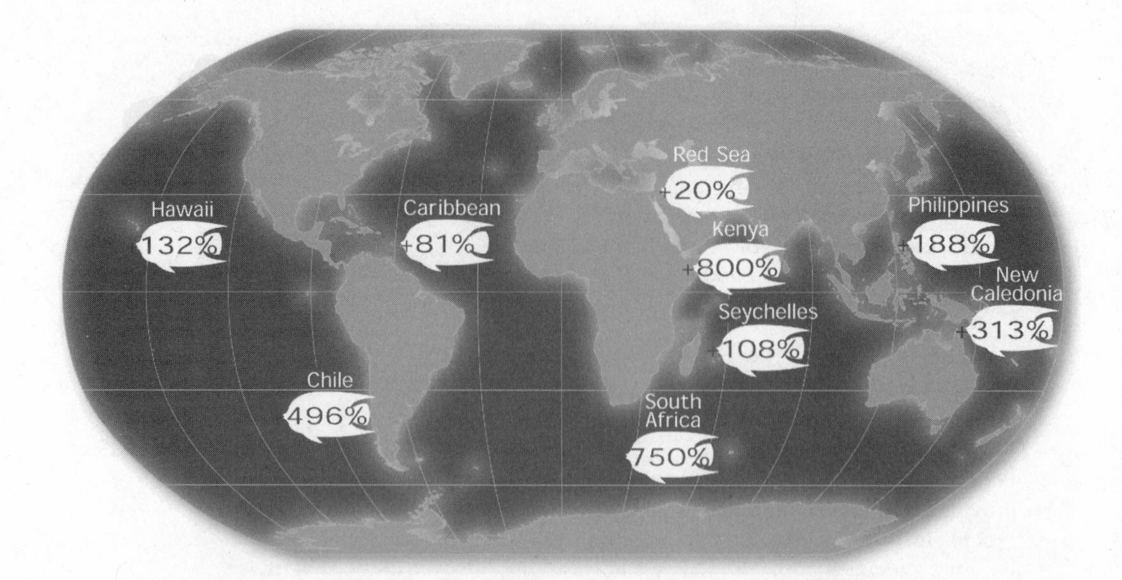


**FIG. 4.2 Average Increase in Fish Biomass within Marine Reserve Borders by Location.** Around the world, marine reserves have demonstrated the ability to increase fish biomass inside their boundaries. The numbers on this map show average increases of fish biomass inside marine reserves after the reserves were established for nine locations based on quantitative analysis of 32 peer-reviewed marine reserve studies. Source: Jerome N. Cookson based on data from Halpern and Warner (2002).

abundance or density before and after the implementation of reserve protection may be confounded by other factors. In this case, factors that vary over time in both reserve and nonreserve areas (e.g., recruitment or the arrival of new individuals to an area) may cause the observed changes in density, rather than the protection from exploitation within the reserve. Even with such limitations, the collective weight of evidence from the large number of studies demonstrating such increases is compelling (Dugan and Davis 1993a; Halpern and Warner 2002; NRC 2001; Roberts and Hawkins 2000; Roberts et al. 1995; Ward et al. 2001).

Studies in which density of an exploited species greatly increased over time within a reserve but did not increase substantially outside of the reserve support the case that the protection offered by the reserve is responsible for observed increases in density. For example, in a reserve that encompasses rocky intertidal habitats along the coast of Chile, the density of a harvested gastropod, the loco (*Concholepas concholepas*), increased steadily after an area was fenced off to create a marine reserve. The increase in this predatory gastropod led to a trophic cascade discussed later in this section that transformed the entire local community. In areas outside the reserve, however, there were virtually no increases in loco density (See later this section and chapter 11; Castilla et al. 1994; Durán and Castilla 1989).

Similarly, studies that show reversible changes in abundance following the creation of marine reserves and their subsequent reopening to fishing strengthen the case that observed changes in abundance are due to reserve protection. In the Philippines, this pattern is evident in the abundance of exploited fish over time following the creation of two marine reserves and the subsequent reopening and closing of one of the reserves to fishing (Fig. 4.3 and chapter 11). In the reserve that was continuously closed, there was a steady 12-fold increase in the density and a 17-fold increase in the biomass of large predators over a ten-year period, and increases appeared to be continuing at a steady rate (Russ and Alcala 1996, 1998, 1999, 2003). Clearly, such a shift in the predator guild would likely also have impacts on community structure, predation, competition, energy flow, and other ecosystem properties.

In addition to increasing the abundance or density of exploited species, marine reserves have often been shown to increase the average size of exploited species or to increase the proportion of large individuals in the population. For example, a study of spiny lobster (*Jasus edwardsii*) populations inside and outside a New Zealand marine reserve found that both density and mean size of lobsters increased within the marine reserve for several years, and that the largest lobsters and most lobsters beyond the legal size limit were found within the marine reserve. Female density and size continued to increase throughout the study, though male increases leveled off, probably due to migrations of large males outside of the reserve where they were subject to fishing (MacDiarmid and Breen 1992). Another study of spiny lobster in four New Zealand marine reserves compared populations within the reserves to similar sites nearby, and conservatively estimated increases in density of 3.9 percent per year and 9.5 percent per year in shallow and deep water sites, respectively, with a similar but somewhat higher rate of increase for biomass and egg production (Fig. 4.4; Babcock et al. 1999; Kelly et al. 2000).

Given the ecosystem-focused spiny lobster discussion in chapter 1, how did these changes impact the surrounding ecosystem? Babcock et al. (1999) studied this question and found that, in addition to the changes in the lobster population,

1. abundances of the most common demersal predatory fish, the New Zealand snapper (*Pagrus auratus*) (also discussed in chapter 3), large enough to feed on urchins were six to nine times more abundant within two of the marine reserves than outside;

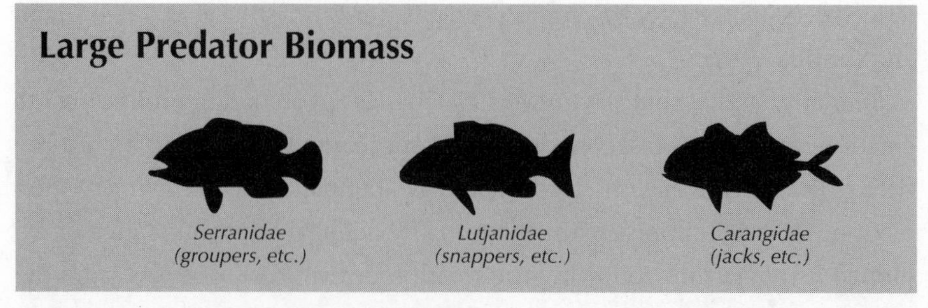


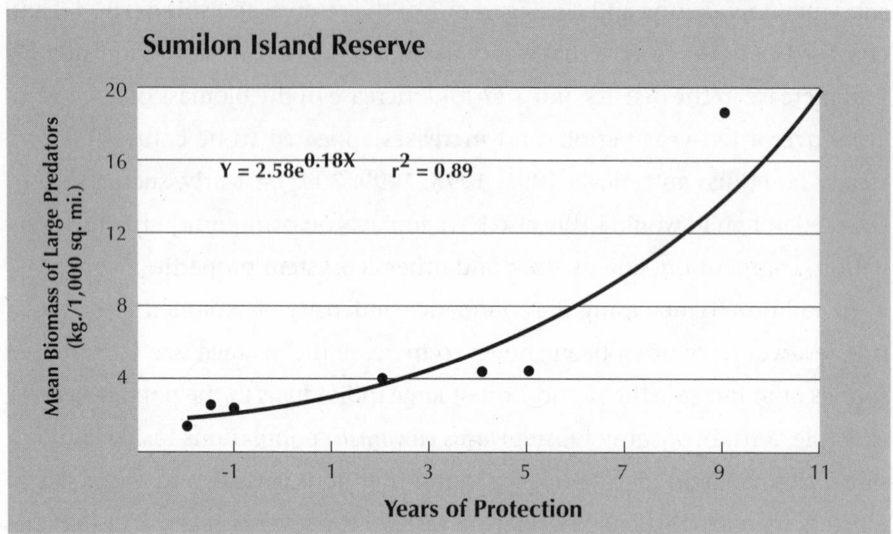


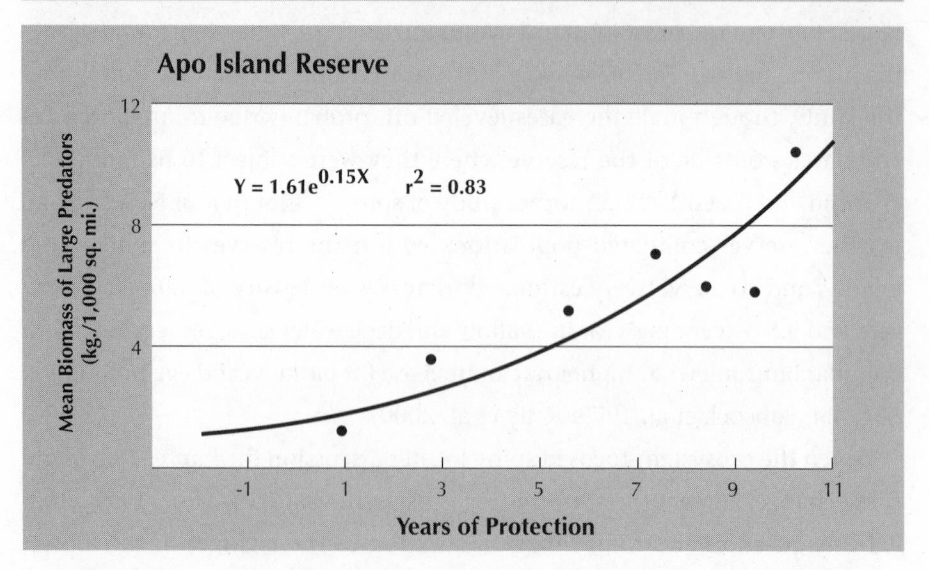


**FIG. 4.3 Changes in Biomass of Large Predators within Two Philippine Marine Reserves (Sumilon Island and Apo Island Marine Reserves) Following Closures versus Length of Time Protected.** Large predator biomass continues to increase exponentially in both reserves and had increased more then 17-fold at Apo following 18 years of protection. Source: Adapted with Permission from Russ and Alcala 2003.

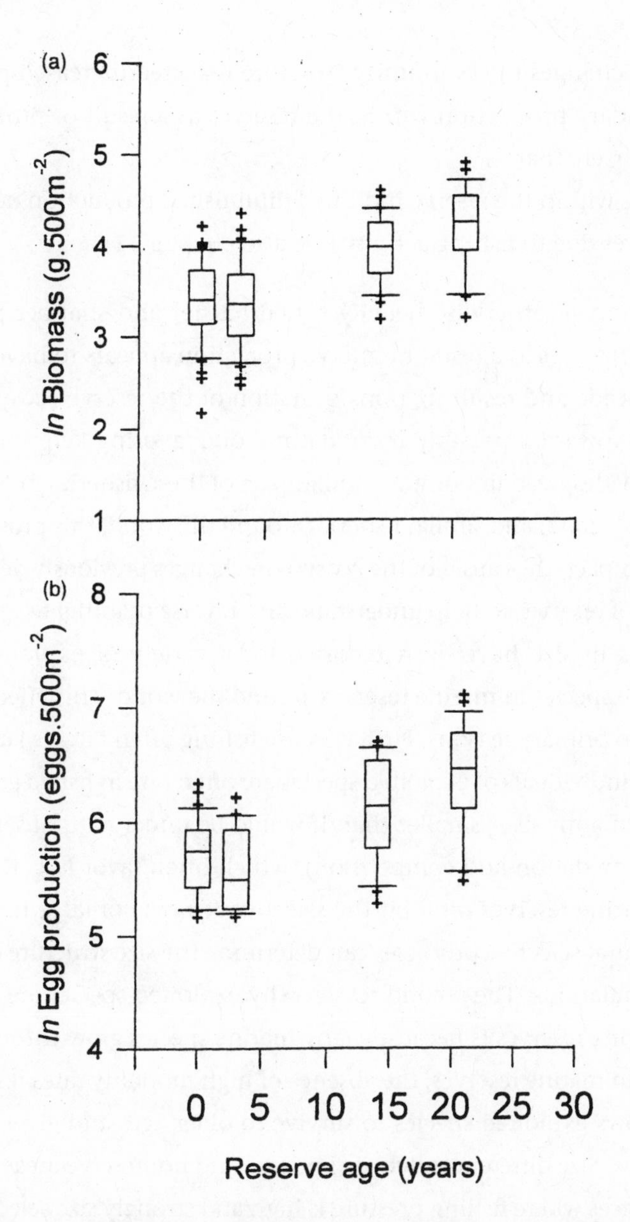


**FIG. 4.4 Changes in Spiny Lobster Biomass and Egg Production in New Zealand Marine Reserves Following Protection.** Boxplot using natural log of (A) biomass and of (B) egg production versus reserve age to compare marine reserves of different ages. Source: Kelly et al. 2000, reprinted with permission from *Biological Conservation.*

2. sea urchin (*Evechinus chloroticus*) abundance had declined in areas frequented by it in one of the reserves from 4.9 to 1.4/m$^2$ since its creation;
3. kelp forest cover had increased within the reserves since their creation;
4. urchin dominated barrens covered only 14 percent of available reef area in the reserve compared to 40 percent outside; and concluded that

5. these changes in community structure reflected increased primary and secondary production within the reserves as a result of protection; and conversely that
6. trends within the reserve indicated diminished production outside of the reserves due to fishing activity (see also chapter 11).

Restoration of otherwise heavily fished lobster and snapper populations within the reserves as a result of reserve protection appears to have triggered a trophic cascade and resulting transformation of the reserve ecosystem, shifting it back toward a possibly more natural one favoring kelp forest habitat. Given the widespread importance and impact of these fisheries in New Zealand (Hauser et al. 2002) and similar fisheries around the world, this provides insight into their impact, the causes of the ecosystem changes previously described, and the ability of reserves to help understand and reverse or mitigate such changes.

Increases in size have been reported for a variety of exploited fish and invertebrate species in marine reserves around the world. This effect on size results for two primary reasons. First, because fishing often targets larger individuals, larger individuals of exploited species are often rare in fished areas, and the mean size of animals is smaller than it would be under natural selective pressures (e.g., predation and competition), which often favor larger individuals. Because marine reserves prohibit the selective removal of large individuals by fishers, natural selective processes can determine the size structure of exploited species populations. The second reason why exploited species are often larger within marine reserves is because many marine species grow throughout their entire life. In marine reserves, the absence of high mortality rates resulting from fishing allows exploited species to survive to older ages and gives them more time to grow. Size differences between reserve and nonreserve areas are likely to occur in places where fishing pressure is high and strongly size selective outside of reserves, but effectively excluded from within the reserve (Chapman and Kramer 1999; Roberts and Polunin 1992). It is also important to note that a reserve's effect on size may take time to become noticeable because species within the reserve must outlive and consequently outgrow their conspecifics in nonreserve areas, and may be confounded by the effects of recruitment and emigration.

Although it is not surprising that the density and sizes of exploited species increase within reserves and are higher than in surrounding areas subject to fishing, these effects have important implications for the ability of reserves to provide specific benefits, including ones related to improving fisheries, protecting ecosystems, and conserving biodiversity. Reproductive output is often size dependent for many exploited fish and invertebrate species, with the number of eggs produced by a female often increasing as a third power function of her length. Red snapper (*Lutjanus campechanus*) provide a good example of this: a single 61 cm female produces the same number of eggs as 212 42 cm females (Figure 4.5; NMFS 1990). Consequently, if large individuals are more common inside a reserve, a relatively small number of such large individuals within a reserve can contribute a disproportionately large amount to reproduction, which may be essential to the viability and recovery of the species.

As discussed in chapter 1, exploitation has reduced the abundance of several species to levels where they are threatened or endangered with extinction. Some of these species are those that suffer from Allee effects, in which reproductive output declines greatly or ceases altogether below a particular density threshold. Many invertebrate species with limited mobility such as abalone, sea urchins, and conch, for example, exhibit Allee effects at low densities (e.g., Levitan et al. 1992; Stoner and Ray-Culp 2000; Tegner 1993). In fished areas, the densities of these species are often reduced below the level required for reproductive success or below some minimum viable population size, reducing their chance of recovery. Within reserves, however, densities of these species may be maintained at a high level, and reproduction within reserves may be essential for the short-term preservation and long-term recovery of the species.

Marine reserves can also preserve other natural aspects of the population structure, life history characteristics, and even genetic composition of populations that have been altered by selective fishing pressure. For example, in several exploited fish and invertebrate species, males and females may grow to different sizes. Because fishing often targets larger individuals, the sex ratio or ratio of males to females may be altered dramatically. For species in which females grow to larger sizes than males, the effect of fishing that removes larger individuals can greatly reduce the number of eggs produced by the population and limit its ability to replenish itself. The same problem occurs for species in which males grow to larger sizes than females and males are selectively removed from the population by fishing. In several important fisheries species fertilization success is directly related to sperm density and motility, both of which may be lower for smaller males. Moreover, smaller males may not be as successful in mating as large males. Thus severe changes in the ratio of males to females may result in a shortage of males and, subsequently, reduced reproductive success (Trippel et al. 1997).

For several exploited species in which males are larger than females, the females can actually change sex and become males as they get older and larger.

## One Large Red Snapper Equals the Egg Production from Nearly Two Hundred and Twelve Smaller Ones

**Size comparison:**

*One large 61 centimeter female weighs 12.5 kilograms and produces 9,520,000 eggs.*

*One small 42 centimeter female weighs 1.1 kilograms and produces 44,000 eggs.*

*Put another way, one large (61 cm) female red snapper can lay 9,520,000 eggs – the equivalent of nearly two hundred and twelve smaller (42 cm) females.*

=

**FIG. 4.5 All Fish Are Not the Same: The Importance of Large Individuals.** A single large female red snapper (*Lutjanus campechanus*) produces as many eggs as 212 smaller ones. Large females of many other species are similarly important. For other species, including certain groupers and spiny lobster, large males can also be important for successful fertilization. Current research is also elucidating the importance of large individuals with respect to quality of eggs and larvae in addition to quantity. Source: Data from NMFS PDT (1990).

This is common for groupers and other reef fish. As fishing selectively removes larger males, females may begin changing to males at smaller sizes. This can severely reduce egg production as the average size of females is reduced, and may reduce the ability of males to fertilize eggs as the average size of males is reduced. Thus fishing that selectively removes larger males may reduce reproductive rates no matter how the population responds. Within reserves, however, protection from exploitation that allows exploited species to survive to larger sizes can maintain sex ratios at natural levels and allow changes in sex to occur at natural sizes. For example, a spawning aggregation of red hind (*Epinephelus guttatus*) was fished to such an extent that the more natural ratio of females to males, approximately 4:1 (which may also have been distorted due to earlier lower levels of fishing), was skewed to as many as 15:1. Closing the spawning area to fishing returned sex ratios to more natural levels after about seven years of protection (Beets and Friedlander 1999). Unfortunately, the more vulnerable Nassau grouper (*Epinephelus striatus*) once shared the same spawning site but has not yet made a similar comeback despite the closure of the area to fishing.

By increasing the abundance and size of exploited species and preserving natural characteristics of their population structure and life history, marine reserves can further protect these species from harmful human impacts. For example, if any characteristics that are altered by fishing are genetically based, changes in characteristics of the population may reflect changes in the genetic composition of the population or a reduction of the genetic variability of the population. Such reduced genetic variability may reduce the resistance or resilience of the species to changes in factors that influence the population. By maintaining natural population structure and life history characteristics, marine reserves may serve as a repository for natural genetic characteristics of the species and may be critical for maintaining the long-term viability of the population.

The galjoen (*Coracinus capensis*) endemic to southern Africa and among its most popular and exploited recreational fish species provides another documented example of human alteration to the genetics of a wild fish population, discovered via a marine reserve. Two separate populations of galjoen exist with apparently limited movement between them. Mark recapture studies conducted within and surrounding the Hoop Marine Reserve strongly suggest that observed dispersal behavior is best explained by genetic differences resulting alternatively in resident versus nomadic behavior. Further, the emigration estimated from the reserve suggests that the 50 km reserve is not only protecting the population's genetic characteristics but is also contributing to this recreational fishery by providing an unassailable source of mature fish to nearby and distant exploited areas (Attwood and Bennett 1994).

Even when the characteristics affected by fishing are not genetically based, the preservation of the natural characteristics of exploited species by marine reserves can still provide numerous benefits. Increases in abundance and size as well as preservation of sex ratios and life history characteristics within marine reserves may all contribute to preserving the reproductive capacity of these species and their ability to replenish their populations. Similarly, marine reserves may help to buffer exploited species against population fluctuations that, when coupled with high levels of exploitation, may threaten the species. Variable environmental conditions and other factors may make replenishment of exploited species infrequent. A single cohort or year-class may constitute the majority of a population, whose long-term viability may depend on reproductive output from this single year-class. In fished areas, strong year-classes may be rapidly reduced or even wiped out entirely as soon as they are large enough to be caught in the fishery, limiting or eliminating their contribution to the population's reproductive output. Within marine reserves, however, strong year-classes are preserved in the population's structure (e.g., Ferreira and Russ 1995; Russ et al. 1998), and can continue to replenish the population for a number of years until another strong year-class can also contribute to the population. Thus marine reserves can help to stabilize population fluctuations and ensure against population crashes that may occur in fished areas.

Marine reserves may also affect nonexploited species and the structure and function of ecological communities and entire ecosystems. For example, fish and invertebrate assemblages within marine reserves often have higher species richness or diversity than do fished areas (e.g., Jennings et al. 1996, 1995; McClanahan 1994; Wantiez et al. 1997). In the absence of fishing, processes such as predation and competition for limited resources can limit or regulate the distribution and abundance of species and the organization of species into ecological communities. In most areas of the sea, these processes may still operate, but their effects are often masked or overshadowed by the dominant effects of fishing, which removes certain species and alters the balance of marine ecosystems. Often, species selectively removed by fishing play an important role in structuring the ecosystem.

In marine reserves however, all components of the ecosystem are preserved, and natural processes are allowed to dominate how the ecosystem is structured. In the rocky intertidal zone of Chile, a predatory gastropod called the loco is commonly harvested for food. In most areas, where locos are fished, their abundances are low. As a result, their preferred prey, mussels, are able to cover up to 100 percent of the intertidal zone because they are competitively dominant over barnacles, kelp, and other organisms for the limited space available on the rocks. Within a marine reserve, however, loco abundance or density is much greater, which increases their impacts as predators and reduces populations of mussels. The reduction in mussels on the rocks opens up space that can be colonized by barnacles, kelp, and other inferior competitors. The resulting dramatic transformation of this rocky intertidal landscape provides a whole new and different vision of its natural condition. Thus the increase in abundance of an exploited species that serves as a keystone predator, in this case, the loco, directly and indirectly affects other species in the community at lower levels of the food chain. (See also chapter 11; Castilla et al. 1994; Durán and Castilla 1989).

Nonexploited or nontarget species that may benefit the most from protection within marine reserves are those species that are taken as bycatch outside reserves. Because fishing is prohibited in marine reserves, the incidental take of nontarget species or bycatch is also eliminated within them. For species that suffer greatly from being taken as bycatch, like the barndoor skate, marine reserves may be the only areas where viable populations still exist (Casey and Myers 1998).

Nonexploited and exploited species alike may also benefit from the ability of reserves to protect habitats from destructive fishing gear and other human impacts. As mentioned in the previous chapter, marine habitats face multiple human impacts, including threats from fishing. Because habitat characteristics often have an important influence on the distribution and abundance of fish and invertebrates, habitat preservation by reserves may affect both exploited and nonexploited species. In a marine reserve in New Caledonia, for example, the species richness of reef fish increased by 67 percent in marine reserves but did not increase in fished areas (Wantiez et al. 1997). As the distribution and abundance of species change in response to habitat protection, the composition and structure of the ecological community may change to a more natural state.

By protecting both exploited and nonexploited species from various human impacts, marine reserves provide numerous benefits for ecosystem conservation. By protecting all components of the ecosystem and removing the dominant effects of exploitation, marine reserves allow "natural" processes to determine community and ecosystem structure and function. Unlike the case of fishing where it may be profitable to harvest every last individual of a particular species, many of these natural processes are driven by the density of various species or components of the ecosystem (e.g., the density of predators, prey, or competitors) and operate through negative feedback loops. Thus they may help to stabilize the ecosystem, making it resistant and resilient to change. Allowing ecosystems to be self-regulating or function under their own natural controls is one of the fundamental principles of ecosystem-based management.

## IMPROVING FISHERY YIELDS

Although effective fisheries management requires that ecosystem structure, function, and integrity are maintained to some extent, it also requires that marine resources be optimally utilized for long-term sustainability (Dayton et al. 1995). We have already indicated how marine reserves can maintain ecosystem structure, function, and integrity, but can marine reserves also contribute to the optimal use of marine resources? It may seem counterintuitive to some that closing an area to fishing and thereby reducing the proportion of fish available to the fishery can actually enhance landings and allow the optimal use of marine resources, but marine reserves can do this in several ways.

One of the guiding principles for optimizing the sustainable use of marine resources is that the harvest of marine stocks cannot go unchecked. Limits must be placed on catches so that they do not exceed the ability of a species to replenish itself. In chapter 3, we briefly discussed traditional fishery management approaches to ensure this, their shortcomings and poor track record, and the use of marine reserves as a supplementary tool. Marine reserves can help prevent overfishing by providing a spatial refuge from fishing mortality. Such protection from overfishing may result from reduced overall fishing effort, but it can also arise with no overall reduction in fishing effort. As the difference in spawning stock biomass within and outside the reserve grows, even the displacement of fishing effort from reserve to nonreserve areas reduces average fishing mortality (especially for older and larger fish residing disproportionately in the reserve areas), provides increased protection against overfishing, and can provide a net benefit to fisheries.

Most modern fisheries management plans have target levels for maintaining spawning stock biomass, usually at levels on the order of at least 20 to 30 percent of the estimated unexploited spawning stock biomass. Because marine reserves contain higher abundances and larger individuals than surrounding areas, they can contribute significantly to conserving spawning stock biomass and sound fishery management. However, the size required for a reserve or reserve network alone to protect adequate spawning stock biomass may be quite large—20 to 80 percent of an area (e.g., Dahlgren and Sobel 2000). In some cases, proper placement of reserves may enhance their contribution disproportionally to their size. Even so, the utility of reserves for protecting spawning stock biomass may be greatest as a complement to other management tools and as insurance against the failure of those tools to protect sufficient spawning stock biomass.

In addition to simply safeguarding biomass against exploitation, marine reserves can actually support or enhance fisheries landings in several ways. For this to happen, the increase in abundance or biomass of exploited species within the reserve must be accompanied by a net export of individuals from the reserve to areas where they can be caught in the fishery. Moreover, for reserves to actually enhance fisheries productivity, increases in target species biomass outside the reserve must be greater than the biomass that would have been available to the fishery in the reserve area if it were not protected.

Reserves may serve to boost productivity in nonreserve areas in two ways: the net movement of juvenile or adult animals from the reserve to fished areas, often referred to as spillover, increases recruitment of older juveniles and adults into the fishery; and the net export of larvae from reserves to nonreserve areas increases settlement and larval recruitment initially, and recruitment into the fishery later. The terms *spillover* or *leakage* are often used synonymously to describe movement of older juvenile and adult fish and invertebrates out of a reserve and into fished areas, but it may be more appropriate to use the second of these terms to differentiate between how such animals move out of a reserve into the fishery.

Spillover is expected to result from the effects that reserves have on exploited species, such as the increase in density, size, or biomass within the reserve. For example, individuals may move or spill over from the reserve if their increase in abundance or biomass within the reserve results in a decrease in the per capita availability of resources, such as food or refuges. In this case, as the carrying capacity of a particular species is approached within the reserve, animals respond to crowding and the subsequent decrease in resource availability by moving to areas with lower abundance or biomass. Thus they may leave the reserve and enter the fishery. In addition to increases in density within reserves, increases in body size of exploited species within reserves can lead to increases in movement rates or home range size that can cause spillover. Because spillover results directly from the positive effects of marine reserves on the density, biomass, or size of exploited species, such movement from the reserve is expected to result in enhanced landings in fished areas adjacent to the reserve.

Furthermore, the reserve is also expected to produce numerous other benefits, such as preserving natural ecological community structure and protecting spawning stock biomass of exploited species.

In contrast, the term *leakage* can be used to refer to the case where movement out of the reserve is independent of reserve effects on abundance, size, or biomass. Ontogenetic habitat shifts, periodic migrations, home ranges that overlap with reserve boundaries, or other types of movements that may occur independently of the density or size of individuals may cause leakage from the reserve into the fishery. Because leakage refers to movement that is largely independent of any effects produced by marine reserves and animals "leaked" by the reserve may be caught in the fishery, it can prevent populations from approaching their maximum biomass, so it may limit other reserve benefits. Just like a leaky container doesn't fill with water, a leaky reserve may not accumulate biomass of exploited species. Although the overall benefits produced by leaky reserves may be limited, leakage may still allow some increases in biomass of exploited species within the reserve and the export of that biomass to fished areas. Thus landings adjacent to the reserve may be somewhat elevated, but not to the extent that a reserve with high spillover rates would. Similarly, even small increases in size and abundance within a reserve may provide some additional benefits to protecting ecosystem structure, function, and integrity. Although both spillover and leakage may be present to some degree for any species in any reserve, the different mechanisms underlying these movements may have implications on the effects or benefits of this movement.

Indirect evidence that reserves can support fisheries outside their borders via emigration comes from studies of the movement rates and distribution patterns of important fishery species. For example, mark and recapture experiments in reserves and adjacent exploited areas show that marked fish released within reserves were frequently caught outside the reserves (Attwood and Bennett 1994; Johnson et al. 1999; Munro 1998; Roberts et al. 2001). Although this movement by itself does not demonstrate that fisheries outside reserves were enhanced by reserves, it suggests that, as abundance and biomass of fishery species increase within reserves, emigration from reserves has the potential to enhance fisheries.

Examination of population distribution patterns provides further indirect evidence that juveniles and adults from reserves may support fisheries in exploited areas outside of reserves. Fish and invertebrates with home ranges in the center of a reserve are less likely to emigrate from the reserve and be captured in the fishery than those at the edge of the reserve. Therefore, population distribution characterized by decreasing density with increasing distance from the center of the reserve may indicate emigration from the reserve to exploited areas, which supports local fisheries, especially if such decreasing density continues outside of the reserve boundaries (Kramer and Chapman 1999; Rakitin and Kramer 1996). However, a decreasing trend only within the reserve itself could also be confounded by a poaching gradient.

Nassau grouper (*Epinephelus striatus*) within and outside of the Exuma Cays Land and Sea Park, a large Bahamian marine reserve, show a peak at the center of the reserve and decreasing density as distance from the center increases, including a continuing decline within fished areas outside the reserve (Fig. 4.6; Sluka et al. 1996; see also chapter 9, Bahamas case study). Such a pattern is consistent with spillover or leakage from the park into adjacent fished waters, but not with a poaching gradient alone. A poaching gradient could explain such a decrease within the park, but not beyond its borders.

Identifying direct and indirect evidence of emigration from reserves is an important step in testing hypotheses related to the potential of marine reserves to provide spillover benefits to fisheries but does not directly test whether this movement enhances fisheries. Net emigration out of the reserve is indicative of spillover and a prerequisite for such benefits. Resulting increased landings or catch per unit effort (CPUE) attributable to such movements are better evidence of fishery enhancement. Correlating such movement with the occurrence of increased landings near a reserve, provides even stronger evidence that movement out of the reserve is actually enhancing fisheries.

In lagoonal marine reserve zones within Florida's Merritt Island National Wildlife Refuge, originally protected for security reasons due to its proximity to the Cape Kennedy Space Center, increased abundance and mean size of recreationally targeted fish species inside the reserve, movement of tagged individuals out of the reserve and their capture in the fishery, and a proliferation of world record fish caught in adjacent areas, suggest that the reserve is supporting an important recreational trophy fishery in surrounding waters. (Fig. 4.7; Johnson et al. 1999; Roberts et al. 2001).

Further evidence of fishery enhancement by emigration from a marine reserve comes from the Sumilon Island Marine Reserve example discussed earlier and its adjacent reef fish fishery in the Philippines. In this case, a reserve was established for a number of years before it was opened to fishing. Catch per unit effort of local fishers was high when 25 percent of a reef was protected within a marine reserve, and decreased dramatically shortly after the entire reef was open to fishing (Alcala 1988; Alcala and Russ 1990; Russ and Alcala 1996).

The authors believed this was the result of spillover effects, but could also not rule out larval export as a factor. In Kenya, a large reserve encompassing over 60 percent of fishing grounds dramatically increased catch per unit effort in nearby fisheries (McClanahan and Kuanda-Arara 1996). Similarly, artisanal fisher catches adjacent to a network of five small reserves in St. Lucia increased from 46 to 90 percent, depending on gear type, within five years of their creation (Roberts et al. 2001). These three examples from St. Lucia, the Philippines, and Kenya demonstrate that emigration from reserves to fished areas can be important in supporting local fisheries; in fact landings actually increased when a significant portion of the fishing grounds was closed to fishing.

Whether or not emigration from a reserve can support or enhance fisheries, however, may depend on the underlying mechanism or process driving the movement. Although movement out of the reserve may provide a mechanism for transferring the increase in biomass (size and/or abundance) of exploited species from the reserve to the fishery, frequent movement out of the reserve and into the fishery may result in high capture rates of animals from the reserve and prevent biomass increases from occurring within the reserve.

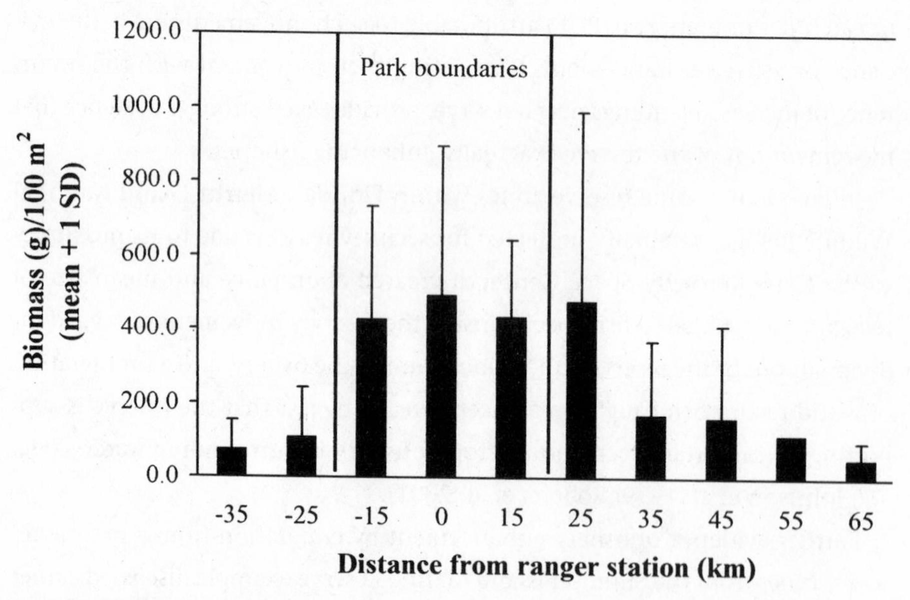


**FIG. 4.6 Nassau Grouper Density Distribution within and outside Bahamian Marine Reserve.** Graph shows relative density (biomass/area) of Nassau grouper versus distance from the ranger station located near the center of the Exuma Cays Land and Sea Park. Densities are much higher within the park and taper off away as you move away from it, consistent with emigration and net export out of it. Source: Adapted from Sluka, et al. 1997. Data provided by Mark Chiappone.

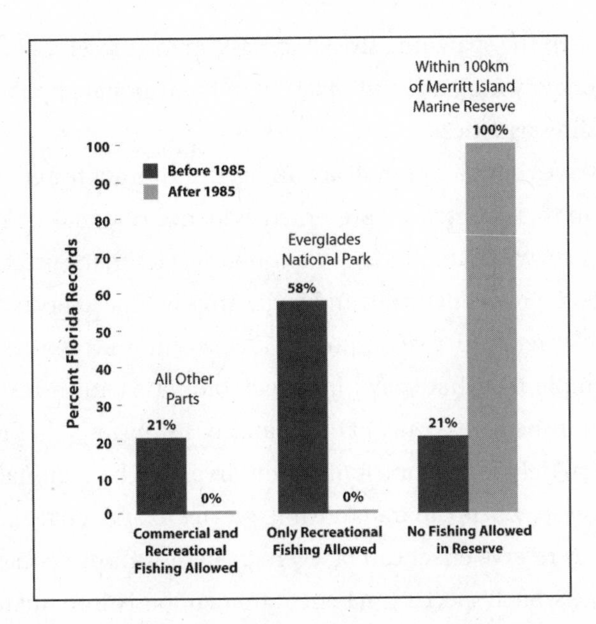


**FIG. 4.7 Merritt Island Florida Marine Reserve and Adjacent World Record Catches.** This graph compares the percent of Florida's world record catches coming from waters adjacent to the Merritt Island reserve to that of Everglades National Park and to all other parts of the state, both before and after 1985, when the reserve was established. Prior to 1985, Everglades National Park, where only recreational fishing is allowed produced most of the records. Since then, waters adjacent to Merritt Island reserve have produced nearly all of them. Source: Adapted and reproduced with permission from PISCO 2003.

Even though spillover, and to a lesser extent leakage, from marine reserves can enhance fisheries on local scales, marine reserves may provide more important benefits to fisheries on even greater spatial scales via larval export or replenishment. Unfortunately, it is much more difficult to demonstrate the fisheries enhancement benefits of marine reserves via enhanced larval supply because it is difficult to directly link the production of larvae within a marine reserve to enhanced landings elsewhere. Nevertheless, there are several lines of evidence suggesting that the export of larvae from marine reserves can support or enhance fisheries. In addition, developing techniques and technology are enhancing our capacity to directly link within-reserve production to landings outside (see chapter 7).

Most fisheries management tools operate on the premise that preserving a certain amount of spawning stock biomass in the sea is necessary for populations of exploited species to produce enough larvae to replenish themselves and support fisheries. Because the spawning stock biomass of exploited species is often higher in marine reserves than in fished areas, and larval production is correlated with the spawning stock biomass, marine reserves may contribute a disproportionately high amount of larvae to the regional population and support surrounding fisheries.

Marine reserves have additional advantages over other fishery management tools. Earlier in this chapter, we presented evidence that marine reserves often harbor greater spawning biomass of exploited species than surrounding fished areas due to both greater densities and larger sizes within reserves and discussed the likely greater reproductive output per area within the reserves than outside them. The simple fact that spawning stock biomass within reserves is often greater than in fished areas on a per unit area basis suggests that reserves make a disproportionately large contribution of larvae to the regional population. More and larger spawners can translate into greater reproductive output for most species, but this reserve effect can be even greater for those species that experience Allee effects. Such species can benefit tremendously from increased density within marine reserves, especially when heavily fished or depleted outside. Similarly, by permitting exploited species to grow to larger sizes than in fished areas where other management tools are used, other aspects of the natural population structure (e.g., male to female ratio) of exploited species are preserved, which can increase reproduction success. For any increase in reproductive output from reserves to enhance fisheries, the increased output must be transported to areas where larvae can settle and grow until they enter the fishery.

Although transport mechanisms can retain larvae locally within a reserve, currents are likely to advect or diffuse the majority of larvae for most marine species out of the reserve and replenish exploited stocks downstream unless the reserve is large. Because many reserves are only a few kilometers long, even if larvae only travel or mix over tens of kilometers before the settle from the plankton, it is likely that they will be transported out of the reserve and into areas where they may grow to enter the fishery.

Even when present, documenting fisheries enhancement by marine reserves through either spillover or larval replenishment may be difficult in practice. Russ (2002) and Jones et al. (1992) provide good overviews of the inherent difficulties of demonstrating such reserve effects. Consequently, various mathematical models have been used to predict the effect of marine reserves on fisheries. For example, a model developed by Hastings and Botsford (1999) found that marine reserves can produce yields similar to other fisheries management tools that are more commonly used. Similarly, several models suggest that larval export from marine reserves can support or enhance fisheries, particularly for species that are heavily exploited outside of reserves (Holland and Brazee 1996; Sladek Nowlis and

Roberts 1997, 1999; Sladek Nowlis and Yoklavich 1998). In addition, other models indicate that larval dispersal from reserves can reduce variability or uncertainty in fisheries management (Guenette et al. 2000; Lauck et al. 1998), or maintain genetic variability that may be reduced by fishing (Trexler and Travis 2000).

In addition to mathematical models indicating that the advection of larvae from a reserve can support or enhance fisheries, there is some direct evidence that increased larval production within a reserve can enhance recruitment of larvae in areas that are accessible to fishers. An example of this comes from the Bahamas, where the queen conch (*Strombus gigas*) is an important fishery species. The density of queen conch within a marine reserve there is several times higher than nearby areas that are fished. As a result, the density of early stage queen conch larvae was measured to be about ten times higher within and around a marine reserve than in fished areas further away. Late stage larvae (several weeks old), however, were not concentrated around the marine reserve but were dispersed throughout the region and concentrated in areas where hydrographic features retained them. Thus larval production within the park is likely to contribute substantially to supporting the conch fishery throughout the region (Stoner et al. 1998; Stoner and Ray 1996). As we gain a better understanding of larval transport and develop new techniques for determining where larvae and recruits were spawned, we will be able to further investigate this important benefit of marine reserves.

Marine reserves can also contribute to improved fisheries management and fisheries yields in other pragmatic ways. Enforcing other fishery management tools is often difficult since enforcement of these tools requires complex inspection and tracking systems. Management tools that limit catch or landings in the fishery require that all landings be inspected and tracked, often on the basis of an individual fisher. Management tools that limit or restrict fishing effort require tracking fishing effort in terms of the number of fishers, boats, or boat days fished, and/or the number and type of fishing gear used. In either case, to determine if a fisher or fishing vessel is operating illegally, enforcement officers must often board the vessel and inspect the fishing gear or catch to see if there are any violations. Within marine reserves, however, enforcement is simplified; because no fishing is allowed within marine reserves, all enforcement officers must do is determine whether a boat within a marine reserve is fishing or not. This benefit of marine reserves is very attractive in areas where there is limited capacity for fisheries enforcement.

At the cusp of improving fisheries and expanding our knowledge and understanding of marine ecosystems, marine reserves can play a critically important and frequently overlooked role in providing reliable stock assessment and other information necessary for sustainable fisheries and other purposes. Schroeter et al. (2001) used information gleaned from a long-term fisheries independent monitoring program that included reserve and nonreserve sites to assess a new dive fishery for warty sea cucumbers (*Parastichopus parvimensis*) in California's Channel Islands. The long-term data showed a decrease in abundance throughout the Channel Islands within three to six years after the onset of the fishery. Using before–after, control–impact (BACI) analyses, they implicated fishing mortality as responsible for a 33 to 83 percent decline in stocks. By contrast, traditional CPUE data showed no declines and even an increase for one island.

## EXPAND KNOWLEDGE AND UNDERSTANDING OF MARINE ECOSYSTEMS

Several marine reserves have been created in areas near marine laboratories for scientific reasons (e.g., Leigh Marine Reserve [Ballantine 1994] and Hopkins Marine Reserve [Paddack and Estes 2000]). Marine reserves can enhance scientific research and education in several ways. The simple fact that marine reserves may be used to separate different uses of marine resources contributes to this benefit of marine reserves. It is difficult for scientists to conduct controlled experiments in a marine environment when components of their experiments are being captured and eaten or otherwise disrupted by human impacts. By setting aside areas of the sea that are free from many human impacts, scientists may be able to conduct more effective experiments and learn more about the marine environment. The understanding gained from such experiments is essential for the effective management of marine environments.

Similarly, by setting aside areas of the sea that are free from fishing and other human impacts, scientists have "control" areas to study, compare, and contrast with areas that are subject to fishing and other human impacts. This allows scientists to examine what effects exploitation has on living marine resources and marine ecosystems. In chapter 1 and previous sections of this chapter, we discussed how fishing alters marine ecosystems and how populations of marine species and ecological communities are often structured differently in areas that are subject to exploitation when compared to areas in which natural processes dominate. Because human impacts have spread throughout the global marine environment, such studies are impossible without the creation of marine reserves, as McClanahan (1990) has eloquently stated (see chapter 1).

Studies of species interactions and ecological processes within reserves can provide critical insights to determine how marine ecosystems are regulated by natural processes and how they respond to various natural and human induced stresses. Comparisons between populations, communities, and ecological processes operating in marine reserves and nonreserve areas allows scientists not only to determine the impact of exploitation and other impacts but also to evaluate the efficacy of other management tools and strategies. As the idea of ecosystem-based management of marine environments is increasingly put into practice, use of marine reserves as a baseline reference or guide will allow managers to more effectively manage the marine environment and adapt management measures as necessary to meet the goals of maximizing sustainable productivity of fishery resources while preserving ecosystem integrity.

Earlier in this chapter, we recognized that the four broad categories of marine reserve benefits should be considered overlapping rather than discrete. Several of the examples discussed earlier under protecting ecosystems and improving fisheries also clearly demonstrate how reserves can provide critical insights about the ocean that might otherwise remain hidden and are worth mentioning here. The Chilean reserve example involving the loco, its recovery, and its role in transforming its rocky intertidal community greatly expanded our knowledge and understanding of this ecosystem, how it functions, and what is natural (Castilla et al. 1994). The Kenyan reserve example provided similar information regarding Kenyan coral reefs; the ecological roles and relationships involving predatory fish, herbivorous urchins, coral, and algae; and their roles in the stability and resilience of this ecosystem (McClanahan et al. 1999). Likewise, the examples from reserves in Tasmania and New Zealand both provided valuable insights into the functioning of kelp forest ecosystems, the roles of fish and lobster in them, and their response to fishing activities (Ballantine 1994; Edgar and Barrett 1999). The Channel Islands warty sea cucumber fishery example clearly demonstrates how reserves expand our knowledge about fishing impacts by detecting trends that might otherwise go unnoticed (Schroeter et al. 2001).

The list of examples documenting the role of reserves in expanding our knowledge and understanding of the ocean realm is a long one, but two more from the North Pacific are worth mentioning here. The first is discussed in more detail later in this book and involves the use of reserves and de facto reserves in British Columbia to study large adult pinto abalone (*Haliotis kamtschatkana*), which were nearly impossible to find elsewhere (Wallace 1999), and lingcod (*Ophiodon elongatus*) mating behavior and related movements that were difficult to unravel in fished waters. On a much grander scale, a U.S. National Research Council panel recently recommended that the U.S. government run a decade-long test involving the creation of two large marine reserves and two control areas off Alaska to help unravel the mystery surrounding the thirty-year decline of Stellar sea lions (*Eumetopius jubatus*), determine the role of fishing in it, and settle a high-stakes dispute concerning regulations for one of the world's most valuable fisheries. The panel concluded that the experiment "is the only approach that directly tests the role of fishing in the decline" and controls for other factors including climate change (Malakoff 2002).

Finally, marine reserves provide a natural classroom for marine education. Just as terrestrial parks and wildlife refuges serve as a center for people to visit in order to appreciate and learn about the natural environment, marine reserves serve as such centers, simply due to the fact that they are protected. This is perhaps one of the most important aspects of marine reserves because it helps to foster an understanding of marine environments—wild ecosystems with their full diversity and abundance of marine life—and a conservation ethic across a wide range of people.

## ENHANCE NONCONSUMPTIVE OPPORTUNITIES

Although they function by preserving natural marine ecosystems and permit ecological processes to operate in the absence of human impacts, marine reserves are created to meet human social, economic, and cultural needs. Earlier in this chapter, we saw examples of how marine reserves enhance fisheries through emigration from reserves and larval replenishment. But their economic benefits are not limited to enhancing consumptive uses of marine resources. Enhancing scientific research and marine education are among the many nonconsumptive opportunities enhanced by marine reserves. There are a number of other ways that marine reserves enhance nonconsumptive opportunities that provide economic, social, and cultural benefits. We discuss some of these benefits here and also later in this book.

Tourism is now the world's leading industry. As with terrestrial parks, marine reserves are often areas people visit to experience nature. In many cases, the simple designation of an area as a marine park or reserve has caused an increase in visitation (Ballantine 1991). Although heavy visitation of these areas may be incompatible with reserve protection intended to minimize human impacts, some degree of visitation is desirable to promote marine conservation education and to provide economic benefits to communities around the reserve in the form of ecotourism. Ecotourism has proven to be a viable economic activity in a number of marine settings. Tourism based on scuba diving, sea kayaking, whale watching, and other recreational activities may cause minimal impact to the marine environment but provide a source of income to local communities. The creation of marine reserves may enhance opportunities for these nonconsumptive uses of the sea. In the Florida Keys, the socioeconomics of these and related activities have been extensively documented in connection with marine protected area (MPA) and reserve development (Johns et al. 2001), and small marine reserves created around several popular dive sites have been promoted as "extra strength" dive sites because of their protected status and the likely increase in fish biomass that will develop there (Fig. 4.8).

Ecotourism can be particularly important to communities surrounding marine reserves both as a local economic benefit and as an alternative source of income for fishermen if the creation of the reserve is intended to reduce overall fishing effort on a regional scale. Ecotourism and other economic opportunities that result from marine reserves may not only be a more lucrative opportunity than fishing but may also be a more dependable source of income than fishing. Such tourism, however, must be monitored and if necessary regulated to ensure that these human impacts do not degrade the marine environment in the manner of more consumptive uses. Rudd and Tupper (2002) conducted a preliminary quantitative assessment of the impact of Nassau grouper size and abundance on scuba diver site selection and marine reserve economics in the Turks and Caicos Islands. They concluded that Nassau groupers provide nonextractive economic value to divers. This value would increase under reserve management as the size and number of fish increased, and it could affect the economic viability of such reserves.

There are also many social benefits to marine reserves. The simple fact that they can support the livelihood of fishermen and provide alternative sources of income is of great social benefit. In many places where fisheries have crashed, fishermen have lost their jobs, and capital invested in the fishery has either been wasted or financed by the government. Society therefore must pay a heavy price to support fishermen who can no longer fish and may not have alternative sources of income available to them. Although they are by no means the entire solution to this problem, marine reserves may contribute toward avoiding this costly situation.

Marine reserves may also provide some important cultural benefits. By supporting fisheries, marine reserves can support traditional ways of life that may be lost forever if fisheries crash. Many cultural values may be associated with

**FIG. 4.8 Florida Keys Tourism Promotion Featuring "Extra Strength" Marine Reserve.** Photo and ad courtesy of photographer Stephen Frink.

these traditions and may be lost with them. By preserving marine resources, marine reserves can contribute to preserving these cultural values and traditions. Similarly, in parts of the world where there is a conservation ethic, even among a minority of the population, marine reserves can provide peace of mind to those who are concerned about the health of the marine environment. The social and cultural dimensions of marine reserves are discussed in more detail in chapter 6.

## WHAT MARINE RESERVES CAN'T DO

Marine reserves may be thought of as an essential vaccine, insurance, and wellness program for the oceans, not a panacea for their growing ills. Even among the strongest reserve proponents, few consider them to be a silver bullet. Most supporters consider them to be a vital tool and necessary part of a comprehensive ocean protection and management approach, but not adequate by themselves. Some reserve opponents have tried to frame the debate as a choice between marine reserves and sound resource management, but this is a false dichotomy. Effective ocean resource management increasingly requires the development of effective marine reserves and marine reserve networks among its most important tools.

Sound public policy requires meeting and balancing multiple societal goals and values (see also chapter 6). Most coastal communities, societies, and nations want to both protect the oceans and utilize them for multiple purposes, including recreation and food production. Treating the entire ocean as a marine reserve might go a long way to achieving ocean protection, but would clearly not meet all other goals. Size and boundary considerations, political realities, and other societal needs and desires require that marine reserves be used in conjunction with other tools to achieve all of their and society's other goals and to maximize reserve benefits.

Both protecting the oceans generally and maximizing the effectiveness of individual reserves and reserve networks of any size also requires addressing threats and impacts generated from activities external to reserves. Reserve design and regulations can mitigate such impacts and sometimes provide a limited degree of protection against some of them, but frequently they operate outside of reserve jurisdiction or on a scale where they are best dealt with using additional approaches. Pollution, coastal development, and other threats to water quality are among the most critical of these. Global change, operating on an even larger scale, is another. Even fishing-related impacts, given likely societal choices, will best be addressed through a combination of marine reserves and other management tools. The following passage summarizes these points well:

> Reserves will be essential for conservation efforts because they can provide unique protection for critical areas, they can provide a spatial escape for intensely exploited species, and they can potentially act as buffers against some management miscalculations and unforeseen or unusual conditions. . . . Reserves are insufficient protection alone, however, because they are not isolated from all critical impacts . . . with-

> out adequate protection of species and ecosystems outside of reserves, effectiveness of reserves will be severely compromised. (Allison et al. 1998)

## CONCLUSION

Despite scientific evidence that marine reserves provide and are capable of providing a multitude of benefits and can make significant contributions to marine conservation and fisheries production, many still ask, "Just because a reserve works there, will it work here?" when a reserve is proposed in a new area. Although a particular reserve may not produce all of the benefits that reserves are capable of, and it is often difficult to predict what benefits are likely for a particular reserve, the large number of studies showing similar reserve effects from a variety of diverse systems suggests that the response of marine populations and communities to reserve protection is fairly robust to unique features of individual reserves. Many of the reserve effects and benefits discussed in this chapter are supported by examples from around the globe in a variety of marine systems. Thus we can expect reserves to achieve many of their proposed goals much of the time. To answer the simple question, Do reserves work? the answer is yes. The efficacy for a particular reserve to achieve specific management objectives, however, may depend on a variety of factors, including the species that the reserve is intended to protect, public compliance with reserve protection, and design characteristics of the reserve.

However, this evidence is not always ironclad or accepted for all potential benefits in the context of a dispute over designation of a particular reserve. Here one must ask, Can anyone provide similarly conclusive, empirical evidence, using similar standards, that other (nonmarine reserve) fishery management tools have been directly responsible for improving long-term fishery yields in a tropical reef or other system? Have other tools been proven to prevent the collapse, extirpation, and extinction of the more vulnerable fish populations? Can they prevent degradation of reefs and maintain natural diversity and intact systems? The point is not to question whether these other tools can be effective. Many are widely accepted. But it is clear that, by these criteria, marine reserves can provide benefits likely unavailable via traditional management tools.

People often view the marine environment as being invulnerable or at least resilient to human impacts. This may be reinforced by the fact that most people, although they may be reliant on the sea in many ways, do not have an appreciation for the sea and the diversity of life that inhabits marine environments. Because changes in the marine environment may often occur over generations, even fishers and others who make their living from the sea may not realize the changes that result from human impacts until these impacts cause severe disruptions to their livelihood. By having marine reserves as a reference, people may better appreciate how different the marine environment is in surrounding areas where human impacts dominate. This understanding may help to dispel the myth that the sea is too vast to be affected by humans, and instill the notion that conservation and effective management of marine systems are necessary to preserve them.

The strong track record of marine reserves to date combined with their sound theoretical underpinnings provides compelling arguments to greatly expand their use in order to address the problems discussed in chapters 1 through 3, reverse related trends, restore natural baselines and the abundance and diversity of marine life, and protect the oceans' vitality and integrity. A preponderance of evidence strongly supports the expectation that an appropriate system or network of reserves can provide a full suite of benefits falling under the four general categories of protecting ecosystems, improving fishery yields, expanding our knowledge and understanding of the oceans, and enhancing nonconsumptive opportunities.

## REFERENCES

Alcala, A. C. 1988. Effects of marine reserves on coral fish abundances and yields of Philippine coral reefs. *Ambio* Stockholm 17(3):194–199.

Alcala, A. C., and G. R. Russ. 1990. A direct test of the effects of protective management on abundance and yield of tropical marine resources. *Journal Du Conseil International Pour L'Exploration De La Mer* 47:40–47.

Allison, G. W., J. Lubchenco, and M. H. Carr. 1998. Marine reserves are necessary but not sufficient for marine conservation. *Ecological Applications* 8(1, suppl):S79–S92.

Attwood, C. G., and B. A. Bennett. 1994. Variation in dispersal of galjoen (*Coracinus capensis*) (Teleostei: Coracinidae) from a marine reserve. *Canadian Journal of Fisheries and Aquatic Sciences* 51(6):1247–1257.

Babcock, R. C., S. Kelly, N. T. Shears, J. W. Walker, and T. J. Willis. 1999. Changes in community structure in temperate marine reserves. *Marine Ecology Progress Series* 189:125–134.

Ballantine, W. J. 1991. *Marine Reserves for New Zealand.* Auckland: University of New Zealand.

———. 1994. The practicality and benefits of a marine reserve network. In Gimbel, K. L. *Limiting Access to Marine Fisheries: Keeping the Focus on Conservation,* 205–223. Washington, DC: Center for Marine Conservation and World Wildlife Fund US.

Beets, J., and A. Friedlander. 1999. Evaluation of a conservation strategy: A Spawning aggregation closure for red hind, *Epinephelus guttatus,* in the U.S. Virgin Islands. *Environmental Biology of Fishes* 55:91–98.

Bohnsack, J. A. 1998a. Application of marine reserves to reef fisheries management. *Australian Journal of Ecology* 23(3):298–304.

———. 1998b. Ecosystem management, marine reserves, and the art of airplane maintenance. In Creswell, R. L., ed. *Proceedings of the Gulf and Caribbean Fisheries Institute,* no. 50, p. 304.

Casey, J. M., and R. A. Myers, R. A. 1998. Near extinction of a large, widely distributed fish. *Science* 281(5377):690–692.

Castilla, J. C., G. M. Branch, and A. Barkai. 1985. Human exclusion from the rocky intertidal zone of central Chile: The effects on *Concholepas concholepas* (Gastropoda). *Oikos* 45(3):391–399.

———. 1994. *Exploitation of Two Critical Predators: The Gastropod,* Concholepus concholepus, *and the Rock Lobster,* Jasus lalandii. Heidelberg, Germany: Springer-Verlag.

Chapman, M. R., and D. L. Kramer. 1999. Gradients in coral reef fish density and size across the Barbados Marine Reserve boundary: Effects of reserve protection and habitat characteristics. *Marine Ecology Progress Series* 181:81–96.

Dahlgren, C. P., and J. Sobel. 2000. Designing a Dry Tortugas ecological reserve: How big is big enough? To do what? *Bulletin of Marine Science* 66(3):707–719.

Dayton, P. K. 1998. Reversal of the burden of proof in fisheries management. *Science* 279(5352):821–822.

Dayton, P. K., S. F. Thrush, M. T. Agardy, and R. J. Hofman. 1995. Environmental effects of marine fishing. *Aquatic Conservation: Marine and Freshwater Ecosystems* 5(3):205–232.

Dugan, J. E., and G. E. Davis. 1993a. Applications of marine refugia to coastal fisheries management. *Canadian Journal of Fisheries and Aquatic Sciences* 50:2029–2042.

———. 1993b. Introduction to the International Symposium on Marine Harvest Refugia. Dugan, J. E., and G. E. Davis, eds. 1993.

Durán, L. R., and J. C. Castilla. 1989. Variation and persistence of the middle rocky intertidal community of central Chile, with and without human harvesting. *Marine Biology* 103(4):555–562.

Edgar, G. J., and N. S. Barrett. 1999. Effects of the declaration of marine reserves on Tasmanian reef fishes, invertebrates and plants. *Journal of Experimental Marine Biology and Ecology* 242(1):107–144.

Ferreira, B. P., and G. R. Russ. 1995. Population structure of the leopard coralgrouper, *Plectropomus leopardus,* on fished and unfished reefs off Townsville, central Great Barrier Reef, Australia. *Fishery Bulletin* 93(4):629–642.

Godoy, C. and C. Moreno. 1989. Indirect effects of human exclusion from the rocky intertidal in southern Chile: a case of cross-linkage between herbivores. *Oikos* 54[1]:101–106.

Guenette, S., T. J. Pitcher, and C. J. Walters. 2000. The potential of marine reserves for the management of northern cod in Newfoundland. *Bulletin of Marine Science* 66(3):831–852.

Halpern, B. S. 2003. The impact of marine reserves: Do reserves work and does reserve size matter? *Ecological Applications* 13(1, suppl):S117–S137.

——— and R. R. Warner. 2002. Marine Reserves Have Rapid and Lasting Effects. *Ecology Letters* 5:361–366.

Hastings, A., and L. W. Botsford. 1999. Equivalence in yield from marine reserves and traditional fisheries management. *Science* 284:1537–1538.

Hauser, L., G. J. Adcock, P. J. Smith, J. H. B. Ramirez, and G. R. Carvalho. 2002. Loss of microsatellite diversity and low effective population size in an overexploited population of New Zealand snapper (*Pagrus auratus*). *Proceedings of the National Academy of Sciences, USA* 99(18):11742–11747.

Holland, D. S., and R. J. Brazee. 1996. Marine reserves for fisheries management. *Marine Resource Economics* 11(3):157–171.

Jennings, S., E. M. Grandcourt, and N. V. C. Polunin. 1995. The effects of fishing on the diversity, biomass and trophic structure of Seychelles' reef fish communities. *Coral Reefs* 14(4):225–235.

Jennings, S., S. S. Marshall, and N. V. C. Polunin. 1996. Seychelles' marine protected areas: Comparative structure and status of reef fish communities. *Biological Conservation* 75(3):201–209.

Johns, Grace M., Vernon R. Leeworthy, Frederick W. Bell, Mark A. Bonn. 2001. *Socioeconomic Study of Reefs in Southeast Florida, Final Report.* Hazen and Sawyer Environmental Engineers & Scientists in association with Florida State University and NOAA.

Johnson, D. R., N. A. Funicelli, and J. A. Bohnsack. 1999. Effectiveness of an existing estuarine no-take fish sanctuary within the Kennedy Space Center, Florida. *North American Journal of Fisheries Management* 19(2):436–453.

Jones, G. P., R. C. Cole, and C. N. Battershill. 1992. Marine Reserves: Do They Work? In C. N. Battershill et al. Proceedings of the Second International Temperate Reef Fish Symposium. Wellington, NIWA Marine.

Kelly, S., D. Scott, A. B. MacDiarmid, and R. C. Babcock. 2000. Spiny lobster, *Jasus edwardsii,* recovery in New Zealand marine reserves. *Biological Conservation* 92:359–369.

Kramer, D. L., and M. R. Chapman. 1999. Implications of fish home range size and relocation for marine reserve function. *Environmental Biology of Fishes*. 55:65–79.

Lauck, T., C. W. Clark, M. Mangel, and G. R. Munro. 1998. Implementing the precautionary principle in fisheries management through marine reserves. *Ecological Applications* 8:72–78.

Levitan, D. R., M. A. Sewell, and C. Fu-Shiang. 1992. How distribution and abundance influence fertilization success in the sea urchin *Strongylocentrotus franciscanus. Ecology* 73(1):248–254.

Ludwig, D., R. Hilborn, and C. Walters. 1993. Uncertainty, resource exploitation, and conservation. *Science* 260:17–18.

MacDiarmid, A. B., and P. A. Breen. 1992. Spiny lobster population changes in a marine reserve. In Battershill, C. N., et al. *Proceedings of the 2nd International Temperate Reef Symposium,* 47–56. Wellington, New Zealand: NIWA Marine.

Malakoff, D. 2002. Report Seeks Answers to Marine Mystery. *Science* (298):2110–2111, December 13.

McClanahan, T. R. 1990. Are conservationists fish bigots? *Bioscience* 40(1):2.

———. 1994. Kenyan coral reef lagoon fish: Effects of fishing, substrate complexity, and sea urchins. *Coral Reefs* 13:231–241.

McClanahan, T. R., and B. Kaunda-Arara. 1996. Fishery recovery in a coral-reef marine park and its effect on the adjacent fishery. *Conservation Biology* 10:1187–1199.

McClanahan, T. R., N. A. Muthiga, A. T. Kamukuru, H. Machano, and R. W. Kiambo. 1999. The effects of marine parks and fishing on coral reefs of northern Tanzania. *Biological Conservation* 89(2):161–182.

Munro, J. L. 1998. Comment on the paper by J. B. C. Jackson: Reefs since Columbus. In *Coral Reefs* 16(suppl):S23–S32. *Coral Reefs* 17:191–192.

Murray, S. N., R. F. Ambrose, J. A. Bohnsack, L. W. Botsford, M. H. Carr, G. E. Davis, P.

K. Dayton, D. Gotshall, D. R. Gunderson, M. A. Hixon, J. Lubchenco, M. Mangel, A. MacCall, D. A. McArdle, et al. 1999. No-take Reserve Networks: Sustaining Fishery Populations and Marine Ecosystems. *Fisheries* 24(11):11–25.

National Marine Fisheries Service (NMFS) Plan Development Team (PDT). 1990. *The Potential of Marine Fishery Reserves for Reef Fish Management in the U.S. Southern Atlantic.* NMFS.

National Research Council (NRC). 2001. *Marine Protected Areas: Tools for Sustaining Ocean Ecosystem.* Washington, DC: National Academy.

Paddack, M. J., and J. A. Estes. 2000. Kelp forest fish populations in marine reserves and adjacent exploited areas of central California. *Ecological Applications* 10:855–870.

Palumbi, S. R. 2002. *Marine Reserves: A Tool for Ecosystem Management and Conservation.* 45 pages. Arlington, Virginia, Pew Oceans Commission.

Pauly, D., M. L. Palomares, R. Froese, P. Sa-a, M. Vakily, D. Preikshot, and S. Wallace. 2001. Fishing down Canadian aquatic food webs. *Can. J. Fish. Aquat. Sci./J. Can. Sci. Halieut. Aquat.* 58(1):51–62.

Rakitin, A., and D. L. Kramer. 1996. Effect of a marine reserve on the distribution of coral reef fishes in Barbados. *Marine Ecology Progress Series* 131(1–3):97–113.

Roberts, C. M., and J. P. Hawkins. 2000. *Fully Protected Marine Reserves: A Guide.* Washington, DC: WWF Endangered Seas Campaign.

Roberts, C. M., and N. V. C. Polunin. 1992. Effects of marine reserve protection on northern Red Sea fish populations. *Proceedings of the Seventh International Coral Reef Symposium* 2:969–977.

Roberts, C. M., and N. V. C. Polunin. 1991. Are Marine Reserves Effective in Management of Reef Fisheries? *Reviews in Fish Biology and Fisheries* 1:65–91.

Roberts, C. M., W. J. Ballantine, C. D. Buxton, P. Dayton, L. B. Crowder, W. Milon, M. K. Orbach, D. Pauly, and J. Trexler. 1995. *Review of the Use of Marine Fishery Reserves in the U.S. Southeastern Atlantic.* National Marine Fisheries Service, Southeast Fisheries Center.

Roberts, C. M., J. A. Bohnsack, F. Gell, J. P. Hawkins, and R. Goodridge. 2001. Effects of marine reserves on adjacent fisheries. *Science* 294(5548):1920–1923.

Rowley, R. J. 1994. Case studies and reviews: marine reserves in fisheries management. *Aquatic Conservation: Marine and Freshwater Ecosystems* 5:233–254

Rudd, M. A., and M. H. Tupper. 2002. The impact of Nassau grouper size and abundance on scuba diver site selection and MPA economics. *Coastal Management* 30(2):133–151.

Russ, G.R. 2002. Yet Another Review of Marine Reserves in Reef Fishery Management. Tools. In Peter Sale, Ed. *Coral Reef Fishes; Dynamics and Diversity in a Complex Ecosystem,* 421–444. Academic Press: London, UK.

Russ, G. R., and A. C. Alcala. 1996. Marine reserves: Rates and patterns of recovery and decline of large predatory fish. *Ecological Applications* 6(3):947–961.

———. 1998. Natural fishing experiments in marine reserves 1983–1993: Community and trophic responses. *Coral Reefs* 17(4):383–397.

———. 1999. Management histories of Sumilon and Apo Marine Reserves, Philippines, and their influence on national marine resource policy. *Coral Reefs* 18(4):307–319.

Russ, G.R. and A.C. Alcala 2003. Marine Reserves: Rates and Patterns of Recovery and Decline of Predatory Fish, 1983–2000. *Ecological Applications.* 13(6):1553–1565.

Russ, G. R., D. C. Lou, J. B. Higgs, and B. P. 1998. Ferreira. Mortality rate of a cohort of the coral trout, *Plectropomus leopardus,* in zones of the Great Barrier Reef Marine Park closed to fishing. *Marine and Freshwater Research* 49(6):507–511.

Schroeter, S. C., D. C. Reed, D. J. Kushner, J. A. Estes, and D. S. Ono. 2001. The use of marine reserves in evaluating the dive fishery for the warty sea cucumber (*Parasticho-pus parvimensis*) in California, U.S.A. *Canadian Journal of Fisheries and Aquatic Sciences* 58:1773–1781.

Sladek Nowlis, J., and C. M. Roberts. 1997. Theoretical approaches to marine reserve design. *Proceedings of the 8th International Coral Reef Symposium* 2:1907–1910.

Sladek Nowlis, J., and C. M. Roberts. 1999. Fisheries benefits and optimal design of marine reserves. *Fishery Bulletin* 97(3):604–616.

Sladek Nowlis, J., M. M. Yoklavich, and Puerto Rico Sea Grant. 1998. Design criteria for rockfish harvest refugia from models of fish transport. In Yoklavich, M. M. *Marine Harvest Refugia for West Coast Rockfish: A Workshop*. NOAA Technical Memorandum: NOAA-TM-NMFS-SWFSC-255. Pacific Grove, CA, Department of Commerce.

Sluka, R. D., M. Chiappone, and K. M. Sullivan. 1996. Habitat preferences of groupers in the Exuma Cays. *Bahamas Journal of Science* 4(1):8–14.

Sobel, J. 1996. Marine Reserves: Necessary Tools for Biodiversity Conservation? *Global Biodiversity* 6(1): 8–17.

Stoner, A. W., and M. Ray. 1996. Queen conch, *Strombus gigas,* in fished and unfished locations of the Bahamas: Effects of a marine fishery reserve on adults, juveniles, and larval production. *Fishery Bulletin* 94:551–556.

Stoner, A. W., and M. Ray-Culp. 2000. Evidence for Allee effects in an over-harvested marine gastropod: Density-dependent mating and egg production. *Marine Ecology Progress Series* 202:297–302.

Stoner, A. W., N. Mehta, and M. Ray-Culp. 1998. Mesoscale distribution patterns of queen conch (*Strombus gigas* Linne) in Exuma Sound, Bahamas: Links in recruitment from larvae to fishery yields. *Journal of Shellfish Research* 17(4 Dec).

Tegner, M. J. 1993. Southern California abalones: Can stocks be rebuilt using marine harvest refugia? *Canadian Journal of Fisheries and Aquatic Sciences* 50:2010–2018.

Trexler, J. C., and J. Travis, J. 2000. Can marine protected areas restore and conserve stock attributes of reef fishes? *Bulletin of Marine Science* 66(3):853–873.

Trippel, E. A., O. S. Kjesbu, and P. Solemdal. 1997. Effects of adult age and size structure on reproductive output in marine fishes. In Chambers, R. C., and E. A. Trippel, eds. *Early Life History and Recruitment in Fish Populations,* 31–62. London: Chapman and Hall.

Wallace, S. S. 1999. Evaluating the effects of three forms of marine reserve on northern abalone populations in British Columbia, Canada. *Conservation Biology* 13(4):882–887.

Wantiez, L., P. Thollot, and M. Kulbicki. 1997. Effects of marine reserves on coral reef fish communities from five islands in New Caledonia. *Coral Reefs* 16(4):215–224.

Ward, Trevor J., D. Heinemann, and N. Evans. 2001. *The Role of Marine Reserves as Fisheries Management Tools: A Review of Concepts, Evidence, and International Experience.* Canberra, Australia: Bureau of Rural Sciences.

FIVE

# Design and Designation of Marine Reserves

JOSHUA SLADEK NOWLIS AND
ALAN FRIEDLANDER

Marine reserves represent a complex biological and social phenomenon. Any given marine reserve is likely to encompass a unique mixture of species, habitats, and ecosystems, and these combinations as well as their broader distributions will help to determine the best design principles. The social and political atmosphere will also vary from one reserve to the next, with different levels of support, sense of need, and even goals and objectives for the reserve or reserve network. An effective marine reserve designation process should follow advice based on local biological and ecological characteristics, but should also provide flexibility to address the social and political atmosphere. This process should include plenty of opportunity for interaction between technical advisers and the general public, with an open invitation to members of the public to propose designs and a chance for scientists to inform and review these proposals.

Marine reserves provide a valuable and powerful tool to help meet the multiple goals and address the complex challenges facing managers in many areas, but are clearly not the best tool for all purposes. In assessing their use, it is important to keep in mind what makes reserves different from other management tools, and ultimately what their relative strengths and weaknesses are. Four properties make reserves stand out from other management measures. First, reserve boundaries are simpler to enforce and thus more difficult to circumvent than other regulations. It is easier to see whether a boat is fishing in an area than to board it and examine its catch or gear for compliance. Second, reserves allow fish to grow large and realize their full reproductive potential. Third, reserves protect an entire area from many major human impacts, letting nature flourish and effectively manage itself. In this manner, they represent a form of effective ecosystem-based management even when we do not understand the details of how the ecosystems work (Buck 1993). Finally, reserves provide lightly affected areas that we can use as a reference to understand the effects of human activities and how to control them in the future (see also chapters 4, 8, and 10 for examples and discussion).

## DESIGN CRITERIA

Ecological considerations are important in marine reserve design, but it is imperative to view them in the broader context of a designation process. We begin with this broader view, then discuss some general design principles and conclude with a detailed discussion of several design elements.

### Ecological Dimensions in Context

Ecological considerations are an important part, but only one part, of an effective process to establish marine reserves (Table 5.1). The ocean is a public resource and as such its management should reflect the desires of society, whether those desires are for sustainable fishing, conservation, nonconsumptive uses, or some combination of these. Enforcement considerations should also influence reserve design to ensure that goals are actually met. These considerations can usually be addressed through a set of general criteria that may be satisfied with a wide range of possible designs.

If no-fishing zones are regularly violated, reserves will only benefit the poachers. Enforcement can be aided through the selection of appropriate shapes, sizes, and locations for reserves. However, the most important factor in achieving compliance is often broad community-level support, including acceptance by fishers—the people most likely to be excluded from marine reserves (Proulx 1998). If they are involved early on in the process, exposed to and educated about scientific deliberations, and ultimately held responsible for proposing and modifying reserve designs, local communities and fishers are more likely to take ownership of reserves and assist in both compliance and enforcement. Where reserves have been established without broad public support, they may be vulnerable to dismantling when politics shift (e.g., Russ and Alcala 1999; see also chapters 4 and 8) or in danger of never being created in the first place.

Stakeholder input is also a means to collect and consider invaluable in-

Table 5.1 Recommended Marine Reserve Designation Process

1. Goals
   a. Representatives of the general public specify goals for the management area.
   b. Scientific and enforcement advisers work with these representatives to clarify goals and specify measurable objectives.
2. Design criteria
   a. Scientific and enforcement advisers determine appropriate design criteria for a reserve or reserve network in the management area—including recommendations for overall percent inclusion in the reserve; habitat types to consider; critical areas for inclusion; and the size, shape, and configuration of individual reserves within a network—based on the management goals.
   b. If requested, advisers illustrate their criteria by drawing examples on maps.
3. Drafting alternatives
   a. Representatives of the general public draft alternative network designs, taking socioeconomic and cultural impacts into consideration.
   b. Scientific and enforcement advisers work with the general public representatives to ensure their alternatives reflect the general design criteria.
   c. The public submits alternatives along with descriptions of the socioeconomic and cultural ramifications to the government agency or working group. Advisers provide formal reviews of each alternative and its capacity to meet stated goals.
4. Selecting alternatives
   a. The government group drafts a final list of alternatives, taking into account the public's values, scientific and enforcement reviews, and any potential short-term socioeconomic or cultural impacts.
   b. The public comments on draft alternatives.
   c. The government group chooses a preferred alternative.
   d. More public comments are made.
   e. The government group makes a final choice.

formation about the biology and socioeconomic properties of ocean use (Johannes 1997). Some cultures have studied and fished local waters for centuries, and even shorter-lived fishing traditions can provide a wealth of knowledge for effective reserve design, as discussed further in chapter 7. Stakeholder input will be best represented and included if stakeholders are exposed to the development of scientific and enforcement criteria and then encouraged to develop reserve design proposals that meet these criteria.

### Act Now? or Study the Problem?

One key question in developing marine reserves is whether to act quickly to establish reserves or wait for more study. Several authors have made the point that reserves seem to work for most species even when they are set up with little or no scientific guidance and, consequently, we needn't wait for more study (e.g., Roberts 1998). Other authors have used simple models to illustrate the potential pitfalls of establishing reserves in poor locations (e.g., Crowder et al. 2000). *Sources* are areas that produce more fish than they contain, some of which move to other areas at some point in their life cycle. By contrast, *sinks* are areas that produce fewer fish than live there and are only sustained because they are supplemented from source areas. Crowder et al. developed a model in which areas were either sources or sinks and showed that managers could actually do more harm than good if they created a marine reserve network that encompassed more sinks than sources.

A major flaw with the preceding argument, though, is the assumption that sources and sinks are static—that they do not change with the establishment of marine reserves. In fact, strong evidence supports the fact that reserves usually become sources by creating an area with many fish producing lots of offspring (Appeldoorn 2001). As a result, it is an incorrect simplification of reality to assume that all sinks will remain so if they are designated as marine reserves. Very strong sinks may not be overcome by marine reserves, but these areas are likely to be poor enough in quality that they aren't identified as useful areas from the start. More subtle sinks, on the other hand, are more likely to become sources after reserves result in the buildup of high abundance and reproductive potential within them. In this manner, the concern over sinks is similar to that of poor habitat in general. Reserves may not function well if placed in poor quality habitats, whether because of unrecoverable degradation, source–sink issues, or low natural productivity.

To avoid these potential pitfalls, there are some pieces of information that may be worth taking the time to collect, depending on how readily they might be available. Traditional knowledge of crucial areas—spawning grounds for example—and other key life history traits may be readily available from experienced fishers in the area (Johannes 1978). It also may be possible, depending on the ecosystems and budgets involved, to make at least rough maps showing habitat distributions throughout the management area. Habitat mapping is more achievable than ever with the advent of technologies that can discern habitats remotely, as discussed in chapter 7. Finally, there may be a wealth of scientific information already collected from the area that can be instructive once it is compiled. These sorts of information may add substantially to the effectiveness of marine reserve design without long delays. For most other types of information, though, the benefits gained by learning them would not be worth the time it takes to do so. If important discoveries are made during or after reserve implementation, they may justify a reevaluation of the reserve design. For example, a black grouper spawning aggregation was discovered less than 100 meters outside a newly designated marine reserve in the Florida Keys (Eklund et al. 2000). It is a challenge to provide sufficient flexibility to allow rapid response to discoveries such as these while providing enough process to maintain public support.

## The Designation Process

The process for designating marine reserves will be more effective if it is driven by well-defined goals (Ballantine 1997), which could include the conservation of healthy natural ecosystems, insurance of fisheries against collapse due to management errors, or many other possibilities. The process should clearly specify how it will address public values, ecological, socioeconomic, and enforcement considerations, and the input of fishing communities and other stakeholders. General processes have already been proposed for developing marine reserve networks (Hockey and Branch 1997; Roberts et al. 2003a). These processes score potential reserve areas relevant to the specified goals, engineer a set of biologically adequate alternatives, and select among those according to socioeconomic criteria.

We contend that the most successful site selection and designation processes will rely on the same basic philosophy but be driven less by government agencies or scientists and more by the public (see Table 5.1). While a top down approach may work well in some settings, a bottom up approach is ultimately the most likely to produce long-lasting site designations. However, top-down interest and pressure can provide a broader context for individual site designations and a cohesive national policy can provide a framework for developing a stronger network or system approach, especially if it includes bottom-up input. Regardless, it is valuable to set up several working groups that interact extensively from the start. One should represent the public at large, another should consist of informed and objective scientists with relevant biological and social expertise, a third should consist of enforcement experts, and in situations where multiple agencies have overlapping jurisdiction, a fourth may be required consisting of representatives of those agencies.

Goal setting is a crucial first step. Marine reserves can help to achieve many societal goals, ranging from fisheries enhancements to conservation of natural environments for economic and intrinsic reasons (NRC 2001). Since no single reserve design will satisfy all goals equally it is important to specify goals as the first step in a designation process (Murray et al. 1999). At the same time, efforts should be made to ensure that the reserve design is capable of meeting a range of goals (Roberts et al. 2003a). These goals should incorporate the desires of local people, and their enumeration is an opportunity to involve stakeholders early in the reserve creation process. When establishing goals, it should be clear to what area they apply (Roberts et al. 2003a). It is important that these goals are communicated clearly to a group of scientific advisers in a manner that is amenable to their asking relevant questions. To facilitate this process, we recommend that draft goals be shared with the science group who can then provide feedback to the public group, who may choose to revise the stated goals for clarity, specificity, or both.

Once goals are clearly defined, the science and enforcement groups can develop relevant design criteria. The criteria should state acceptable ranges for several design elements and highlight how choices of one element may affect how other elements are addressed. Scientific criteria will most constructively address several elements:

- The total size of the reserve or reserve network
- The habitat types to consider
- Critical areas for inclusion
- The size, shape, and configuration of individual reserves within a network

Enforcement criteria will most constructively address:

- The size, shape, and location of individual reserves

If requested, advisers should illustrate criteria by drawing examples on maps. In our experience, examples are better received when several quite different options are presented. Single examples tend to be construed as detailed recommendations even if they are meant only to serve as an illustration of how to achieve general design criteria.

Extensive communication with the public helps to inform them of the design criteria and prepare them to draft alternatives for the design of the reserve or reserve network. Although all members of the public deserve the opportunity to present alternatives, special attention should be paid to those stakeholders that spend the most time on or in the water, including fishers. The opinions of stakeholders with extensive on-the-water experience are especially important because of their knowledge of local ocean life and fisheries, their vulnerability to short-term negative impacts if they are displaced by marine reserves, and the crucial role their opinions play in achieving compliance and assisting with enforcement. Scientific and enforcement experts should provide constructive critiques of proposed alternatives, including suggestions of how to make the alternatives fit better with the general design criteria. This step can be formal or informal but will have the greatest impact if it is done in an interactive manner. Modified alternatives should then be submitted to the government agency or working group along with descriptions of how the alternatives were devised, what their short- and long-term socioeconomic and cultural ramifications may be, and the rationale for excluding specific other areas. Scientific and enforcement advisers can then provide analysis of how effective each alternative is likely to be at meeting the stated goals. The government agency or group then drafts a final list of alternatives, taking into account the public's values, ecological and enforcement reviews, and any potential socioeconomic or cultural impacts. They should aim to include a suite of alternatives that covers the range of realistic possibilities and addresses the desires and concerns of a broad cross section of the public. The public should have a chance to comment on the alternatives to help inform the government's choice of a draft preferred alternative, and be granted another chance for comment before a final selection is made.

This process may be more time consuming than a more autocratic one and less rigorous than a more scientifically driven one. But, it is worth the time investment and loss of some scientific rigor to involve the general public deeply in the process. If the general design criteria are done well, they should ensure that the reserve or reserve network will be effective enough. This process emphasizes public involvement because there are far greater dangers for failure due to lack of public acceptance than due to poor design.

## Goals

Marine reserves can help to achieve many societal goals, which can generally be lumped into four categories: (1) ecosystem protection, (2) improved fisheries, (3) expanded knowledge and understanding of marine systems, and (4) better nonconsumptive opportunities (NRC 2001). There has been a disconnect in the ways in which design criteria have been addressed for each goal. Whereas optimality models—which predict the design that will maximize reserve performance—have been the norm when addressing fishing benefits, especially the maximization of yields (NRC 2001), risk minimization models and other methods to predict the minimum necessary design have been the norm for other goals. The reason for this disconnect is simple: the optimal solution for addressing conservation and nonconsumptive goals would be to close the entire ocean to fishing and other major human impacts, while our scientific

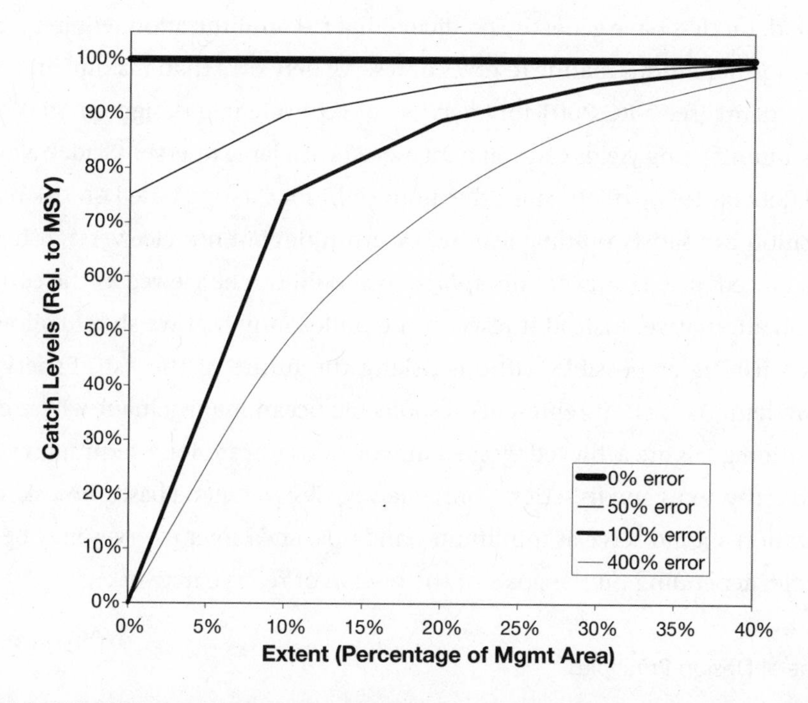


**FIG. 5.1 Risk Minimization Overrides Other Design Considerations.** This graph shows sustained catch levels, relative to maximum sustainable yields (MSY), as a function of reserve size and error level. Marine reserves, or another method of setting aside a protected population, maintain productive fisheries even if managers are making large errors. Reserves increase catches under all circumstances when management errors lead to overfishing, with more dramatic increases under larger errors. Note that maximum sustainable yields can be achieved with large reserve networks even if errors are large. Also note that the use of reserves does not necessarily cost any yield—without errors, maximum sustainable yields can be achieved with or without reserves. Source: Adapted from Sladek Nowlis and Bollermann 2002.

understanding would be best served by having an extensive reserve network with smaller scale experimental fishing areas.

How do we find balance given this dichotomy? It depends on societal values but, more often than not, marine reserves should be viewed in terms of risk minimization for all benefits rather than optimality for fishing benefits. This conclusion follows two important observations. First, risk minimization is an underlying goal for all four of the categories. Unless we conserve the entire ecological and socioeconomic system with some minimal network of marine reserves, none of the other goals can be assured (Appeldoorn and Recksiek 2000). Research has shown that setting aside populations free from risk of fishing makes the populations as a whole much more robust and resilient to inadvertent management errors (Fig. 5.1; Sladek Nowlis and Bollermann 2002).

Second, models have generally predicted that risk minimization requires larger and more ambitious marine reserves or reserve networks than maximizing fishing benefits (see NRC 2001 for overview of goal-oriented design criteria), and maximum fishing yields can still be achieved with large reserves (Sladek Nowlis and Roberts 1995, 1999). Since the more ambitious design based on risk minimization can satisfy optimal fishing opportunities but not vice versa, it has to take precedence. However, this approach should not be viewed as the conservation alternative. Instead it rests on the philosophy that we should allow as much fishing as possible without risking the future of the fish, fishery, or ecosystem. As such, it represents responsible ocean management where conservation goals are achieved by stacking the odds toward long-term success for both ecosystems and fisheries. Consequently, design criteria based on risk minimization should serve as minimum standards, and larger reserves may be desirable, depending on the goals of the reserve or reserve network.

## General Design Principles

A substantial effort has gone into developing design principles for conservation areas on land, and much can be learned about marine reserve design from them. These efforts have examined and illuminated a wide range of concepts, including (1) minimum viable population sizes, (2) effective population sizes, (3) biodiversity hotspots, and (4) landscape processes. The first two of these concepts are aimed at ensuring any protected area is sufficiently large to contain a viable population. For many land species and a few ocean ones, an area may need to be very large to maintain a viable population, often considered to require at minimum an effective population size of 200 (Gilpin and Soulé 1986).

The latter two concepts are aimed at identifying priority areas for protection and represent two very different approaches. Under the hotspot approach, scientists map out the ranges of any and all species of interest. They then analyze those maps to identify hotspots that contain particularly large numbers of species. Ideally, areas should be chosen so that all species are represented in at least one conservation area. This approach offers the potential to find and use complementary areas to achieve broader conservation goals, but it also raises some concerns. Species ranges are not static and may change with developmental stages, seasons, and ecological succession—the natural process of recovery of an area to natural or human disturbance. Moreover, concern has been raised that areas of high species diversity may in fact be poor quality habitat for many of them (Araújo and Williams 2001). In contrast, the landscape process approach looks at systems in a more dynamic way. Including an entire watershed as part of a protected area (marine or terrestrial) is one simple illustration of this sort of thinking since upstream activities can impact a downstream conservation area (Pickett et al. 1997). On land, this approach has led to such ideas as providing corridors to link networks of protected areas (Noss 1987; Simberloff and Cox 1987).

We should heed the lessons learned on land, but with care to avoid overgeneralizing them. One fundamental difference between ecosystems on land and in the sea is their status. Although marine ecosystems have suffered badly (Jackson et al. 2001; Myers and Worm 2003), they are generally not as badly degraded as those on land. One important implication of this difference is that we still rely on the ocean to a greater degree for wild-caught food. The design of parks on land, where much habitat has already been urbanized or farmed, often focuses on keeping as much of the biota inside as possible—an island of nature amidst human development. The greatest threat to land-based biodiversity has gradually shifted from overexploitation to habitat destruction, in part because many of the great natural food sources have been hunted to unproductive levels. In the sea, there are still widespread habitats that have the potential to perform their natural ecological functions if managed appropriately. As a result, we can think of marine reserves in the context of wild fish production to sustain outside fishing areas (NRC 2001). Another difference that contributes to the capacity of marine reserves to produce food is the relative openness of ocean systems. The fluidity of the ocean makes it inevitable that many species will disperse beyond reserve boundaries. These differences all favor the use of the landscape process concept over hotspots as a guiding principle for designing networks of marine reserves.

Scientists have already identified some guiding principles for designing marine reserves or reserve networks and we will build on these. Ballantine (1995, 1997) identified three important concepts:

- Representation of all habitats
- Replication of reserve units to avoid losing too much from the occasional poor quality area
- Networking the reserve units in a self-sustaining manner

He suggested that the network should encompass 20 to 30 percent of the total management area. Roberts et al. (2003b) added a few additional rules of thumb. They recommended prioritizing sites to most efficiently achieve the greatest result. Specifically, they recommended preferentially including four site categories:

- Sites that include vulnerable habitats
- Sites that contain vulnerable life history stages
- Sites that are capable of supporting exploited species or rare species
- Sites that provide ecological services

The services include coastal barriers and water purification but might also include places that have special nonconsumptive value, like a popular diving spot (see chapter 4 for more detail on other ecological services). These authors also recommended avoiding sites with very high threats from human or natural disasters. We agree with avoiding sites under high threat of human catastrophes and also those highly vulnerable to natural disasters if reserves are going to be sparse. However, if a reasonably large reserve network is going to meet the other general rules of thumb, then it will be important to include areas that are frequently disturbed naturally. We believe they should simply be treated as one more habitat type to be included in a representative and replicated manner.

This landscape-level approach is being used more frequently. Sala et al. (2002) showed that biodiversity in the Gulf of California, Mexico, was not random. Instead it showed organization corresponding to latitude and depth. Friedlander et al. (2003a) took a similar broad look at the pattern of diversity around Old Providence and Santa Catalina Islands in the Seaflower Biosphere Reserve, Colombia. They found strong similarities in the full assemblages and bottom-dwelling reef communities within habitat types they had previously defined, differences among habitat types generally, and additional differences between sites near the island and those on a long shallow bank that extended to the north. Once these distinctions were recognized, they were incorporated into the design process by ensuring that each distinct type was represented.

### Crucial Factors

Although a number of different factors can influence marine reserve design, two stand out as especially important: the fluidity of the ecosystems involved and the extent of damaging activities outside the reserve. Not coincidentally, these two factors underlie our ability to rely on the sea for wild-caught food. The fluidity is a crucial factor because reserves will be more effective the better they retain adults, although some degree of export of reproduction is desirable (PDT 1990; Sladek Nowlis and Roberts 1999). The extent of damaging activities is important because it determines the extent to which reserves have to accomplish all management objectives. For example, high yields can be achieved from many fisheries in the absence of marine reserves if fishing is at relatively low levels, carefully controlled, or both. In contrast, very large reserves may be necessary to achieve similar fishery yields if fishing activity is high in the remaining fishing grounds (Sladek Nowlis and Roberts 1999).

*Movement to and from Reserves*

Marine reserve design will be influenced a great deal by the degree of interaction across space in the ocean. Most marine species, particularly those targeted by fishing efforts, have planktonic larvae, which generally spend from a week to several months in the water column (Boehlert 1996). These lengthy larval periods provide great potential for long-distance dispersal. It has been demonstrated that, if larvae drift passively on surface currents, they may move hundreds of kilometers during their larval phase (Roberts 1997). Reinforcing this evidence of high dispersal potential, genetic studies have shown surprisingly high homogeneity in marine populations across entire ocean basins, suggesting that populations mix genetically over these broad ranges (e.g., Lacson 1992). However, work in population genetics has also shown that very little mixing, on the order of a few individuals per generation, need take place to maintain this homogeneity among otherwise distinct populations (Slatkin 1987).

Not all organisms can disperse so far. Tunicates, for example, generally produce large larvae with limited dispersal ability. Larvae of the tunicate *Lissoclinum potella* are visible to the naked eye in field conditions and can be followed from parent to settlement (e.g., Olson and McPherson 1987). The lack of dispersal capability can influence tunicate biodiversity patterns, with extremely limited distributions for some species (e.g., the Chilean tunicate *Pyura praeputialis,* Clarke et al. 1999). Though tunicates are not a traditional target for exploitation most places in the world, there is growing interest in previously nonexploited groups, including tunicates, by the aquarium and pharmaceutical industries. Other exploited species also show limited dispersal patterns (e.g., the bull kelp *Durvillea antarctica,* Castilla and Bustamente 1989), but these species are exceptions to the general rule of high dispersal potential among exploited species.

However, larvae might not disperse as far as surface currents on the open ocean would suggest. Small-scale coastal oceanography can play a major role in larval dispersal and recruitment (Caselle and Warner 1996) and can lead to local retention of larvae (Black 1993; Wolanski and Sarsenski 1997). Larval behavior can also play an important role in dispersal (e.g., Jenkins et al. 1999; Katz et al. 1994). Together, local oceanography and larval behavior can lead to significant amounts of local retention (e.g., Cowen et al. 2000). These patterns can be accentuated by differential survival of newly settled individuals. Even if larvae manage to move far away from parents, the chances of discovering suitable habitat decrease with increasing distance. If they do not find suitable habitat, they are more likely to die—making the dispersal ultimately ineffective.

These discoveries are backed by growing evidence that there are genetic gradients, which indicate relatively low exchange through dispersal (Palumbi 2003). Collectively, the body of work on larval dispersal suggests that many coastal marine populations retain larvae locally but also allow some larvae to disperse large distances. Perhaps not coincidentally, this strategy of mixed dispersal ranges has been shown to have profoundly positive ecological and evolutionary benefits in terrestrial and freshwater systems (Cohen and Levin 1987). The interactions among larger-scale oceanographic processes, smaller-scale coastal oceanography, and larval behavior are still poorly understood and remain a great mystery of marine ecology.

Reserves provide an opportunity to learn more about larval dispersal patterns. Reserves create a buildup of biomass, and therefore potential reproductive output, within their borders. To the extent that larvae are retained locally, this phenomenon should be observable as gradients of larval abundance, decreasing as one moves away from the reserve. Only a few studies have attempted to document a larval gradient, or any phenomenon, outside of reserve boundaries, but they offer promising results. In one reservelike experiment, Tegner (1992) reintroduced green abalone (*Haliotis fulgens*) into an area off of California. Green abalone were depleted to very low levels at the time of reintroduction from heavy overfishing, so the reintroduced abalone served as a reserve of sorts. The author demonstrated higher than expected recruitment in the general area of the reintroduction, with apparent recruitment enhancement up to 8 km away. Additional studies like this one offer potential to gain a better understanding of larval dispersal and the links between adult biomass and new recruitment into a population.

Like larval dispersal, postsettlement movements by juveniles and adults may contribute to the openness of marine systems. Marine species can be classified into three general categories: benthic, pelagic, and demersal. Benthic species are associated with bottom substrate as adults, and thus have extremely limited adult dispersal capabilities. There are some benthic species that have been shown to move substantial distances as adults (e.g., spiny lobsters, Acosta 1999) but most are physically attached to the bottom with minimal movement capabilities. Incidentally, benthic species may be most capable of benefiting from reserve protection because their limited movement capabilities can make them susceptible to reproductive failure if density is too low (Levitan 1991). Pelagic species form the other extreme, being associated almost exclusively with the water column. Demersal species fall in between, typically associated with bottom structure but with capabilities to swim in the water column. Reserve studies and reserve efforts have generally focused on benthic and especially demersal species (Halpern 2003) because pelagic species may be less likely to benefit from reserves (Bohnsack 1996; but see Guenette and Pitcher 1999).

Although postsettlement movement in demersal species is easier to study than larval dispersal, it is nearly as poorly understood. Studies have examined movement patterns of both tropical and temperate demersal species both indirectly and directly, albeit with a bias toward reef-associated fish.

Reserves provide the majority of indirect evidence about postsettlement movement. For example, Russ and Alcala (1996) showed higher densities of adult fish close to the border of a reserve on Apo Island, Philippines, than farther from it, suggesting the possibility that fish move across the border but not too far. Similar results were found at the border of a marine reserve on Barbados (Rakitin and Kramer 1996). Johnson et al. (1999) demonstrated that, in addition to a buildup of biomass within a Florida reserve, some fish moved in and out, and a number of world record trophy fish were caught in the area. Additional direct evidence comes from observable phenomena like spawning aggregations where high abundances concentrated in space and time could only be explained by movements to and from the aggregation.

Direct evidence consists of tagging experiments, where tagged fish are retrieved by fishers or by underwater visual observation, and tracking experiments, where fish are equipped with an acoustic device that can be tracked using a hydrophone from the surface. Tagging experiments provide a coarse-scale picture of movement, showing the limits over longer periods of time. Tracking experiments provide the fine-scale picture, showing detailed movement patterns over short time frames.

Although juvenile and adult fish movement patterns are still poorly understood, a picture is emerging that includes specialized movements tied to particular life history events with less frequent and often habitat-limited movement patterns at other times.

Many species of fish and marine invertebrates utilize different habitats as they grow and mature, typically moving from shallow inshore habitats, to deeper offshore habitats (Roberts 1996). In some cases, these movement patterns through development will vary depending on whether appropriate habitats are adjacent, and may be inhibited by habitat types that act as a barrier to dispersal (e.g., Acosta 1999). These sorts of movements can also occur on a daily basis. Juvenile grunts in the Caribbean undergo predictable movements between daytime resting and nighttime feeding areas (Ogden and Ehrlich 1977).

Numerous species of fish aggregate to spawn, requiring long-distance movements for some. Tropical species like groupers and snappers are particularly well known for this phenomenon. Larger groupers and snappers can migrate great distances to specific sites and form spawning aggregations of hundreds or thousands of individuals at specific times of the year (Domeier and Colin 1997). Long-term persistence of these aggregations at specific sites (e.g., Colin and Clavijo 1978) makes these species extremely susceptible to fishing pressure (Sadovy 1993). Several grouper and snapper species have been greatly overfished throughout the world, largely due to extreme exploitation of spawning aggregations. Despite the recognition of this behavior and great importance to conservation, only a small, but growing, number of spawning aggregations have been closed to fishing, and a paucity of scientific information exists on the details of spawning aggregations, especially the potentially important habitat characteristics of spawning aggregation sites (Fig. 5.2; and see chapters 9 and 10). It is not surprising that fish seek this complex habitat while spawning because the complexity offers shelter for the large number of fish that gather, and it may also influence the dispersal of offspring so that some are retained within the complex habitat structure whereas others are effectively dispersed longer distances (Wolanski and Sarenski 1997).

Many temperate adult fishes show similar patterns (Cushing 1995). Fish tend to be found in higher concentrations on spawning grounds compared to feeding grounds and thus are more susceptible to fishing pressure in these locations. The spawning grounds for plaice in the Southern Bight in the southern North Sea have remained in the same location since the grounds were discovered in 1921 (Harding et al. 1978). Between 1921 and 1967, the mean peak date of spawning for these fish was January 19 with a standard deviation of less than one week (Cushing 1969). Plaice spawning therefore exhibits a high degree of predictability in both space and time. Cod tagged in five regions on the Canadian Shelf tended to return to their grounds of first spawning, with a low emigration rate (distant recaptures/all recaptures) of 0.0375, which would

**FIG. 5.2 Complex Habitat at a Spawning Aggregation Site (NOAA/NOS/NCCOS/CCMA-Biogeography Program unpublished data).** The Red Hind Bank Marine Conservation District, St. Thomas, U.S. Virgin Islands, was established to protect spawning aggregations of red hind (*Epinephelus guttatus*), a Caribbean reef fish, and other species. Not coincidentally, red hind and other groupers utilized a site with complex habitat, providing refuge for spawners (see Beets and Friedlander 1999 for more detail).

promote a relatively high rate of gene flow between each group (Thompson 1943) but relatively little mixing from a standpoint of stock productivity.

Despite the ability to migrate large distances and the propensity of some species to do so at specific stages of their life history, most reef-associated fishes appear to be rather sedentary and possess relatively small home ranges. Holland et al. (1993) found that a population of *weke* (white goatfish, *Mulloidichthys flavolineatus*) in Hawaii showed high site fidelity, with 93 percent of recaptures occurring at the release site. In a trapping and tagging study conducted in Hanalei Bay, Kauai, Friedlander et al. (1997) recaptured or resighted 85 percent of all tagged individuals of twenty-three species within 50 meters of their release site. The limited range of dispersal of recaptured *omilu* (blue trevally, *Caranx melampygus*) (75.5 percent within 0.5 km of the release site) and strong site fidelity observed from sonically tagged fish suggest that dispersal is much less than might be predicted for a highly mobile, piscivorous species (Holland et al. 1996). *Kumu* (whitesaddle goatfish [*Parupeneus porphyreus*]), a Hawaiian endemic goatfish and important fisheries species, were acoustically tracked around the Coconut Island refuge for periods up to ninety-three hours (Meyer et al. 2000). The home ranges of all fish were within the boundaries of the Coconut Island reserve. This small reserve (less than 1 $km^2$) was capable of protecting both large juveniles and some spawning size individuals (Meyer et al. 2000). *Kala* (blue spined unicornfish [*Naso unicornis*]) were acoustically tracked for periods of up to twenty two days in the shallow high-energy fringing reef habitat in the Waikiki Marine Life Conservation District (Meyer and Holland 2001). The home ranges of all of the *kala* tracked were completely encompassed by the boundaries of the Waikiki Marine Life Conservation District. However, more mobile species, such as jacks and goatfishes, ranged over an area slightly larger than 1 $km^2$ (Meyer 2003). Even this limited movement exceeded the 0.32 $km^2$ size of the Waikiki Marine Conservation District and left these fish vulnerable to fishing. Caribbean reef fish showed similarly restricted movements in and near a marine reserve on Barbados, with most species rarely showing much movement away from the site of first capture (Chapman and Kramer 2000). However, some species (e.g., horse-eye jacks [*Caranx latus*] and bar jacks [*C. ruber*]) did appear to move frequently from the study area. Moreover, even sedentary species showed relatively more movement when the reef habitat was uninterrupted than when it was fragmented.

These results suggest considerable site fidelity on the part of a number of species. They also suggest that an association exists with a particular locality of rather limited size. Short-term (e.g., day-to-day) movements may be common, but tag recovery data and telemetry data indicate that if these fishes make such movements, most of them return to the home locality. Although different species have widely different movement patterns these results are generally consistent with existing ideas about the limited normal range of movements of many demersal, reef habitat–associated species.

Nevertheless, some evidence suggests that marine systems are open and thus vulnerable to impacts unless marine protected areas are very large. Friedlander and DeMartini (2002) identified twice the fish biomass in the large, remote, and lightly fished Northwestern Hawaiian Islands than in small, fully protected marine reserves in the main Hawaiian Islands (Sladek Nowlis and Friedlander, 2004). Large apex predators made up the majority of fish by weight in the Northwestern Hawaiian Islands but were virtually absent from reefs in the main Hawaiian Islands, even those protected from fishing. The marine reserves themselves contained more than twice the fish biomass of areas that received partial or no special protections within the main Hawaiian Islands (Friedlander et al. 2003b). Nevertheless, the reserve effect was not adequate to reestablish fully functioning ecosystems, most likely because of the small extent of marine reserves as a whole and the heavy fishing pressure surrounding them. Although many benefits can be reaped from small reserves, larger ones may be necessary to sustain fully functioning ocean ecosystems.

*Outside Impacts*

The second crucial marine reserves design factor is the degree of outside impacts. For example, models have shown consistently that maximum yields can be obtained over a range of reserve sizes depending on the intensity of fishing outside of the reserve (Sladek Nowlis and Roberts 1995, 1999). If impacts are light and strictly controlled outside of reserves, marine reserves may not be necessary. However, experience suggests that fishing rates rarely stay light and are even less frequently under strict control of managers (e.g., Myers and Worm 2003).

In addition to modeling reserve design based on the scales of outside impacts, managers may wish to reduce the magnitude of outside impacts on reserves. One way in which managers can do so is by integrating reserves with coastal zone, ecosystem, or broader ocean zoning management plans. Another way is through the creation of linked land–sea protected areas that protect adjacent terrestrial and marine areas. Reserve designations can help to protect the designated area from fishing, point-source pollution, and other directed human impacts, but may not offer protection from non–point source runoff. As a result, it may be desirable to locate marine reserves downstream from terrestrial protected areas, or at least to enact stricter controls on upstream development if a reserve is put in place. In some cases, though, it may be necessary to scale up the size of marine reserves to account for major disturbances such as oil spills or hurricanes (Allison et al. 2003).

### Total Extent of Reserves or Reserve Network Coverage

The hottest debate regarding marine reserves usually surrounds their extent of coverage through the management area in question. A recommendation of 20 percent (PDT 1990) created uproar along the southeastern Atlantic coast of the United States, whereas a recommendation of 30 to 50 percent (Airamé et al. 2003) caused a similar stir in the Channel Islands of California. Total extent is a key design consideration because it is arguably the most important for achieving goals but also has the greatest influence on what the short-term costs are likely to be for displaced stakeholders (Fig. 5.3). Costs can be significant if large marine networks are created (Sladek Nowlis and Roberts 1997). Fortu-

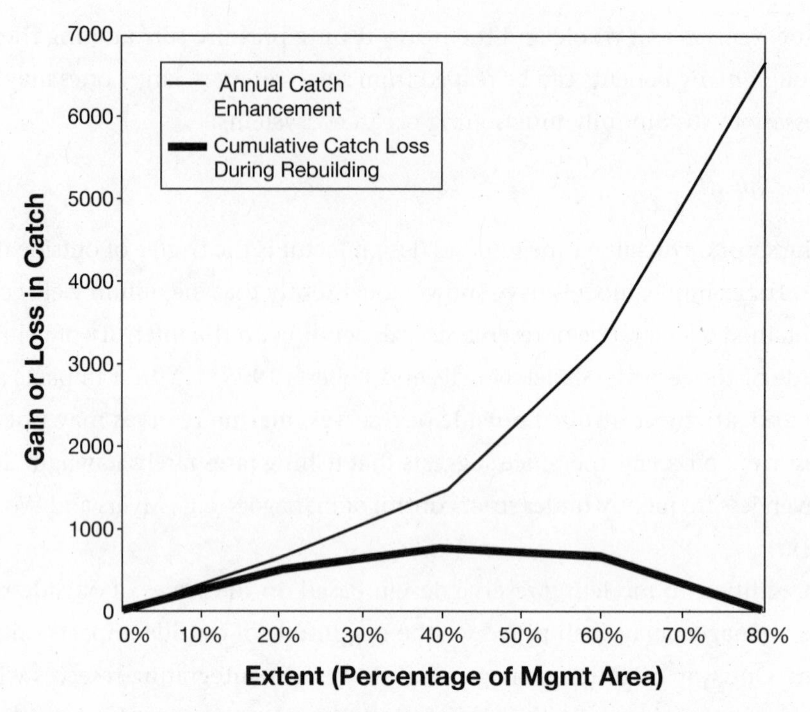


**FIG. 5.3 Extent of Reserves: More Benefits, More Costs.** A population model of the white grunt (*Haemulon plumieri*), a Caribbean reef fish, experiencing heavy overfishing initially. The creation of a marine reserve or reserve network enhances catches in the long term but also generates short-term costs. Source: Adapted from Sladek Nowlis 2000.

nately, those costs are expected to be offset by reserve benefits clearly and quickly when fisheries are most depleted, and can be deferred under any circumstances by phasing in reserves (Sladek Nowlis and Roberts 1997). It is also worth noting that for most overfished fisheries—ones that have been depleted to the point where they have lost productivity potential—reserves often impose smaller opportunity costs than other more conventional management techniques while offering the greatest chance of achieving rebuilding to more productive levels (Sladek Nowlis 2000).

A number of scientists have examined the question of how much area to include in a marine reserve or reserve network (reviewed in NRC 2001). The recommendations depend on a number of factors, foremost among them being the goals and objectives for the reserve or reserve network. A number of studies have examined the extent of reserve that will maximize fishery yields (e.g., Sladek Nowlis and Roberts 1999) and even these studies found a wide range of possible extents, from none to sizable, depending on several circumstances. On the other hand, there are also many nonconsumptive benefits one might achieve from marine reserves, and most of these benefits will be maximized by having very large extents. Consequently, the percentages discussed following here should be viewed as minimum standards rather than optimal recommendations. To move forward on this complex issue, we focus on risk minimization. While this is not the only goal one might have for a marine reserve or reserve network, it is fundamentally important because without it managers are gambling with the future of the ecosystem and the people that rely on it for sustenance.

Scientific results suggest that reserve networks need to protect a population consisting of 30 to 50 percent of its pristine size to ensure against collapses (Mangel 1998; Sladek Nowlis and Bollermann 2002). This level of insurance reduces the tremendous uncertainties that surround fisheries management. For example: in the United States, where more resources are available than almost anywhere else on the planet for fisheries management and science, over three quarters of all federally managed fish populations (some species are managed as separate populations across their range) were of unknown abundance, had unknown levels of fishing pressure, or both, in the year 2000 (NMFS 2001). State managed fisheries may do even worse. A 2002 California report indicated that 85 percent of all nearshore species (the ones most likely to be State managed) were of unknown status (CDFG 2002). If we cannot even identify the status of fish, we surely cannot manage them with certainty. Scientific studies clearly indicate that uncertainty can be countered most effectively by maintaining a portion of a fished population as off limits from all fishing (e.g., Sladek Nowlis and Bollermann 2002).

Closing 30 percent of an area will not necessarily protect 30 percent of the fished population from fishing, nor will it necessarily reduce catches by 30 percent. If especially productive areas are chosen, the effects may be amplified. On the other hand, two factors will cause a smaller portion of the population to be protected than reserves might suggest.

1. Marine systems vary tremendously in their openness, and some reserve benefits will be diluted the more individuals move across reserve boundaries. Openness can be minimized by using relatively fewer large reserve units within a network because large reserves will leak less than smaller ones (Diamond 1975). However, it is likely that systems with high fluidity will need a greater extent of reserve coverage to counteract this fluidity. As such, reserves are a more obvious choice for less open systems—including most coastal ocean ecosystems—than more fluid systems, like open-ocean pelagic ecosystems. It is also possible to address this issue by selecting areas with high habitat diver-

sity since many species move among different habitats daily (Holland et al. 1993) or throughout their life cycles, and may need to move shorter distances if several habitat types are in close proximity (Appeldoorn et al. 1997, 2003).

2. Ecological disasters will also tend to increase the extent of reserve coverage necessary to minimize risks. Natural and human-caused ecological disasters may disrupt equilibrium processes inside reserves and in doing so make them less effective at meeting management goals. However, natural ecological disasters are part of natural cycles and we should not harbor a static view of the world. When habitats are disturbed, the natural process of succession gradually restores mature living communities; for example, old-growth forests or coral reefs. Succession is an important process in ecology. It is through this process that early successional, fast-growing, and widely dispersing organisms coexist with the hardier, slower-growing species that ultimately outcompete them in an undisturbed site. Retaining the natural cycles of disturbance is important for maintaining ecological balance. When damaging human activities are added to natural forms of disturbance, the results depend on their relative magnitudes and frequencies. In an environment with little natural disturbance, even small amounts of human impacts can disrupt the ecosystem. By contrast, in an environment with high rates of natural disturbance, relatively high rates of human impacts may not noticeably affect the ecosystem. If human impacts do degrade certain areas more than the natural cycle of disturbance, reserve coverage will need to be scaled up to account for the degradation. For example, it has been determined that along the California coastline reserve coverage needs to be scaled up by 20 to 80 percent to address oil spills (Allison et al. 2003).

A number of scientists have identified 20 percent reserve coverage as a minimum societal goal (e.g., Ballantine 1997; PDT 1990). This percentage was originally proposed based on overfishing definitions that suggested fished populations should be maintained at levels that on average allowed individuals to achieve 20 percent of their expected reproductive output (Goodyear 1993). Since that time, overfishing definitions have been overhauled, often to far more conservative levels (e.g., 40 percent for rockfish, Clark 1993). Recent recommendations have focused on broad concepts of insurance rather than on single-species management. These approaches have identified that reserve coverage of as little as 10 to 20 percent can help sustain a fishery, whereas 30 to 50 percent may be necessary to ensure high, long-term abundance and catch levels (Sladek Nowlis and Bollermann 2002). Depending on the scale of impacts outside, the openness of the system, and the rate and extent of ecological disasters, reserve extents may need to be modified to encompass the desired proportion of the unfished abundance. And these recommendations should be viewed as minimum standards rather than optimal recommendations because there are many nonconsumptive benefits that may also be important in the designation of marine reserves.

### Size and Shape of Individual Reserves

Scientists have actively debated the size and shape of conservation areas on land since the 1970s and developed some general principles. Their debate was based on the presumption that resources would limit the total coverage of land conservation areas and, as such, a key decision would be how to parcel out the coverage. Coined the single-large-or-several-small (SLOSS) debate, a general consensus emerged that few larger reserves were generally better than several small because they were more likely to contain functional ecosystems within their borders and to suffer less from outside effects (Diamond 1975).

In the sea, the size and shape of individual reserves can have important effects on ecological and socioeconomic performance as mediated by the fluidity of the system and scale of impacts in outside areas. Whether goals are to enhance fishing opportunities or conservation of natural ecosystems, it will be desirable to design marine reserves so that adults stay inside them while some of their offspring disperse out (PDT 1990; Sladek Nowlis and Roberts 1999). There is one fairly minor exception to this rule. In a few cases (e.g., recreational trophy fisheries) there will be far greater value placed on the catch of a few very large individuals. In these cases, it may be desirable to have a small amount of adult movement across reserve boundaries (Johnson et al. 1999). But even this minor motivation to have leaky reserves should be tempered for a couple of reasons. First, too much leakiness will prohibit fish from growing large enough to provide the trophy opportunities. Second, because of the wide range of movement patterns exhibited by different species and sometimes even different members of the same species (as discussed following here), most sizes and shapes that maintain most adults within reserve borders will also foster enough spillover of some species to provide trophy fishing opportunities.

Keeping adults in marine reserves will be easier in some systems than in others. Given new studies that show relatively low rates of adult movement in coastal environments (e.g., Attwood and Bennett 1994; Holland et al. 1996; Fig. 5.4), relatively small marine reserves may adequately protect adults near

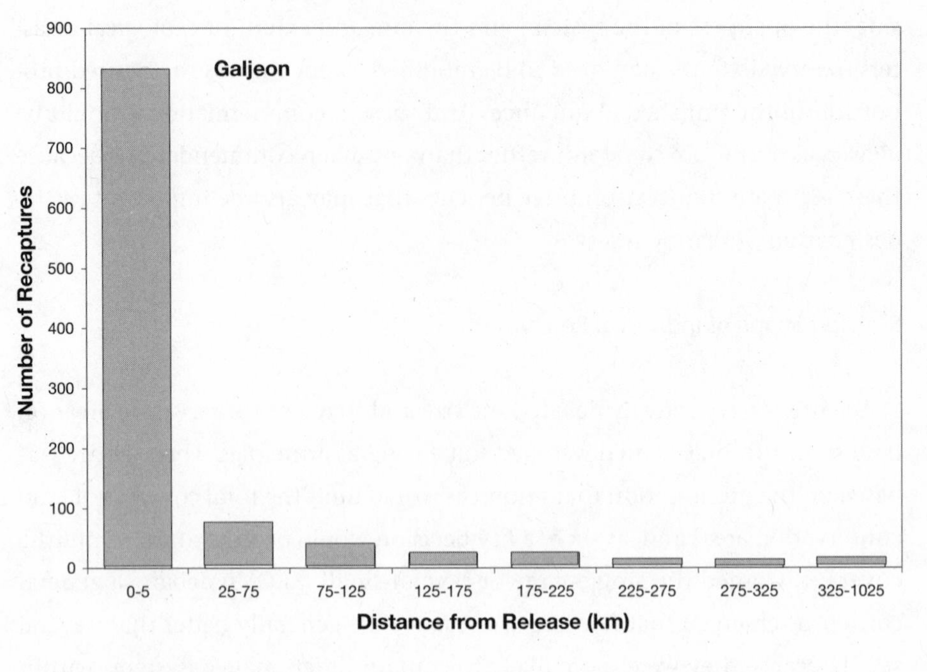


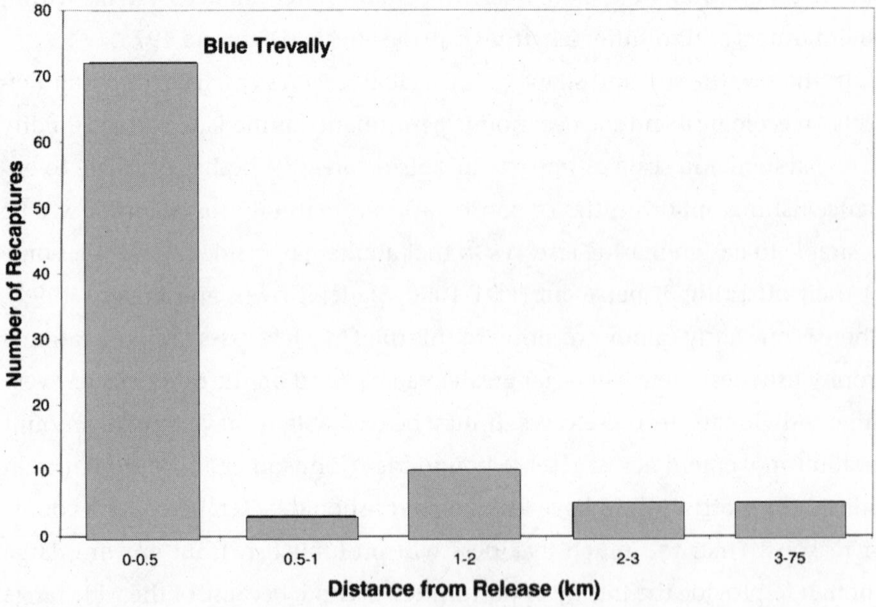


**FIG. 5.4 Movement Tendencies and Capabilities.** Both the (a) galjeon, a southern African shoreline fish, and (b) blue trevally, a Hawaiian shallow-water predator, showed remarkable dispersal potential, yet the vast majority of individuals stayed very close to home (see Attwood and Bennett 1994 and Holland et al. 1996 for more detail).

the coast. In fact one reserve on the Caribbean island of St. Lucia was highly effective despite measuring only 150 by 175 meters (Roberts and Hawkins 1997). The shape of a reserve could help if it accounts for connectivity among habitat types. Since fish may use more than one habitat type throughout the day or their life cycle, it is usually preferable to include multiple habitat types within the same reserve (Appeldoorn et al. 1997). There is no single best shape for doing so, although swaths stretching from shore into deep water are more likely to contain a diversity of habitats than reserves without as much depth range (e.g., PDT 1990), and may encompass common natural migrations from shallower, land-associated to deeper habitats (Appeldoorn et al. 2003; Davis and Dodrill 1989; Love 1996).

For highly mobile species like the bluefin tuna, reserves might need to be extremely large if they are going to protect a substantial portion of adults. Instead, these species might gain more realistic protection from reserves located in areas where large groups of animals come together to feed or reproduce (Hyrenbach et al. 2000). Complementary regulations will be especially important for highly mobile species to address uncertainty. These could include size limits (Myers and Mertz 1998) or quota systems (Sladek Nowlis and Bollermann 2002). Regardless, we find relatively sedentary species in virtually every part of the ocean, particularly associated with bottom habitats. Given increasing fishing efforts targeting deepwater species and the impacts this can have on bottom habitats (Dayton et al. 1995), it makes sense to consider all areas of the ocean when designing networks of marine reserves.

Enforcement can also be enhanced through the shape of individual reserves. Both enforcement and compliance will be greatly aided if reserve borders are straight lines running north–south and east–west or utilizing other obvious navigational reference points. Enforcement will also generally be easier if there are relatively few large rather than many small reserves. However, the size issue has enough ecological and socioeconomic implications that enforcement considerations may be secondary in this respect.

### Site Selection

Marine reserve networks have the greatest chance of including all species, life stages, and ecological linkages if they encompass representative portions of all ecologically relevant habitat types in a replicated manner (Ballantine 1995, 1997). Studies indicate that habitats are a good surrogate for species, so that a system of protected areas that incorporates all habitat types is also likely to provide refuge for most species. In fact, habitats are generally a better focus for protected area design than species because they are easier to map and are more closely tied to the ecological processes whose conservation should be the ultimate goal. As a precursor to including habitat types in a protected area, it will be necessary to define habitat types in an ecologically relevant manner and map out their distributions. Let us consider an example.

We have performed extensive surveys of the coastal ocean environments around Old Providence and Santa Catalina islands (in close proximity to each other) in the Archipelago of San Andrés, Old Providence, and Santa Catalina, Colombia, which were recently designated the Seaflower Biosphere Reserve (Friedlander et al. 2003a). We used preexisting habitat maps (Díaz et al. 1996) and surveyed a wide range of habitat types for ecological differences as identified by distinct fish assemblages and communities of organisms living on the seafloor. Based on these surveys, we were able to lump several habitat types into a few simple yet distinct categories and identify ecological connections among these habitats. We also identified a major difference between otherwise similar appearing habitats based on their proximity to land (Appeldoorn et al. 2003). The shelf extends approximately 20 kilometers north of the islands but only a short distance south. Our surveys showed distinct differences in fish assemblages for all habitat types depending on whether the survey site was close to the island or on the northern bank. This finding confirmed the importance of links between coral reefs and other nearshore habitats, like mangrove lagoons and sea grass beds (Ogden 1988), which serve as nursery grounds for a number of species. The bank habitats were nevertheless valuable and worthy of protection because of their differences (some species thrive in the absence of the ones that start life in mangroves or sea grass beds), but it was helpful to identify them as different from otherwise similar looking habitats near the islands.

It is important to represent all habitats, but some may have greater conservation value than others. It is especially important to identify limiting habitat types and ensure that these are preferentially included in no-fishing or no-entry zones. These habitats fall into three categories: rare habitats, especially vulnerable habitats, and habitats where fish are especially vulnerable to overfishing. Habitats may be rare because they only develop under limited ecological conditions or because they have been disproportionately impacted by previous activity. For example, coastal wetlands exist in a narrow band at the water's edge and have been targeted heavily by coastal development (Rosenberg et al. 2000). Vulnerable habitats would include those that are especially likely to be impacted by fishing activity. Structurally complex habitats are especially vulnerable to the impacts of bottom trawling, for example (Dayton et al. 1995; Watling and Norse 1998). Habitats where fish are especially vulnerable primarily include places where fish gather to feed or reproduce. When they are

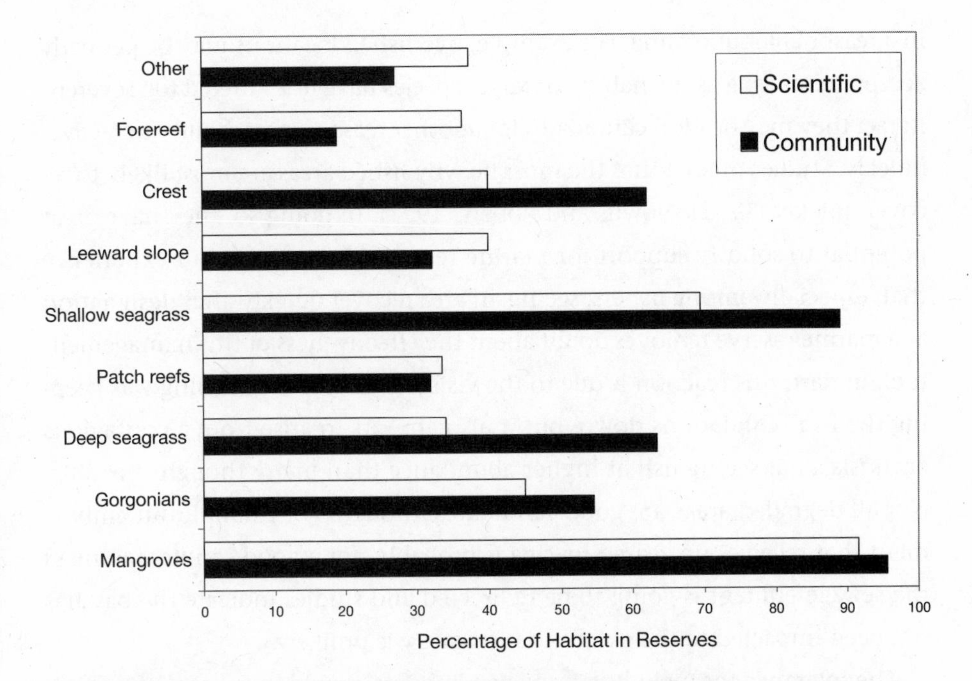


**FIG. 5.5 Overrepresentation of Sensitive Habitats in Reserve Networks.** Scientists working in the Seaflower Biosphere Reserve, San Andresés Archipelago, Colombia, developed recommendations for the percent of each habitat type to include in well-designed marine reserve networks around Old Providence Island. These targets were 36 percent inclusion for most habitat types, but higher for those habitats known to be rare and vulnerable, including mangrove lagoons and spawning aggregations (see Friedlander et al., 2003a, for more detail). The scientists also provided examples of how reserves might be designed to achieve these target percentages. The percent of each habitat type included in the scientists' examples (white bars) is contrasted here with the percent of each habitat type included in proposals made by stakeholders for such a reserve network (black bars).

gathered together, fish are easier to catch en masse. Moreover, they may selectively choose structurally complex, and thus vulnerable, habitat for shelter during these aggregations (Beets and Friedlander 1999). At the same time, there may be some areas that are of particular value because they contain or are in close proximity to all the habitats necessary to support a productive ecosystem. These areas, along with those that are rare or vulnerable, such as spawning aggregation sites, should receive particular attention (Fig. 5.5).

Another question with respect to habitat inclusion is whether it is acceptable or desirable to include degraded habitats within a reserve. From a socioeconomic perspective, a site may be desirable if it was impacted at some point in the past and is now no longer of great utility. Sites like this could be acceptable reserves if the source of degradation is identified and eliminated, and if the system appears to have the capacity to recover its former productivity in a reasonable timeframe. For example, overfished locations may be perfectly acceptable as long as the habitat or target species have not suffered too severely. In fact they may be ideal candidates for another reason—their ability to recover quickly. Studies predict that the most heavily fished areas are most likely to recover quickly (Sladek Nowlis and Roberts 1997). In doing so, they have great potential to solidify support for marine reserves. It has been our experience that, especially among fishers, seeing an area recover quickly after designation as a marine reserve removes doubt about the effectiveness of this management tool. In part, this reaction is due to the visible evidence that fishing was keeping the fish populations down, but it also appears to arise from a contagious enthusiasm at seeing fish at higher abundance than many thought possible. Not all degraded areas are good candidates, though. For example, an embayment that receives untreated sewage is probably not a good candidate unless the sewage effluent is going to be redirected and studies indicate the bay has not been impacted to the extent that recovery is unlikely.

The tolerance for including impacted habitats should increase with greater extent of reserve coverage. If reserves will encompass a substantial portion of an entire management area, it might actually be desirable to include habitats in varying conditions (while avoiding severely degraded habitats) to learn whether and to what extent habitats can recover. However, greater damage will require higher coverage to ensure that reserves minimize risk of collapse since it will take some time before they are capable of providing resilience to the region's marine ecosystems.

We therefore recommend that stakeholders lead site selection with support and guidance from technical advisers. The ecological, socioeconomic, and enforcement design criteria should be conveyed to stakeholders, with continuing discussion and feedback as stakeholder groups create and collaborate on proposals. To the extent that stakeholder groups can agree on a single proposal that meets the basic scientific criteria, we recommend that such a proposal be adopted.

### Regulations

A common question when designating marine reserves is whether a little fishing is acceptable if conducted by some group that might cause relatively light impacts. On the one hand, there are some examples where certain kinds of fishing are likely to have little effect on the species of concern. In one studied case, pelagic fishing (for fish in the water column) seemed compatible with the recovery of a bottom-dwelling scallop and some groundfish populations on Georges Bank off the northeastern United States (Murawski et al. 2000). However, this result should not be taken as typical. This region has greater management capacity than almost any other place on the planet, and enforcement was carried out using satellite-based vessel monitoring systems. It would be difficult to replicate this scenario in most places in the world.

On the other hand, allowing some fishing in an area can open up a host of enforcement and ecological difficulties. In contrast to the Georges Bank example, most studies of partially protected marine protected areas (MPAs) show they fare poorly, most likely because of the difficulty in enforcing the ban on certain kinds of fishing when other fishing activity with similar appearance is allowed (Reed 2002; see also New Zealand Poor Knight's discussion in chapter 11). Wallace (1999) examined northern abalone abundance in several sites around southern Vancouver Island, British Columbia, Canada. Abalone collection was banned in all of the study sites and throughout the region, but sites varied in their other regulations. Five sites were open to other types of fishing, and all had extremely low abalone abundance—suggesting poaching. Three other sites were closed to all fishing. One was a designated ecological reserve, a second was a prison site where fishing was prohibited close to shore, and the third was near a military installation. The military site had the highest abundance of abalone, probably owing to high levels of enforcement. The prison site and the ecological reserve had fewer abalone, suggesting that some poaching may have taken place, but both had substantially more abalone than the five partially protected sites. More impressively, the ecological reserve had as many abalone as the prison site, suggesting that compliance within the reserve matched the prison site. These results indicate that allowing some fishing can create a real enforcement challenge.

In a larger study, Friedlander et al. (2003b) surveyed sixty sites around the main Hawaiian Islands. These sites varied in their fishing regulations. Some sites were open to all forms of fishing, others were MPAs that allowed some forms of fishing but not others, a few were managed using traditional Hawaiian methods and light overall fishing pressure, and still others were no-take marine reserves. Results indicated that open and partially protected sites were quite similar both in their fish assemblages and in the amount of fish they contained. These sites were distinct in both respects from marine reserves and from lightly fished sites managed using traditional Hawaiian practices, which had similar characteristics to each other. These results would indicate that most exceptions were deleterious to any benefits partial protection might provide, with the possible exception of light fishing using traditional methods in a culture where these methods have evolved over centuries and management responsibility was delegated to the local community.

Allowing some forms of fishing also threatens an area with ecological effects that cascade through an ecosystem. Certain species have been identified as key players in ecosystems. Their removal can trigger changes throughout the entire ecosystem. For example, the loco (*Concholepas concholepas*), a predatory snail, plays a central role in keeping mussels from dominating the Chilean rocky intertidal zone. It is also a prized food source and has been depleted throughout Chile. Protected areas have not only allowed the loco to recover, but at more natural abundance levels it restores the Chilean rocky intertidal environments to more natural conditions (Castilla and Durán 1985; Durán and Castilla 1989). There are many other examples of similar key species in marine environments (Pinnegar et al. 2000; see also examples in chapters 4, 8, and 11). If an MPA were closed to all other forms of fishing but allowed collection of these important species, the MPA would not contain an ecosystem in natural ecological balance. These types of ecological interactions are especially important to be aware of because we know relatively little about ecological interactions in the ocean, and these types do appear fairly often.

To assure that a number of fishery and conservation goals are achieved, a core network of true marine reserves is necessary. However, these reserves could fit naturally within a broader zoning plan that includes a full range of MPAs and other management zones. Such a zoning approach would also enable managers to address a broad range of threats to and conflicts about protection and use of marine resources. Conservationists and fishers do not monopolize conflict about use of the ocean. There are also rifts between commercial, recreational, and subsistence fishing; between motorized and nonmotorized water sports; and between fishing and oil drilling, just to name a few. Zoning provides an opportunity to reduce all of these conflicts by designating areas where each activity is allowed.

In a broader zoning approach marine reserves should lie at the core of larger marine protected areas (NRC 2001). Doing so can buffer the reserves from outside impacts and reduce the impact of leaky reserves. Buffer zones might be compatible with such uses as light fishing using traditional practices and other forms of tightly regulated commercial and recreational fishing.

## CONCLUSIONS

Reserves should be designed using a process that clearly defines the role of the general public, scientific and enforcement advisers, and fishing communities.

The aim of the process should not be one of compromising scientific advice with the will of fishers, but instead achieving mutually agreed upon goals related to sustaining fish, fisheries, and ocean ecosystems into the foreseeable future. Toward this end, scientific and enforcement advice should play a key role in shaping designs, but the task of selecting actual reserve sites is ideally left to a public process that engages key stakeholders, including fishers and the broader public, as long as those sites are compatible with the agreed to and stated goals for the reserve and the expert advice about their ability to meet them.

There are some useful rules of thumb for designing marine reserves or reserve networks. While different goals could result in different designs, it is worth paying special attention to minimizing the risk of collapsing fished populations and the fisheries and ecosystems they support. Given our lack of knowledge, we would need marine reserves or other tools that kept 10 to 20 percent of all fish off limits to fishing to assure persistence, if not health, of these populations. To ensure relatively healthy fisheries and ecosystems would require more—30 to 50 percent of all fish must be protected. Depending on the openness of the systems, the magnitude of impacts outside reserves, the efficacy of other management tools, and the scale and frequency of ecological disasters, reserves may have to be modified, typically scaled up. Reserve networks should be divided into individual reserves capable of supporting viable adult populations of at least bottom-associated species within their boundaries. It is also desirable, and for most species likely, that reproduction will move out across the boundary to fishing grounds and other reserves. All habitats should be represented in a replicated manner, with special emphasis paid to rare, vulnerable, and fish aggregation habitats. Although small amounts of fishing may be compatible with conservation objectives at times, allowing exceptions makes partially protected areas vulnerable to losing all benefits. Especially given the goal of risk minimization, it is important to ensure that marine reserves form the backbone of any marine protected area plan. The reserves may fit naturally as the core of a broader zoning plan designed to reduce a wide range of conflicts surrounding the use of the sea.

We know everything necessary to design effective reserves right now. There are such great needs to ensure against future management mistakes and rebuild depleted fish populations that most reserve designs will prove beneficial even if they are only first steps toward ideal design. As reserves become more common, design choices will become more important for providing real improvements. Fortunately, the process of designating reserves and studies of their performance along the way will provide invaluable information for making the right choices in the future.

## REFERENCES


Acosta, C. A. 1999. Benthic dispersal of Caribbean spiny lobsters among insular habitats: Implications for the conservation of exploited marine species. *Conservation Biology* 13(3):603–612.

Airamé, S., J. E. Dugan, K. D. Lafferty, H. Leslie, D. A. McArdle, and R. R. Warner. 2003. Applying ecological criteria to marine reserve design: A case study from the California Channel Islands. *Ecological Applications* 13(1, suppl):S170–S184.

Allison, G. W., S. Gaines, J. Lubchenco, and H. Possingham. 2003. Ensuring persistence of marine reserves: Catastrophes require adopting an insurance factor. *Ecological Applications.* 13 (l. suppl.):58–524.

Appeldoorn, R. S. 2001. "Do no harm" versus "stop the bleeding" in the establishment of marine reserves for fisheries management. *Proceedings of the Gulf and Caribbean Fisheries Institute* 52:667–673.

Appeldoorn, R. S., and C. W. Recksiek. 2000. Marine fisheries reserves versus marine parks: Unity disguised as conflict. *Proceedings of the Gulf and Caribbean Fisheries Institute* 51:471–474.

Appeldoorn, R. S., C. W. Recksiek, R. L. Hill, F. E. Pagan, and G. D. Dennis. 1997. Marine protected areas and reef fish movements: The role of habitat in controlling ontogenetic migration. *Proceedings of the 8th International Coral Reef Symposium* 2:1917–1922.

Appeldoorn, R. S., A. Friedlander, J. Sladek Nowlis, P. Usseglio, and A. Mitchell-Chui. 2003. Habitat connectivity in reef fish communities and marine reserve design in Old Providence–Santa Catalina, Colombia. *Gulf and Caribbean Research* 14(2):61–78.

Araújo, M. B., and P. H. Williams. 2001. The bias of complementarity hotspots towards marginal populations. *Conservation Biology* 15:1710–1720.

Attwood, C. G., and B. A. Bennett. 1994. Variation in dispersal of galjoen (*Coracinus capensis*) (Teleostei: Coracinidae) from a marine reserve. *Canadian Journal of Fisheries and Aquatic Sciences* 51(6):1247–1257.

Ballantine, W. 1995. The practicality and benefits of a marine reserve network. In Gimbel, K., ed. *Limited Access to Marine Fisheries: Keeping the Focus on Conservation,* 205–223. Washington, DC: World Wildlife Federation.

———. 1997. Design principles for systems of "no-take" marine reserves. *Workshop on the Design and Monitoring of Marine Reserves, February 18–20.* Vancouver: Fisheries Centre, University of British Columbia.

Beets, J., and A. Friedlander. 1999. Evaluation of a conservation strategy: A spawning aggregation closure for grouper in the Virgin Islands. *Environmental Biology of Fishes* 55:91–98.

Black, K. P. 1993. The relative importance of local retention and inter-reef dispersal of neutrally buoyant material on coral reefs. *Coral Reefs* 12:43–53.

Boehlert G. W. 1996. Larval dispersal and survival in tropical reef fishes. In Polunin N. V. C., and C. M. Roberts, eds. *Reef Fisheries,* 61–84. London: Chapman and Hall.

Bohnsack, J. A. 1993. Marine reserves: They enhance fisheries, reduce conflicts, and protect resources. *Oceanus* 36:63–71.

———. 1996. Maintenance and recovery of reef fishery productivity. In Polunin, N.V.C. and Roberts, C.M., eds. *Reef Fisheries,* 283–313. London: Chapman and Hall.

Buck, E. H. 1993. *Marine Ecosystem Management.* Washington, DC: Congressional Research Service, The Library of Congress, 12.

California Department of Fish and Game (CDFG). 2002. *Nearshore Fishery Management Plan.* Sacramento, CA: CDFG Marine Region.

Caselle, J. E., and R. R. Warner 1996. Variability in recruitment of coral reef fishes: The importance of habitat on two spatial scales. *Ecology* 77(8):2488–2504.

Castilla, J. C., and R. H. Bustamante. 1989. Human exclusion from rocky intertidal of Las Cruces, central Chile: Effects on *Durvillaea antarctica* (Phaeophyta, Durvilleales). *Marine Ecology Progress Series* 50:203–214.

Castilla J. C., and L. R. Durán 1985. Human exclusion from the rocky intertidal zone of central Chile: The effects on *Concholepas concholepas* (Gastropoda). *Oikos* 45:391–399.

Chapman, M. R., and D. L. Kramer. 2000. Movements of fishes within and among fringing coral reefs in Barbados. *Environmental Biology of Fishes* 57:11–24.

Clark, W. G. 1993. The effect of recruitment variability on the choice of a target level of spawning biomass per recruit. In *Proceedings of the International Symposium on Management Strategies for Exploited Fish Populations,* 233–246. Anchorage: Alaska Sea Grant College Program, AK-SG-93-02.

Clarke, M., V. Ortiz, and J. C. Castilla. 1999. Does early development of the Chilean tunicate *Pyura praeputialis* (Heller, 1878) explain the restricted distribution of the species? *Bulletin of Marine Science* 65(3):745–754.

Cohen, D., and S. A. Levin. 1987. The interaction between dispersal and dormancy strategies in varying and heterogeneous environments. *Lecture Notes in Biomathematics* 71:110–122.

Colin, P. L., and I. E. Clavijo. 1978. Mass spawning by the spotted goatfish, *Pseudopeneus maculatus* (Bloch) (Pisces: Mullidae). *Bulletin of Marine Science* 28:780–782.

Cowen, R. K., K. M. M. Lwiza, S. Sponaugle, C. B. Paris, and D. B. Olson. 2000. Connectivity of marine populations: Open or closed? *Science* 287:857–859.

Crowder L. B., S. J. Lyman, W. F. Figueira, and J. Priddy. 2000. Sink-source population dynamics and the problem of siting marine reserves. *Bulletin of Marine Science* 66(3):799–820.

Cushing, D. 1969. The regularity of the spawning season of some fishes. *J. Cons. Int. Explor. Mer.* 33:81–92.

———. 1995. *Population Production and Regulation in the Sea.* Cambridge, England: Cambridge University Press.

Davis, G. E., and J. W. Dodrill. 1989. Recreational fishery and population dynamics of spiny lobster, *Panulirus argus,* in Florida Bay, Everglades National Park, 1977–1980. *Bulletin of Marine Science* 44:78–88.

Dayton, P. K., S. F. Thrush, M. T. Agardy, and R. J. Hofman. 1995. Environmental effects of marine fishing. *Aquat. Cons.* 5:205–232.

Diamond, J. M. 1975. The island dilemma: Lessons of modern biogeographic studies for the design of natural reserves. *Biological Conservation* 7:129–146.

Díaz, J. M., G. Diaz, J. Garzon-Ferreira, J. Geister, J. A. Sánchez, and S. Zea. 1996. Atlas de los arrecifes coralinos del Caribe colombiano, I: Archipiélago de San Andrés y Providencia. Pub. Esp. INVEMAR.

Domeier, M. L., and P. L. Colin. 1997. Tropical reef fish spawning aggregations: Defined and reviewed. *Bulletin of Marine Science* 60:698–726.

Durán, L. R., and J. C. Castilla. 1989. Variation and persistence of the middle rocky intertidal community of central Chile with and without human harvesting. *Marine Biology* 103:555–562.

Eklund, A. M., D. B. McClellan, and D. E. Harper. 2000. Black grouper aggregations in relation to protected areas within the Florida Keys National Marine Sanctuary. *Bulletin of Marine Science* 66(3):721–728.

Friedlander, A. M., and E. E. DeMartini. 2002. Contrasts in density, size, and biomass of reef fishes between the northwestern and the main Hawaiian Islands: The effects of fishing down apex predators. *Marine Ecology Progress Series* 230:253–264.

Friedlander, A., J. Sladek Nowlis, J. A. Sanchez, R. Appeldoorn, P. Usseglio, C. McCormick, S. Bejarano, and A. Mitchell-Chui. 2003a. Designing effective marine protected areas in Seaflower Biosphere Reserve, Colombia, based on biological and sociological information. *Conservation Biology*. 17:1–16.

Friedlander, A. M., E. K. Brown, P. L. Jokiel, W. R. Smith, and K. S. Rodgers. 2003b. Effects of habitat, wave exposure, and marine protected area status on coral reef fish assemblages in the Hawaiian archipelago. *Coral Reefs* 22: 291–305.

Friedlander, A. M., R. C. DeFelice, J. D. Parrish, and J. L. Frederick. 1997. *Habitat Resources and Recreational Fish Populations at Hanalei Bay, Kauai*. Final report of the Hawaii Cooperative Fishery Research Unit to the State of Hawaii. Honolulu, HI: Department of Land and Natural Resources, Division of Aquatic Resources. 320 pp.

Gilpin, M., and M. E. Soulé. 1986. Minimum viable populations: Processes, of species extinction. In Soulé, M. E., ed. *Conservation Biology: The Science of Scarcity and Diversity,* 19–34. Sunderland, MA: Sinauer.

Goodyear, C. P. 1993. Spawning stock biomass per recruit in fisheries management: Foundation and current use. *Canadian Special Publications in Fisheries and Aquatic Sciences* 120: 25–34.

Guenette, S., and T. J. Pitcher 1999. An age-structured model showing the benefits of marine reserves in controlling overexploitation. *Fisheries Research* 39:295–303.

Halpern, B. 2003. The impact of marine reserves: Do reserves work and does reserve size matter? *Ecological Applications* 13(1, suppl):S117–S137.

Harding, D., J. H. Nicholas, and D. S. Tungate. 1978. The spawning of plaice (*Pleuronectes platessa* L.) in the Southern North Sea and English Channel. *Rapports et Procès-Verbaux des Réunions Conseil International pour l'exploration de la Mer* 172:102–113.

Hockey, P. A. R., and G. M. Branch. 1997. Criteria, objectives and methodology for evaluating marine protected areas in South Africa. *South African Journal of Marine Science* 18:369–383.

Holland, K. N., J. D. Peterson, C. G. Lowe, and B. M. Wetherbee. 1993. Movements, distribution and growth rates of the white goatfish *Mulloides flavolineatus* in a fisheries conservation zone. *Bulletin of Marine Science* 52(3):982–992.

Holland, K. N., C. G. Lowe, and B. M. Wetherbee. 1996. Movements and dispersal patterns of blue trevally (*Cranx melampygus*) in a fisheries conservation zone. *Fisheries Research* 25:279–292.

Hyrenbach, K. D., K. A. Forney, and P. K. Dayton. 2000. Marine protected areas and ocean basin management [Viewpoint]. *Aquatic Conservation: Marine and Freshwater Ecosystems* 10:437–458.

Jackson J. B. C., M. X. Kirby, W. H. Berger, K. A. Bjorndal, L. W. Botsford, B. J. Bourque, R. H. Bradbury, R. Coke, J. Erlandson, J. A. Estes, T. P. Hughes, S. Kidwell, C. B. Lange, H. S.

Lenihan, J. M. Pandolfi, C. H. Peterson, R. S. Steneck, M. J. Tegner, and R. R. Warner. 2001. Historical overfishing and the recent collapse of coastal ecosystems. *Science* 293:629–638.

Jenkins, G. P., K. P. Black, and M. J. Keough. 1999. The role of passive transport and the influence of vertical migration on the presettlement distribution of a temperate, demersal fish: Numerical model predictions compared with field sampling. *Marine Ecology Progress Series* 184:259–271.

Johannes, R. E. 1978. Traditional marine conservation methods in Oceania and their demise. *Annual Reviews in Ecology and Systematics* 9:349–364.

———. 1997. Traditional coral-reef fisheries management. In Birkeland, C., ed. *Life and Death of Coral Reefs,* 380–385. New York: Chapman and Hall.

Johnson, D. R., Funicelli, N. A., and Bohnsack, J. A. 1999. Effectiveness of an existing estuarine no-take fish sanctuary within the Kennedy Space Center, Florida. *North American Journal of Fisheries Management* 19:436–453.

Katz, C. H., J. S. Cobb, and M. Spaulding 1994. Larval behavior, hydrodynamic transport, and potential offshore-to-inshore recruitment in the American lobster *Homarus americanus. Marine Ecology Progress Series* 103:265–273.

Lacson, J. M. 1992. Minimal genetic variation among samples of six species of coral reef fishes collected at La Parguera, Puerto Rico, and Discovery Bay, Jamaica. *Marine Biology* 112:327–331.

Levitan, D. R. 1991. Influence of body size and population density on fertilization success and reproductive output in a free-spawning invertebrate. *Biological Bulletin* 181:261–268.

Love, M. 1996. *Probably More Than You Ever Wanted to Know about West Coast Fishes.* Santa Barbara, CA: Really Big Press.

Mangel, M. 1998. No-take areas for sustainability of harvested species and a conservation invariant for marine reserves. *Ecology Letters* 1:87–90.

Meyer, C. G. 2003. *An Empirical Evaluation of the Design and Function of a Small Marine Reserve (Waikiki Marine Life Conservation District).* Ph.D. diss., Manoa, HI: University of Hawaii.

Meyer, C. G., and K. N. Holland. 2001. A kayak method for tracking fish in very shallow habitats. In J. R. Sibert and J. L. Nielsen, eds. *Electronic Tagging and Tracking in Marine Fisheries,* 289–296. Dordrecht: Kluwer Academic Publishers.

Meyer, C. G, K. N. Holland, B. M. Wetherbee, and C. G. Lowe. 2000. Movement patterns, habitat utilization, home range and site fidelity of whitesaddle goatfish, *Parupeneus poryphyreus,* in a marine reserve. *Environ. Biol. Fish.* 59:235–242.

Murawski, S. A., R. Brown, H.-L. Lai, P. J. Rago, L. Hendrickson. 2000. Large-scale closed areas as a fishery-management tool in temperate marine systems: The Georges Bank experience. *Bulletin of Marine Science* 66(3): 775–798.

Murray, S. N., R. F. Ambrose, J. A. Bohnsack, L. W. Botsford, M. H. Carr, G. E. Davis, P. K. Dayton, D. Gotshall, D. R. Gunderson, M. A. Hixon, J. Lubchenco, M. Mangel, A. MacCall, D. A., McArdle, J. C. Ogden, J. Roughgarden, R. M. Starr, M. J. Tegner, and M. M. Yoklavich. 1999. No-take reserve networks: Protection for fishery populations and marine ecosystems. *Fisheries* 24(11):11–25.

Myers, R. A., and G. Mertz. 1998. The limits of exploitation: A precautionary approach. *Ecological Applications* 8:S165–S169.

Myers, R. A., and B. Worm. 2003. Rapid worldwide depletion of predatory fish communities. *Nature* 423:280–283.

National Marine Fisheries Service (NMFS). 2001. *Report to Congress: Status of the Fisheries of the United States.* Silver Spring, MD: U.S. Department of Commerce.

National Research Council (NRC). 2001. *Marine Protected Areas: Tools for Sustaining Ocean Ecosystems*. Washington, DC: National Academy Press.

Noss, R. F. 1987. Corridors in real landscapes: A reply to Simberloff and Cox. *Conservation Biology* 1:159–164.

Ogden, J. C. 1988. The influence of adjacent systems on the structure and function of coral reefs. *Proceedings of the 6th International Coral Reef Symposium* 1:123–129.

Ogden, J. C., and P. R. Ehrlich. 1977. The behavior of heterotypic resting schools of juvenile grunts (Pomadasyidae). *Marine Biology* 42:273–280.

Olson, R. R., and R. McPherson. 1987. Potential vs. realized larval dispersal: Fish predation on larvae of the ascidian *Lissoclinum patella* (Gotschaldt). *Journal of Experimental Marine Biology and Ecology* 110:245–256.

Palumbi, S. 2003. Population genetics, demographic connectivity, and the design of marine reserves. *Ecological Applications* 13(1, suppl):S146–S158.

Pickett, S. T. A., R. S. Ostfeld, M. Shachak, and G. E. Likens, eds. 1997. *The Ecological Basis of Conservation: Heterogeneity, Ecosystems, and Biodiversity.* New York: Chapman and Hall.

Pinnegar, J. K., N. V. C. Polunin, P. Francour, F. Badalamenti, R. Chemello, M.-L. Harmelin-Vivien, B. Hereu, M. Milazzo, M. Zabala, G. D'Anna, and C. Pipitone. 2000. Trophic cascades in benthic marine ecosystems: Lessons for fisheries and protected-area management. *Envionrmental Conservation* 27:179–200.

Plan Development Team (PDT). 1990. *The Potential of Marine Fishery Reserves for Reef Fish Management in the U.S. Southern Atlantic.* NOAA Technical Memorandum NMFS-SEFC-261. Silver Spring, MD: U.S. Department of Commerce.

Proulx, E. 1998. The role of law enforcement in the creation and management of marine reserves. In Yoklavich, M. M., ed. *Marine Harvest Refugia for West Coast Rockfish: A Workshop,* 74–77. NOAA Technical Memorandum NOAA-TM-NMFS-SWFSC-255. Silver Spring, MD: U.S. Department of Commerce.

Rakitin, A., and D. L. Kramer. 1996. Effect of a marine reserve on the distribution of coral reef fishes in Barbados. *Marine Ecology Progress Series* 131:97–113.

Reed, J. K. 2002. Deep-water Oculina coral reefs of Florida: Biology, impacts, and management. *Hydrobiologia* 471:43–55.

Roberts, C. M. 1996. Settlement and beyond: Population regulation and community structure of reef fishes. In Polunin, N. V. C., and C. M. Roberts. *Reef Fisheries,* 85–112. London: Chapman and Hall.

———. 1997. Connectivity and management of Caribbean coral reefs. *Science* 278: 1454–1457.

———. 1998. Sources, sinks and the design of marine reserve networks. *Fisheries* 23:16–19.

Roberts, C. M., and J. P. Hawkins 1997. How small can a marine reserve be and still be effective? *Coral Reefs* 16:150.

Roberts, C. M., G. Branch, R. H. Bustamente, J. C. Castilla, J. Dugan, B. S. Halpern, K. P. Lafferty, H. Leslie, J. Lubchenco, D. McArdle, M. Ruckleshaus, and R. R. Warner. 2003a. Application of ecological criteria in selecting marine reserves and developing reserve networks. *Ecological Applications* 13(1, suppl): 5215–5228.

Roberts, C. M., S. Andelman, G. Branch, R. H. Bustamente, J. C. Castilla, J. Dugan, B. S. Halpern, K. D. Lafferty, H. Leslie, J. Lubchenco, D. McArdle, H. P. Possingham, M. Ruckleshaus, and R. R. Warner. 2003b. Ecological criteria for evaluating candidate sites for marine reserves. *Ecological Applications* 13(1, suppl): 5199–5215.

Rosenberg, A., T. E. Bigford, S. Leathery, R. L. Hill, and K. Bickers. 2000. Ecosystem approaches to fishery management through essential fish habitat. *Bulletin of Marine Science* 66(3):535–542.

Russ, G. R., and A. C. Alcala. 1996. Do marine reserves export adult fish biomass? Evidence from Apo Island, central Philippines. *Marine Ecology Progress Series* 132:1–9.

———. 1999. Management histories of Sumilon and Apo marine reserves, Philippines, and their influence on national marine resource policy. *Coral Reefs* 18:307–319.

Sadovy, Y. M. 1993. The Nassau grouper, endangered or just unlucky? *Reef Encounters* 13:1–12.

Sala, E., O. Aburto-Oropeza, G. Paredes, I. Parra, J. C. Barrera, and P. K. Dayton. 2002. A general model for designing networks of marine reserves. *Science* 298:1991–1993.

Simberloff, D., and J. Cox. 1987. Consequences and costs of conservation corridors. *Conservation Biology* 1:63–71.

Sladek Nowlis, J. 2000. Short- and long-term effects of three fishery-management tools on depleted fisheries. *Bulletin of Marine Science* 66(3): 651–662.

Sladek Nowlis, J., and B. Bollerman. 2002. Methods for increasing the likelihood of restoring and maintaining productive fisheries. *Bulletin of Marine Science* 70:715–731.

Sladek Nowlis, J., and A. Friedlander. 2004. Marine reserve function and design for fisheries management. In Norse, E. A., and L. B. Crowder, eds. Marine *Conservation Biology: The Science of Maintaining the Sea's Biodiversity.* Washington, DC: Island Press.

Sladek Nowlis, J., and C. M. Roberts. 1995. Quantitative and qualitative predictions of optimal fishery reserve design. In Roberts, C., and W. J. Ballantine, C. D. Buxton, P. Dayton, L. B. Crowder, W. Milon, M. K. Orbach, D. Pauly, J. Trexler, and C. J. Walters. *Review of the Use of Marine Fishery Reserves in the U.S. Southeastern Atlantic,* B-12. NOAA Technical Memorandum, NMFS-SEFSC-376. Miami, FL: U.S. Department of Commerce.

———. 1997. You can have your fish and eat it, too: Theoretical approaches to marine reserve design. *Proceedings of the 8th International Coral Reef Symposium* 2:1907–1910.

———. 1999. Fisheries benefits and optimal design of marine reserves. *Fishery Bulletin* 97:604–616.

Slatkin, M. 1987. Isolation by distance in equilibrium and nonequilibrium populations. *Evolution* 47:264–279.

Tegner, M. J. 1992. Brood stock transplants as an approach to abalone stock enhancement. In Shepherd, S. A., M. J. Tegner, and S. A. Guzman del Proo. *Abalone of the World: Biology, Fisheries and Culture,* 461–473. Oxford: Blackwell Scientific.

Thompson, H. 1943. *A Biological and Economic Study of Cod (*Gadus callarias *L.).* Research Bulletin 14. St. John's, Newfoundland: Department of Natural Resources.

Wallace, S. S. 1999. Evaluating the effects of three forms of marine reserve on northern abalone populations in British Columbia, Canada. *Conservation Biology* 13:882–887.

Watling, L., and E. A. Norse. 1998. Disturbance of the seabed by mobile fishing gear: A comparison to forest clearcutting. *Conservation Biology* 12:1180–1197.

Wolanski E., and J. Sarsenski. 1997. Larvae dispersion in coral reefs and mangroves. *American Scientist* 85:236–243.

SIX

# Social Dimensions of Marine Reserves

MICHAEL B. MASCIA

Social factors—not biological or physical variables—may be the primary determinants of marine reserve design and performance. While it may seem counterintuitive that the foremost influences on the emergence, evolution, and success of an *environmental* policy could be *social*, marine reserves result from *human* decision-making processes and require changes in *human* behavior to succeed. Thus, the social, cultural, political, and economic variables that mold individual choice and behavioral change ultimately shape the development, management, and performance of marine reserves and protected areas. For purposes of brevity, in this chapter the term *social* refers to social, cultural, political, and economic factors collectively, except where otherwise noted. Social factors, for example, fostered establishment and influenced the design of the Fagatele Bay National Marine Sanctuary in American Samoa. The participation of local authorities in the site selection process generated popular support for the sanctuary and, in recognition of Samoan cultural traditions, the proposed sanctuary boundary was revised to correspond with the bounds of the local marine tenure system (Fiske 1992).

Marine reserves are not only the product of social processes, but they also have social ramifications. Marine reserves, like other forms of resource management, allocate access to and use of marine resources among individuals and social groups and, thereby, directly and indirectly shape society. In Belize, for example, establishment of the Hol Chan Marine Reserve had far-reaching social impacts in the adjacent town of San Pedro. Reserve establishment catalyzed the transition of San Pedro from a fishing community to a tourism-based economy. Local men left the fishing industry for the higher wages they could garner working as tour guides for snorkelers and scuba divers in the new marine reserve. The predominantly mestizo community diversified and grew rapidly, as newcomers from throughout Belize, North America, and Europe migrated to the area in search of economic opportunities. The standard of living in San Pedro continued to rise markedly following reserve establishment, as did levels of crime and drug abuse (Mascia 2000; unpublished data).

The relationship between marine reserve design and performance is complex and dynamic; just as reserve design influences performance, reserve performance influences design. This reciprocal relationship is seldom discussed in the scientific literature, but it is critical to understanding of reserve emergence and evolution and to the design of effective reserve policy. Following the early success of the Discovery Bay Fishery Reserve in Jamaica, for example, local fishermen successfully lobbied for expansion of the reserve (Woodley and Sary 2003). In Belize, the perceived socioeconomic success of the Hol Chan Marine Reserve not only prompted expansion of the reserve but also spurred nearby communities to initiate development of additional marine reserves. In Barbados, by contrast, widespread dissatisfaction with the social performance of the Barbados Marine Reserve (also known as the Folkestone Marine Park) contributed to the demise of a proposed network of marine reserves along the south and west coasts of the island.

An understanding of the relationship between marine reserve design and performance is essential to decision makers, who design reserves to achieve specific policy objectives. Though there has been a growing appreciation of the role of the social sciences in marine reserve design, social scientific research on reserve design and performance is limited. As a result, efforts to design marine reserves are still largely based on anecdotal evidence and individual experience rather than social scientific knowledge. Though conventional wisdom and trial-and-error have produced many marine reserve success stories, reserves designed in accordance with rigorous social science–based guidelines would be more likely to achieve social and environmental policy objectives.

This chapter reviews the social dimensions of reserve design and performance, the relationship between these two elements, and the implications of this relationship for marine reserve policy. The first section outlines the principal sociopolitical elements of marine reserve design: decision-making arrangements, resource use rules, monitoring and enforcement systems, and conflict resolution mechanisms. The role of cultural beliefs and values in marine reserve emergence, evolution, and performance is then discussed. The third section of this chapter reviews the social dimensions of reserve performance, with particular attention to the effects of reserve establishment on resource users. After outlining the known relationships between reserve design and performance, the chapter concludes with a discussion of the implications of these design–performance relationships for marine reserve policy.

## SOCIAL ASPECTS OF MARINE RESERVE DESIGN

A marine reserve is, in essence, a set of rules that collectively govern human interactions with a specified portion of the marine environment. Rules define reserve boundaries, the activities that may take place within these boundaries, and who may engage in reserve activities. Rules also specify protocols for monitoring and enforcing reserve rules governing resource use, as well as the mechanisms for resolving conflicts. Most importantly, rules govern the decision-making processes that establish marine reserve boundaries, resource use rights, monitoring and enforcement systems, and conflict resolution mechanisms. Thus, the design of a marine reserve is the specific configuration of rules that defines, explicitly or implicitly, *who* may do *what*—and *where, when,* and *how* they can do it—with respect to the portion of the marine environment designated as a reserve. The design of a marine reserve (i.e., reserve rules) directly and indirectly shapes human behavior, human interactions with the marine environment, and, ultimately, marine reserve performance.

There are four principal sociopolitical elements of marine reserve design: decision-making arrangements, resource use rules, monitoring and enforcement systems, and conflict resolution mechanisms. Each of these elements of reserve design may have formal and informal components with written or unwritten origins. Aspects of marine reserve design may be derived from legal statutes, policy statements, organizational practices, social norms, cultural traditions, or a combination of any or all of these. As a result, the de facto design that *actually* governs a marine reserve often differs sharply from the de jure system legally designated to do so. Commercial fishing continues in Glacier Bay National Park (Alaska, U.S.A.), for example, despite legal prohibitions dating to 1966 (NRC 2001, 156–157).

### Decision-Making Arrangements

The design of decision-making arrangements determines the rights of individuals or groups to make choices regarding other aspects of marine reserve development and management. Decision-making rules determine, for example, who may participate in making decisions and who may not (e.g., government officials, resource users), how decision makers are selected for their positions (e.g., elected or appointed), and how decisions are made (e.g., consensus or majority vote). These political variables are significant because policy preferences often vary among individuals or social groups; the structure of decision-making arrangements determines whose interests, beliefs, and values are represented in decision-making processes and thus manifest in policy and management decisions.

During the development of the Florida Keys National Marine Sanctuary management plan, for example, commercial fishermen shared limited decision-making authority with environmental groups and commercial dive operators, among others. Commercial fishermen generally opposed the establishment of marine reserves as part of the sanctuary management plan, whereas environmental groups and commercial dive operators generally supported widespread reserve establishment (Suman, Shivlani, and Milon 1999). If any of these groups had held exclusive decision-making authority, its policy preferences alone would likely have been reflected in the sanctuary management plan. In practice, the system of shared decision-making authority resulted in a policy compromise—immediate establishment of a system of nearly two dozen relatively small marine reserves and a commitment to develop a larger marine reserve in the Dry Tortugas within a defined time frame.

Marine reserve decision-making arrangements are usually complex. The responsibility and authority for decision making often rests with different (though sometime overlapping) sets of individuals or groups during the six sequential stages of the policy process: initiation, assessment, selection, implementation, evaluation, and termination (Brewer and deLeon 1983). Moreover, actors' decision-making rights are often limited to particular aspects of marine reserve development or management, such as enforcement or conflict resolution. Procedural rules that govern voting, decision-making criteria, and the use of scientific information also vary depending upon the stage in the policy process. At each stage, subtle differences in the rules that govern decision making may have significant impacts upon the design, implementation, evaluation, or reform of marine reserve rules governing resource use, monitoring, enforcement, and conflict resolution.

Marine reserve decision-making arrangements range along a continuum from highly centralized to highly participatory. Centralized decision-making arrangements limit decision-making responsibility and authority to a single individual or a small group, often specialists within a single government agency.

Participatory decision-making arrangements, by contrast, permit sharing of decision-making responsibility and authority among diverse groups: resource users; nongovernmental organizations; local, state, and national government officials; and other stakeholders.[1] Because the amount, diversity, and type of information brought to bear upon decisions depends upon who has the right to participate in decision-making processes (Healy and Ascher 1995), participatory decision-making arrangements generally increase the amount and diversity of information brought to bear upon decisions. Participatory decision-making arrangements thus increase the likelihood that policy decisions will be based upon accurate models of human behavior and environmental dynamics. Participatory decision-making arrangements also tend to enhance the perceived legitimacy of decisions that are made. The proposed boundaries of the Hol Chan Marine Reserve, for example, were revised prior to implementation at the request of local fishermen, which enhanced the legitimacy of the reserve in the eyes of affected individuals (Mascia 2000).

The procedural rules that govern how decision-makers make choices can shape the results of marine reserve decision-making processes. Voting rules shape the balance of power between majority and minority interests. Decision-making by consensus, for example, grants significantly more power to minority interests than decision-making by simple majority. Consensus-based voting rules, therefore, may preclude marine reserve designs acceptable to most decision-makers but strongly opposed by a few. Voting rules also shape perceptions of the legitimacy of decision-making processes among both minority and majority groups. Similarly, the rules and criteria established to govern decisions (e.g., requiring that a given percentage of the coastline must be designated as marine reserves) often shape the outcome of decision-making processes.

### Resource Use Rules

Rules governing resource use are the second principal component of marine reserve design. Resource use rules—including laws, regulations, formal and informal policies, codes of conduct, and social norms—specify the rights (i.e., privileges) of individuals or groups to access and appropriate marine resources.

[1] In this chapter, the term *resource user* refers to individuals who derive consumptive or nonconsumptive benefits from their physical interactions with the marine environment. The term *stakeholder,* which includes but is not limited to resource users, refers to individuals and organizations with a significant interest in the marine environment or its management.

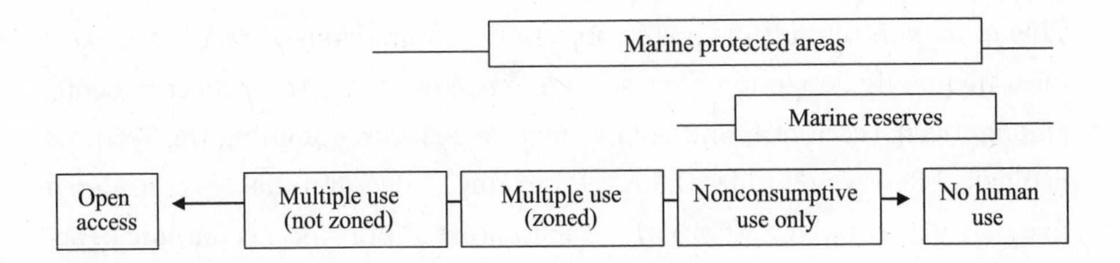


**FIG. 6.1 Relationship of Marine Reserves to Other Systems of Marine Resource Governance, Including MPAs.**

These rights may be held by individuals, groups, organizations, or the state, and are often shared among these actors. Moreover, resource use rights are seldom absolute. The U.S. government, for example, may alter the resource use rights of individuals without compensation when legitimately exercising its public trust authority. Though the right to change the rules governing resource use is generally held by governments, this decision-making authority may be shared with or delegated to resource users or other stakeholders. In the state of Maine, for example, lobster fishermen are governed by formal laws and informal codes of conduct that specify where, when, and how they may fish. The Maine state government recently granted lobstermen limited decision-making authority over resource use rules, including the right to specify trap limits, through the establishment of regional lobstermen-only "councils" (Acheson 2003).

Rules governing resource use thus specify how individuals may interact with each other and the marine environment. Infinite possible configurations of resource use rules exist, ranging along a continuum from "open access" (i.e., no rules) to a complete prohibition on human activities (Fig. 6.1). Marine reserves lie toward the latter end of this continuum but display a great deal of subtle variation in the rules governing resource use. Many reserves limit nonconsumptive recreational, commercial, or scientific activities, in addition to prohibiting all extractive activities (which are forbidden, by definition, in all marine reserves). Marine reserves are also frequently incorporated within larger marine protected areas with multiple regulatory zones, each of which may have a distinct set of rules governing resource use. The Florida Keys National Marine Sanctuary, for example, has several categories of regulatory zones, including three types of marine reserves: ecological reserves, sanctuary preservation areas, and research only special use areas.

Rules governing resource use shape marine reserve performance by establishing use rights that foster specific policy outcomes. Because it is often impossible to maximize multiple policy objectives simultaneously, decision makers must frequently design marine reserve resource use rights that reflect tradeoffs among social, economic, and environmental goals. In designing the Tortugas Ecological Reserve (Florida, U.S.A.), for example, decision makers considered five policy alternatives. Alternative 1 emphasized short-term economic benefits over environmental sustainability by failing to establish a marine reserve in the Tortugas. By contrast, Alternative 5, the most expansive marine reserve proposal, emphasized environmental sustainability at the expense of short-term economic costs. Decision makers ultimately approved Alternative 3, a 151 $nm^2$ marine reserve (U.S. Department of Commerce 2000), which represented the middle-ground balance of economic and environmental outcomes.

The precision and stability of resource use rights mold individual behavior. Precise marine reserve rules specify clear use rights, minimizing conflict among resource users or between resource users and enforcement personnel. In some marine reserves, for example, dive operators may only use particular dive sites at assigned times; such arrangements prevent crowding and conflict among users. Rules foster conflicts when they fail to specify clear resource use rights, raising the costs of resource use and thus dissipating the benefits to resource users. Likewise, imprecise and unstable resource use rights create uncertainty over future opportunities, causing users to discount the future sharply and exploit resources more heavily than they otherwise would. Well-defined resource use rights—precise, stable, easily understood, and easily enforceable—enhance the economic benefits and environmental sustainability of marine reserves by reducing social conflict and creating greater certainty regarding future resource use.

### Monitoring and Enforcement Systems

Marine reserve monitoring systems track changes in the state of reserve-associated social and environmental systems. Reserve monitoring systems vary in what they measure and who does the measuring, as well as where, when, and how measurements are made. Carefully designed monitoring systems—which generally include robust performance indicators, baseline data, and control sites—can provide insights into the changes in social and environmental systems due to reserve establishment. In practice, many marine reserves lack formal systems for monitoring environmental and, especially, social phenomena. As a result, resource users, managers, and other stakeholders often informally monitor environmental and social indicators to assess reserve performance. Monitoring-based assessments of performance can guide future reserve policy and management decisions, as well as enhance confidence in current policies and management practices. In Belize, for example, formal and informal assessment of the social and environmental performance of the Hol Chan Marine Reserve led to widespread support for expansion of the reserve (Mascia 2000; see chapter 10).

Enforcement systems attempt to increase compliance with rules governing resource use by monitoring individual behavior and sanctioning noncompliance. By increasing the severity and likelihood of sanctions and, thus, raising the opportunity cost of noncompliance, enforcement systems act directly upon resource users to foster adherence with established rules. Monitoring user behavior forces would-be poachers to engage in deceptive practices that diminish the benefits of engaging in prohibited activities. Sanctioning noncompliance further diminishes the benefits of engaging in prohibited activities and thus deters malfeasance. The role of enforcement systems has been demonstrated in the Bahamas, where aggressive enforcement of "no fishing" regulations at the Exuma Cays Land and Sea Park dramatically reduced the frequency and extent of fishing within the reserve (Mascia 2000).

Enforcement systems also shape compliance indirectly. By shaping perceptions of the efficacy of enforcement efforts, enforcement systems affect rates of "contingent compliance," where individuals base their decision to comply with rules governing resource use upon the (perceived) rate of compliance by others (Levi 1997). The theory of contingent compliance posits that, because individuals seek to avoid being a "sucker" by obeying the rules while others are not, individuals become increasingly likely to obey the rules as the perceived rate of compliance by others increases. Perceptions of the legitimacy of enforcement systems also shape compliance; both the design of sanction mechanisms and the perceived "fairness" of enforcers shape perceptions of legitimacy. Research suggests that meaningful but graduated and context-dependent sanctions, which ensure that punishment fits the crime, are generally perceived as more legitimate than draconian, one-size-fits-all penalties (Ostrom 1990).

### Conflict Resolution Mechanisms

Conflict resolution mechanisms are formal and informal processes for resolving disputes. Conflict resolution mechanisms permit information exchange, clarification of resource use rights, and adjudication of disputes related to decision making, resource use, monitoring, and enforcement. Critical questions in the design of conflict resolution mechanisms include, Who may participate? and Who adjudicates? Other important design issues include the frequency and location of conflict resolution activities. Readily accessible and low-cost conflict resolution mechanisms enhance regime performance directly by mitigating social conflict and thereby minimizing resource overexploitation and dissipation of reserve benefits (Ostrom 1990). Conflict resolution mechanisms also enhance marine reserve performance by giving voice to aggrieved parties and acknowledging their concerns, which increases the legitimacy of reserve rules and regulations.

## BELIEFS AND VALUES

Underlying marine reserve design and, thus, reserve performance, are human beliefs and values. *Beliefs* are "what people think the world is like," whereas *values* are "guiding principles of what is moral, desirable, or just" (Kempton et al. 1995). Beliefs and values vary among individuals but often display consistent patterns of variation at the level of social or cultural groups. In a study of environmental beliefs and values in the United States, for example, Kempton et al. (1995), found that diverse social groups—loggers, environmentalists, small business owners, and policy makers—shared similar beliefs about the anthropogenic causes of global climate change but differed in the ways that they valued biodiversity. The converse is also possible, where groups may share values but diverge in their beliefs.

Beliefs shape the emergence and evolution of marine reserves. Most conservationists, for example, believe that fishing is the primary threat to marine biodiversity. Based on this shared belief, conservationists generally advocate the establishment of large marine reserves that prohibit fishing and focus less attention on other threats to marine biodiversity, such as land-based sources of pollution, habitat loss, or the introduction of exotic species. Many Caribbean fishermen, by contrast, believe that natural variability and habitat alteration induced by land-based pollution are the principal causes of fish population declines. Consistent with this belief, Caribbean fishermen often argue against the necessity of marine reserves and instead urge more effective coastal zone management (Mascia 2000; Robertson 2002, 197–198). These two belief systems suggest dramatically different approaches to marine conservation in the Caribbean; the policy manifestations of these divergent beliefs are shaped, in large part, by the design of decision-making arrangements.

Beliefs directly and indirectly shape marine reserve performance. Environmental policies based on faulty conceptual models of environmental dynamics or human behavior, for example, have little prospect of achieving their specified objectives. Early efforts to conserve sea turtles through "headstart" programs were ineffective because conservationists believed, incorrectly, that increasing the survivorship of turtle hatchlings was the key to species recovery. In fact, more recent conservation science demonstrates that recovery of sea turtle populations is most sensitive to protection of juveniles, which are now protected from fisheries mortality through legally mandated gear modifications (Crowder et al. 1994; Heppell, Crowder, and Crouse 1996). The efficacy of the gear modification program, however, has been hindered by the belief among fishermen that the modified gear is not actually necessary to conserve turtles and, moreover, reduces catch (Margavio and Forsyth 1996; White 1989). These beliefs have reduced the legitimacy of the gear regulations in the eyes of fishermen and, as a result, have reduced compliance rates—to the detriment of sea turtle populations.

Values also shape marine reserve design. Organizational values (i.e., values shared among members of an organization or agency) shape the manner in which organizations set management priorities and undertake mandated activities. Similarly, decision makers' values limit the bounds of debate over both policy objectives (i.e., ends) and design (i.e., means). Policy objectives and designs that lie outside the bounds of what is considered "good, desirable, or just" are never raised in discussion or are rejected by decision makers. Based on his experience in East Africa, McClanahan (1999, 324) argues that, in any given country, "the types [of] and area in MPAs [marine protected areas, including marine reserves] will depend upon societal values." Similarly, Orbach (1995) notes that the United States has a two-track system of marine wildlife management rooted in American cultural values: U.S. fisheries policy *encourages* direct harvest of marine fishes (to generate "optimum yield"), whereas U.S. marine mammal policy generally *prohibits* direct harvest of all marine mammals (to foster "maximum sustainable populations"). Thus, American values clearly hold that what is "good, desirable, or just" for fish is very different from that which is good, desirable, or just for whales, seals, sea lions, and other marine mammals.

Values shape marine reserve performance through multiple indirect mechanisms. By defining the bounds of policy debate, decision makers' values may preclude consideration of policy approaches that would be effective but are perceived as "unjust," resulting in selection of less effective but more socially acceptable strategies. Differences between stakeholders and decision makers as to what actually constitutes socially acceptable policy can influence the perceived legitimacy of policies, which, in turn, may shape the rate and degree to which agency personnel undertake mandated activities (Mascia 2000). Resource users' value-based perceptions of policy legitimacy also influence compliance rates. In the South Pacific, for example, the efficacy of traditional community-based marine reserves appears correlated with resource user values regarding local customs and traditions—particularly customary law and the authority of traditional village chiefs (Johannes 1978).

## SOCIAL DIMENSIONS OF PERFORMANCE

The social and environmental changes induced by marine reserve establishment can be monitored over time to provide measures of reserve performance. Performance can be measured against implicit and explicit marine reserve policy objectives, as well as using generic standards and criteria for "good" policies, such as social equity, economic efficiency, and environmental sustainability. Evaluations of reserve performance foster accountability, promote social learning, and provide the impetus for replicating successful policies and reforming unsuccessful ones. Most analyses of marine reserve performance to date have focused upon environmental outcomes because these are the primary impetus for reserve establishment and perhaps simpler to measure. The social dimensions of reserve performance (Table 6.1), however, are generally of greater concern to most direct users of marine resources and often the source of contentious debate during reserve development and management. Unfortunately, the dearth of social scientific research on the social and economic performance of marine reserves frequently limits policy discussions to largely conceptual terms.

### Economic

The economic performance of marine reserves can be measured according to both efficiency and equity criteria. The most complete indicator of reserve efficiency, and the most difficult to measure fully, is the relative change in the total economic value (TEV) that society derives from the marine environment following reserve establishment. Because of the difficulty associated with measuring TEV, researchers generally focus on its component parts: use values and nonuse values. Use values are the benefits and costs derived from direct use (e.g., fishery harvests, oil extraction, dive tourism) and indirect use (e.g., fisheries production, shoreline protection, nutrient cycling) of the marine environment. Nonuse values (also known as passive use values) include option, existence, and bequest values. The option value of a marine reserve is the value derived

Table 6.1 Select Social and Economic Performance Indicators[a] for Marine Reserves.

| |
|---|
| Economic Efficiency Indicators |
| Total economic value |
| Direct use value |
| Indirect use value |
| Option value |
| Existence value |
| Bequest value |
| Economic Equity Indicators |
| Income among social groups or subgroups |
| Wealth among social groups or subgroups |
| Wealth disparity among social groups or subgroups |
| Geographic distribution of costs and benefits |
| Sociocultural Indicators[b] |
| Employment levels |
| Crime, domestic violence, or alcoholism rates |
| Gender, ethnicity, age, religious affiliation, or other demographic attributes |
| Perceptions of individual, household, or community well-being |

[a] These indicators may be measured for an affected population in its entirety or for particular groups or subgroups.

[b] Performance indicators measure relative changes in the state of social or environmental systems following reserve establishment.

from the potential future use of the reserve and its components. The existence value of a marine reserve is the value that individuals derive based solely upon the knowledge that the resource exists, whereas the bequest value of a reserve is the value that individuals derive based upon the knowledge that the marine reserve and its components will be available to future generations. Direct use values and some indirect use values accrue in monetary terms and may be directly measured; many indirect use values and all nonuse values provide nonmonetary benefits and costs and therefore cannot be measured directly (Bunce et al. 2000; NRC 2001).

The net effect of marine reserve establishment upon the total economic value of marine resources is not clear. The only study known by the author to measure the TEV of a marine reserve, a survey of the "willingness to pay" of tourists and local residents, estimated the total economic value of the Montego Bay Marine Park, Jamaica, at approximately $20 million (Spash et al. 2000). Clearly, one study is insufficient to make any definitive statements regarding the total economic value of marine reserves, though this research suggests that the total economic value of marine reserves may be quite significant.

The economic literature demonstrates that the direct use costs and benefits of marine reserves may be significant, varying dramatically in accordance with preexisting site-specific resource use patterns and reserve rules governing resource use. Because consumptive direct uses are prohibited within marine reserves, there are usually clear costs associated with reserve establishment and the subsequent discontinuation of consumptive activities. Leeworthy and Wiley (2002) estimate that the "preferred alternative" marine reserve zoning system in the Channel Islands National Marine Sanctuary (California, U.S.A.) will result in maximum annual net costs of $902,000 in forgone benefits from consumptive recreational diving and fishing. Similarly, maximum annual net costs associated with loss of consumptive uses following establishment of the "preferred alternative" Tortugas Ecological Reserve in Florida are estimated at $880,000 in forgone commercial fishing benefits and $126,000 in forgone benefits from consumptive recreational activities (Leeworthy and Wiley 2000). It is worth noting, however, that the economic costs of forgone opportunities within marine reserves may not be incurred by resource users since these individuals may compensate for the loss of access to reserve resources by continuing their activities in nonreserve areas.

The net economic value of nonconsumptive direct uses of marine reserves varies in accordance with the rules governing resource use within reserve boundaries. If nonconsumptive uses such as scuba diving are prohibited, then the direct use costs of reserve establishment will be equivalent to the opportunity cost of forgone activities. If nonconsumptive uses are permitted, however, the net value of nonconsumptive direct uses is likely to be positive; that is, the value of nonconsumptive uses within reserves is likely to increase following marine reserve establishment with enhanced production of ecosystem goods and services. Dixon et al. (1993), and Vogt (1997) both suggest that the economic benefits from the nonconsumptive direct uses of marine reserves exceed the costs of forgone consumptive activities, but both studies' results are based on incomplete cost–benefit analyses. The net economic benefit of the Tortugas Ecological Reserve to nonconsumptive scuba divers at the time of reserve establishment was $25,000 annually (Leeworthy and Wiley 2000), a value that is expected to increase over time.

The indirect use value of marine reserves is not well documented either. Measuring the relative change in indirect use benefits and costs following reserve establishment is a significant challenge that economists have yet to overcome. The most obvious indirect use values of marine reserves, the "spillover" of fishery resources due to greater biological productivity of the protected marine environment, has been estimated using proxy measures in modeling and empirical studies. One bioeconomic fisheries model estimates that optimal establishment of marine reserves worldwide would increase the global harvest value of coral reef fisheries by approximately 5.5 percent ($1 billion) annually

(Pezzey et al. 2000). Using the relative change in catch per unit fishing effort (CPUE) outside marine reserves as a proxy for indirect consumptive use value, studies suggest that the indirect use values of marine reserves vary widely across sites and over time: published values range from no significant change in CPUE to increases of greater than 100 percent (Goodridge et al. 1996; McClanahan and Kaundra-Arara 1996; McClanahan and Mangi 2000; Roberts et al. 2001). In theory, CPUE values could be translated into estimates of the net indirect consumptive use values, though researchers have failed to do so as yet. Researchers comparing total fisheries yields before and after marine reserve establishment (an imperfect proxy for comparing the total direct and indirect use value of fishing) found total yields following reserve establishment to be 30 to 35 percent *less* than yields prior to reserve establishment (McClanahan and Kaunda-Arara 1996; McClanahan and Mangi 2000).

The indirect use value of marine reserves for nonconsumptive activities has not been well measured, though one would predict that the expected indirect value of reserve establishment should be positive. This prediction assumes that there is a positive relationship between the ecological integrity of marine ecosystems and the goods and services that these ecosystems provide to humans. Dive operators, for example, would presumably benefit from the enhanced production of ecosystem goods and services (e.g., spillover of fish) that results from marine reserve establishment. Similarly, coastal residents might benefit from enhanced shoreline protection in coastal areas adjacent to reserves.

The nonuse values of marine reserves appear to be positive, perhaps substantially so, but economists have not yet attempted to measure these costs and benefits directly. Research suggests that these values should vary depending upon the social significance of the marine reserve (Farrow 1996), though economic research has not explicitly tested this hypothesis. Using a thought experiment, Leeworthy and Wiley (2000) estimate the *nonuse value* of a marine reserve within the Florida Keys National Marine Sanctuary at between $3.39 million and $11.3 million annually. Scholz and Fujita (2001, 7), using virtually the same thought experiment, ascribe an identical value to solely the *existence value* of a marine reserve in the Channel Islands National Marine Sanctuary. Assuming a 3 percent discount rate and annual payments of $3 to $10 from 1 percent of U.S. households, the *total nonuse value* of a Florida Keys marine reserve is an estimated $113 to $377 million (Leeworthy and Wiley 2000). Because these estimated values are the product of thought experiments rather than scientific research based on measurement and observation, they must be viewed with caution—actual values could be substantially higher or lower than researchers predict.

The economic performance of marine reserves may also be measured using equity criteria. Indicators of economic equity track the relative changes in monetary and/or nonmonetary benefits and costs that accrue to different social groups as a result of reserve establishment. Measures of relative change in income, wealth, or wealth disparity among specific groups or subgroups (e.g., fishermen and divers, line fishermen and net fishermen), for example, represent useful indicators of the distributive economic effects of reserve establishment. The effect of marine reserves on economic equity may also be measured using indicators that track the net economic effect of reserves on populations of particular concern, such as women, minorities, the poor, the elderly, or traditional cultures. The geographic distribution (e.g., local versus national) of costs and benefits is also a useful indicator of the economic equity of a marine reserve.

The effects of marine reserve establishment on economic equity are perhaps even less well understood and less well studied than reserve effects on efficiency. Among those marine reserves that permit nonconsumptive uses, the general qualitative pattern that follows marine reserve establishment is a transfer of direct use benefits from consumptive resource users such as fishermen to nonconsumptive users such as dive operators and scientists. In Barbados, for example, establishment of the Barbados Marine Reserve shifted the local system of resource use rights from a virtual "open access" system that permitted both consumptive and nonconsumptive uses to an ecotourism and scientific use regime that allowed only nonconsumptive uses (Mascia 2000). Among marine reserves that prohibit both consumptive and nonconsumptive uses, all direct users incur costs associated with the loss of resource use rights within the reserve. In this instance, equity indicators include measures of the relative magnitude or significance of the costs incurred by user groups or populations of particular concern.

Among both consumptive and nonconsumptive users, the distributive economic effects of reserve establishment vary by subgroup. In St. Lucia, for example, establishment of the Soufriere Marine Management Area affected net fishermen and trap fishermen differently (Goodridge et al. 1996). In general, small-scale fishermen, especially those who use fixed gear or fish within informal fishing territories, are more vulnerable to the loss of fishing grounds than larger scale, transient fishermen employing mobile gear. Small-scale and territorial fishermen, when affected by reserve establishment, lose a larger percentage of their fishing grounds than large-scale or transient operators. The latter groups, however, may be more likely to lose a portion of their fishing grounds to marine reserves simply because they fish a larger geographic area. The distributive economic impact of reserve establishment on nonconsumptive users appears correlated with users' degree of economic dependence upon the natural environment. Dive operators, for example, are more likely to benefit from reserve establishment than jet-ski businesses.

### Sociocultural

The extent to which marine reserves achieve sociocultural policy objectives may also be measured and evaluated using performance indicators. Sociocultural performance indicators include relative changes in employment levels, crime rates, domestic violence rates, and alcoholism rates among specific groups, as well as shifts in household relations and modes of production. Demographic indicators include relative changes in the gender, ethnic, age, and religious profile of specific groups (e.g., resource users). Perceptions of individual, household, and community well-being provide measures of aggregate reserve performance—social, economic, and environmental.

The sociocultural dimensions of marine reserve performance have not been well studied. The limited data available suggest that small-scale fishermen may incur significant costs and be fully or partially displaced from the fishing industry by the establishment of marine reserves (Dobrzynski and Nicholson 2003; Goodridge et al. 1996; Mascia 2000; McClanahan and Mangi 2000). McClanahan and Mangi (2000), for example, report a 60 to 80 percent decline in the number of fishermen at the Jomo Kenyatta Beach fish landing site following establishment of the no-take Mombasa Marine Park in Kenya. Many displaced resource users gain full or partial employment in other sectors, such as construction or tourism, but older fishermen, in particular, appear less able to take advantage of alternative economic opportunities. Marine reserves may also induce new migration patterns by restructuring economic opportunities, drawing people to local communities in the case of some reserves and displacing them from adjacent communities in other situations. These shifting migration patterns frequently change the demographic profile of user groups and coastal communities, as was discussed previously with respect to the Hol Chan Marine Reserve. Perceptions of individual, household, and community well-being appear to vary by stakeholder group and depend largely upon the distributive economic impacts of reserves (Mascia 2000). No known research has examined the impact of marine reserve establishment upon social indicators such as rates of crime, domestic violence, or alcoholism, demonstrating the need for further study.

## RELATIONSHIP BETWEEN DESIGN AND PERFORMANCE

The structure of marine reserve decision-making arrangements has a significant effect upon reserve performance. In marine reserves and analogous natural resource governance regimes, research demonstrates that the right of resource users to participate in the design and modification of rules governing resource use is correlated with regime performance—environmental and social (Christie and White 1997; Mascia 2000, 2001; Ostrom 1990; Pollnac et al. 2001). Research also suggests that resource user self-governance rights (i.e., the right to govern the behavior of one's group, independent of external authorities) are correlated with reserve establishment and performance (Mascia 2000; 2001). Selecting basic rules and criteria to govern decision making (i.e., process guidelines) before attempting to make substantive choices about reserve design may help to reduce conflict and facilitate informed decisions among stakeholders with diverse interests, beliefs, and values (Mascia 2001).

Research demonstrates that the clarity and congruence of rules governing resource use influence marine reserve performance. Clearly defined resource and reserve boundaries, as well as clearly defined individual resource use rights, tend to improve the social and environmental performance of marine reserves and other natural resource governance regimes (Ostrom 1990; Mascia 2000, 2001). Rules governing resource use that are explicitly linked to local conditions also tend to enhance reserve performance (Mascia 2000). Research also suggests that the presence of economically congruent resource use rights—where the resource users who benefit most from reserve establishment bear the greatest cost of sustaining reserve benefits, while those who derive the fewest benefits incur the least cost—foster marine reserve performance (Mascia 2000). Among effective marine reserves, research suggests that the rules governing resource use have sufficient scale and scope to address all threats that significantly affect the social or environmental systems of the reserve (Mascia 2000). Finally, the performance of legally designated marine reserves tends to be enhanced when reserve resource use rights are consistent with existing informal or culturally based resource use rights (Fiske 1992; Mascia 2000).

Research on the role of monitoring and enforcement systems in marine reserve performance highlights the importance of accountability, legitimacy, equity, and flexibility. Monitors who actively assess resource conditions and are accountable to resource users (or who are themselves resource users) tend to improve the performance of marine reserves and analogous resource governance regimes (Buhat 1994; Ostrom 1990; Woodley and Sary 2003). Likewise, reserve performance is enhanced by the presence of active and accountable monitors of resource use behavior (Mascia 2000; Roberts 2000; Woodley and Sary 2003). Again, monitors may themselves be resource users. Sanctions for noncompliance must not only be likely and severe enough to raise the cost of noncompliance but also graduated and context-dependent to ensure that punishment fits the crime (Ostrom 1990; Mascia 2000).

The role of conflict resolution mechanisms in marine reserve performance is not yet clear. Available data suggest that low cost, local, and readily accessible conflict resolution mechanisms tend to enhance the performance of marine reserves and analogous natural resource governance regimes (Ostrom 1990; Mascia 2000). Additional research is clearly needed to understand better the role of conflict resolution mechanisms in reserve performance.

## POLICY IMPLICATIONS

The relationships between marine reserve design and performance previously outlined have significant implications for marine reserve policy. Integrating these "lessons learned" into reserve design can contribute to the development of more effective marine reserves, as well as the reform of existing sites. Differences in goals and context make a rigid blueprint design for socially and environmentally effective marine reserves inappropriate, but policy guidelines composed of general principles for reserve design are possible and practical (Box 6.1). The policy guidelines outlined here should be viewed as working hypotheses that are based upon the best available social scientific knowledge, but subject to future revision.

First and foremost, resource users should share responsibility and authority for marine reserve development and management as part of a collaborative management (i.e., comanagement) system. Decision-making arrangements should grant relevant resource user groups "a seat at the table" and an equitable share of voting rights, and, where appropriate, should establish process guidelines that specify basic rules and criteria for decision-making. To ensure that resource user representatives advance group interests when participating in decision-making processes, formal and informal mechanisms (e.g., elections, consultative sessions) should be established to ensure that representatives are accountable to their constituents. As part of marine reserve management systems, mechanisms should be established to encourage and legitimize resource user self-governance initiatives that advance recognized policy objectives. Mechanisms should also be established to facilitate appropriate resource user

### 6.1 Principles for Marine Reserve Design

1. *Share responsibility and authority.* Bringing diverse stakeholder groups, including resource users, into marine reserve decision-making and management processes improves the substance and legitimacy of these decisions, increases management capacity, and enhances the legitimacy of management activities.
2. *Foster accountability.* Accountability mechanisms (e.g., elections, consultative sessions, or open meetings) increase the likelihood that decision makers will further constituents' interests rather than personal interests in decision-making processes. Accountability mechanisms also foster fair and active enforcement of rules governing resources use by enforcement personnel.
3. *Facilitate resource user self-governance.* Resource user self-governance initiatives that are consistent with reserve policy objectives can serve as effective complements to other management efforts.
4. *Clearly define reserve rules and boundaries.* Clear marine reserve boundaries and clear rules governing resource use within reserves foster compliance and simplify enforcement.
5. *Explicitly link rules governing resource use to social and environmental conditions.* Linking reserve rules to the state of social and environmental systems fosters adaptive (and more socially and environmentally sustainable) management of these systems.
6. *Structure reserve rules so that benefits of resource use are roughly proportional to costs of providing these resources.* Reserve rules that allocate resource use benefits to users in rough proportion to the costs that these users incur to provide the same marine reserve resources will likely be perceived as more legitimate, and thus enjoy greater compliance, than rules that allocate benefits disproportionate to their costs.
7. *Build upon informal resource use rights.* Building marine reserves on the foundation of existing systems of informal or customary resource use rights enhances reserve legitimacy and fosters compliance among resource users.
8. *Monitor reserve performance—environmental and social.* Tracking the environmental and social dimensions of marine reserve performance provides the basis for adaptive management.
9. *Make research and monitoring participatory.* Enlisting stakeholders, including resource users, in data collection and analysis educates participants, builds capacity, and fosters trust.
10. *Share monitoring results.* Sharing information regarding the environmental and social performance of marine reserves may enhance reserve legitimacy or provide the impetus for necessary policy reform.
11. *Make punishment fit the crime.* Graduated, context-dependent sanctions enhance compliance by raising the opportunity cost of noncompliance and enhancing the perceived legitimacy of the reserve.
12. *Share information regarding compliance rates and enforcement actions.* Broad dissemination of information regarding compliance rates and enforcement actions can enhance reserve legitimacy and foster contingent compliance.
13. *Establish highly accessible conflict resolution mechanisms.* Highly accessible conflict resolution mechanisms provide a vehicle for resolving disputes that would otherwise increase costs of resource use and, thus, diminish reserve benefits.

Source: Adapted from Mascia 2001.

and other stakeholder participation (formal and informal) in monitoring reserve performance and enforcing rules governing resource use.

Second, resource use rights should be clearly defined and congruent with the local social and environmental context. Reserve boundaries (internal and external) should be designated in a clear and culturally appropriate manner, such as using landmarks or buoys rather than global positioning system (GPS) coordinates. Similarly, resource use rights must be clearly specified (e.g., no-take zone rather than numerous species-specific size limits), so that users and enforcers know what is permissible and what is not. Rules governing resource use should be explicitly linked to and contingent upon the state of site-specific environmental and social conditions. Moreover, the scale and scope of rules governing resource use should be sufficient to address the anthropogenic activities that threaten reserve performance. Resource use rules should also be designed so that the benefits an individual derives from reserve establishment are roughly proportional to the costs he or she incurs to maintain provision of reserve benefits. Finally, rules governing resource use within marine reserves should build upon and reinforce existing informal or customary resource use rights that are consistent with reserve policy objectives.

Third, monitoring and enforcement systems should be active, accountable, and just. Monitoring systems should track the environmental and social aspects of reserve performance. Findings should be disseminated among stakeholders to enhance reserve legitimacy or provide impetus for necessary policy reform. Mechanisms (including the participation of resource users in monitoring efforts) should be established to ensure the accountability of monitors to resource users. Enforcement systems should also include accountability mechanisms, such as participation of resource users in formal and informal enforcement efforts. Information regarding enforcement efforts and compliance rates should be disseminated to increase user confidence in enforcement efforts and encourage contingent compliance with rules governing resource use. Sanctions for noncompliance should be graduated and based upon the seriousness of the offense (as well as other contextual factors) to ensure that punishment is just.

Finally, conflict resolution mechanisms should be established to resolve disputes among resource users, reserve officials, and other stakeholders. These mechanisms should minimize economic or logistical barriers to participation in order to foster rapid resolution of conflicts over resource use rights, enforcement actions, and decision-making processes. This suggests that conflict resolution should be a local, decentralized process that includes both formal and informal mechanisms. Though existing marine reserve research provides little guidance on this point, informal conflict resolution mechanisms might include ad hoc "gripe sessions" (i.e., informal gatherings of stakeholders and reserve personnel to discuss issues of concern) and the informal designation of "elder" reserve personnel and resource users as unofficial arbitrators. More formal conflict resolution mechanisms might include regular consultations between resource users and reserve personnel, as well as nonbinding "stakeholder courts" designed to adjudicate disputes.

## CONCLUSION

The social dimensions of marine reserve design and performance are as complex as the environmental dynamics that have been the focus of marine reserve research to date. More complete consideration of social criteria is critical to both defining and achieving "successful" marine reserves. Decision-making arrangements, resource use rights, monitoring and enforcement systems, and conflict resolution mechanisms all shape marine reserve design and performance by influencing individual choices and human behavior. Beliefs and values, too, shape reserve design and performance by molding individual choices. Reserve establishment impacts not only the state of environmental systems but also that of social, economic, and cultural systems as well. Comprehensive assessments of reserve performance, therefore, include not only measures of changes in fish abundance or species richness but also direct and indirect use values, the distribution of wealth, social relations, and perceptions of well-being. Social scientific research into the relationship between marine reserve design and performance has provided some valuable insights that can serve as working hypotheses for reserve policy, but further study is clearly necessary to enhance our understanding of this relationship and to improve our ability to design socially and environmentally effective marine reserves.

### REFERENCES

Acheson, J. 2003. *Capturing the Commons: Devising Institutions to Manage the Maine Lobster Industry.* Hanover, NH: University Press of New England.

Brewer, G. D., and P. deLeon. 1983. *The Foundations of Policy Analysis.* Homewood, IL: Dorsey.

Buhat, D. 1994. Community-based coral reef and fisheries management, San Salvador Island, Philippines. In White, A. T., L. Z. Hale, Y. Renard, and L. Cortesi, eds. *Collaborative and Community-Based Management of Coral Reefs: Lessons from Experience,* 33–50. West Hartford, CT: Kumarian.

Bunce, L., P. Townsley, R. Pomeroy, and R. Pollnac. 2000. *Socioeconomic Manual for Coral Reef Management.* Townsville: Australian Institute of Marine Science.

Christie, P., and A. T. White. 1997. Trends in development of coastal area management in tropical countries: From central to community orientation. *Coastal Management* 25:155–181.

Crowder, L. B., D. T. Crouse, S. S. Heppell, and T. H. Martin. 1994. Predicting the impact of turtle excluder devices on loggerhead sea turtle populations. *Ecological Applications* 4 (3):437–445.

Dixon, J. A., L. Fallon Scura, and T. van't Hof. 1993. Meeting ecological and economic goals: marine parks in the Caribbean. *Ambio* 22 (2–3):117–125.

Dobrzynski, T., and E. E. Nicholson. 2003. User group perceptions of the short-term impacts of marine reserves in Key West. In Kasim Moosa, M. K., S. Soemodihardjo, A. Nontji, A. Soegiarto, K. Romimohtarto, Sukarno, and Suharsono, eds. *Proceedings of the Ninth International Coral Reef Symposium,* 759–764. Jakarta: Indonesian Institute of Sciences and State Ministry for Environment, Republic of Indonesia.

Farrow, S. 1996. Marine protected areas: Emerging economics. *Marine Policy* 20 (6):439–446.

Fiske, S. J. 1992. Sociocultural aspects of establishing marine protected areas. *Ocean and Coastal Management* 18:25–46.

Goodridge, R., H. A. Oxenford, B. G. Hatcher, and F. Narcisse. 1996. Changes in the shallow reef fishery associated with implementation of a system of fishing priority and marine reserve areas in Soufriere, St. Lucia. In *Proceedings of the 49th Gulf and Caribbean Fisheries Institute, Bridgetown, Barbados, November, 1996.* Ft. Pierce, FL: Gulf and Caribbean Fisheries Institute.

Healy, R. G., and W. Ascher. 1995. Knowledge in the policy process: Incorporating new environmental knowledge in natural resources policy making. *Policy Studies* 28:1–19.

Heppell, S. S., L. B. Crowder, and D. T. Crouse. 1996. Models to evaluate headstarting as a management tool for long-lived turtles. *Ecological Applications* 6(2):556–565.

Johannes, R. E. 1978. Traditional marine conservation methods in oceania and their demise. *Annual Review of Ecology and Systematics* 9:349–364.

Kelleher, G., and C. Recchia. 1998. Lessons from marine protected areas around the world. *Parks* 8(2):1–4.

Kempton, W., J. S. Boster, and J. A. Hartley. 1995. *Environmental Values in American Culture.* Boston: MIT Press.

Leeworthy, V. R., and P. C. Wiley. 2000. *Proposed Tortugas 2000 Ecological Reserve: Final Socioeconomic Impact Analysis of Alternatives.* Silver Spring, MD: U.S. Department of Commerce, National Oceanic and Atmospheric Administration.

———. 2002. *Socioeconomic Impact Analysis of Marine Reserve Alternatives for the Channel Islands National Marine Sanctuary.* Silver Spring, MD: U.S. Department of Commerce, National Oceanic and Atmospheric Administration.

Levi, M. 1997. *Consent, Dissent, and Patriotism.* Cambridge: Cambridge University Press.

Margavio, A., and C. Forsyth. 1996. *Caught in the Net: The Conflict between Shrimpers and Conservationists.* College Station: Texas A&M University Press.

Mascia, M. B. 2000. *Institutional Emergence, Evolution, and Performance in Complex Common Pool Resource Systems: Marine Protected Areas in the Wider Caribbean.* Ph.D. diss., Department of the Environment, Duke University, Durham, NC.

———. 2001. *Designing Effective Coral Reef Marine Protected Areas: A Synthesis Report Based on Presentations at the 9th International Coral Reef Symposium.* Washington, DC: IUCN World Commission on Protected Areas–Marine.

McClanahan, T. R. 1999. Is there a future for coral reef parks in poor tropical countries? *Coral Reefs* 18:321–325.

McClanahan, T. R., and B. Kaunda-Arara. 1996. Fishery recovery in a coral-reef marine park and its effect on the adjacent fishery. *Conservation Biology* 10(4):1187–1199.

McClanahan, T.R., and S. Mangi. 2000. Spillover of exploitable fishes from a marine park and its effect on the adjacent fishery. *Ecological Applications* 10(6):1792–1805.

National Research Council (NRC). 2001. *Marine Protected Areas: Tools for Sustaining Ocean Ecosystems*. Washington, DC: National Academy.

Orbach, M. K. 1995. Ecology and public policy. In Simpson, R. D., and N. L. Christensen, eds. *Ecosystem Function and Human Activities: Reconciling Economics and Ecology* 255–271. New York: Chapman and Hall.

Ostrom, E. 1990. *Governing the Commons: The Evolution of Institutions for Collective Action*. Cambridge: Cambridge University Press.

Pezzey, J. C. V., C. M. Roberts, and B. T. Urdal. 2000. A simple bioeconomic model of a marine reserve. *Ecological Economics* 33:77–91.

Pollnac, R. B., B. R. Crawford, and M. L.G. Gorospe. 2001. Discovering factors that influence the success of community-based marine protected areas in the Visayas, Philippines. *Ocean and Coastal Management* 44:683–710.

Roberts, C. M. 2000. Selecting marine reserve locations: Optimality versus opportunism. *Bulletin of Marine Science* 66(3):581–592.

Roberts, C. M., J. A. Bohnsack, F. Gell, J. P. Hawkins, and R. Goodridge. 2001. Effects of marine reserves on adjacent fisheries. *Science* 294:1920–1923.

Robertson, L. F. 2002. *Dealing in Self-Ownership: The Pursuit of Money and Personal Autonomy in Urban Jamaica*. Ph.D. diss., Department of Social Anthropology, University of Edinburgh, Edinburgh, Scotland.

Scholz, A. J., and R. M. Fujita. 2001. *Supplementary Report: Social and Economic Implications of a Channel Islands Marine Reserve Network*. Oakland: Environmental Defense.

Spash, C. L. , J. D. van der Werff ten Bosch, S. Westmacott, and J. Ruitenbeek. 2000. Lexiconographic preferences and the contingent valuation of coral reefs in Curacao and Jamaica. In Gustavson, K., R. Huber, and J. Ruitenbeek, eds. *Integrated Coastal Zone Management of Coral Reefs: Decision Support Modeling*. Washington, DC: The World Bank.

Suman, D., M. Shivlani, and J. W. Milon. 1999. Perception and attitudes regarding marine reserves: A comparison of stakeholder groups in the Florida Keys National Marine Sanctuary. *Ocean and Coastal Management* 42(12):1019–1040.

U.S. Department of Commerce. 2000. *Tortugas Ecological Reserve: Final Supplemental Impact Statement/Final Supplemental Management Plan*. Washington, DC: U.S. Department of Commerce.

Vogt, H. P. 1997. The economic benefits of tourism in the marine reserve of Apo Island, Philippines. In Lessios, H. A., and I. G. Macintyre, eds. *Proceedings of the 8th International Coral Reef Symposium*. Panama City, Panama: Smithsonian Tropical Research Institute.

White, D. 1989. Sea turtles and resistance to TEDs among shrimp fishermen of the U.S. Gulf Coast. *Maritime Anthropological Studies* 2(1):69–79.

Woodley, J. D., and Z. Sary. 2003. Development of a locally managed fisheries reserve at Discovery Bay, Jamaica. In Kasim Moosa, M. K., S. Soemodihardjo, A. Nontji, A. Soegiarto, K. Romimohtarto, Sukarno, and Suharsono, eds. *Proceedings of the Ninth International Coral Reef Symposium,* 627–634. Jakarta, Indonesia: Indonesian Institute of Sciences and State Ministry for Environment, Republic of Indonesia.

SEVEN

# Research Priorities and Techniques

JOSHUA SLADEK NOWLIS AND
ALAN FRIEDLANDER

The ocean is a complex place with a wide diversity of life forms, habitats, and ecosystems. Far from a static environment, oceans change both cyclically and unpredictably. Tides ebb and flow and seasons change, both having profound effects on ocean life. Longer cycles affect oceans and the life they contain as well; an example is the El Niño Southern Oscillation (Bakun 1993). These cycles are only one reason why oceans are so complex and difficult to predict.

Deep ocean environments are more remote and less visited (by people) than the dark side of the moon (Earle 1995). With the burgeoning popularity of scuba diving in the 1950s, people began to explore coastal waters. Yet scuba only allows intensive study of shallow waters, with physiological limitations making such studies difficult at depths greater than 20 meters (65 feet). More recently, submersible technology has increased the depths at which people can work at length. However, these technologies are expensive and have only scratched the surface of the deep ocean. The unknown status of most deep environments and many shallower ones helps explain our inability to fully understand or predict the ocean.

Especially given the many unknowns surrounding the ocean, marine reserves provide an unparalleled opportunity to study marine ecosystems in the absence of fishing pressure and other major human impacts. As such, they offer the potential to teach us a great deal about how we affect marine ecosystems, how those systems behave in the absence of major human impacts, and how we can best minimize those impacts outside of reserves. We have already learned a great deal despite the current limited size and extent of reserves. In any debate about whether to establish marine reserves, their ability to teach us about the influences we have on the ocean should be a big point in their favor. Thus marine reserves are both an important subject for future research and a key tool to enhance our understanding of marine species and ecosystems. This chapter examines research needs for the designation and management of marine reserves. This research will, in turn, enhance our understanding of the oceans.

## WHAT DO WE NEED TO KNOW ABOUT MARINE RESERVES?

We know a great deal about how to design effective marine reserves and the outcomes of doing so (see chapters 4 and 5 for further discussion). Nevertheless, there are some important unknowns surrounding marine reserve effectiveness. First and foremost, we are unable to precisely predict the full suite of responses of organisms inside and outside of newly created reserves, or how people will respond to these changes. Monitoring is the key to tracking these issues and is discussed at length in this chapter. Movement of adults, larvae, and eggs is a key area where we understand relatively little and the implications are large for reserve design and function. When habitat-based these movements form one component of habitat connectivity, another relatively poorly understood phenomenon with significant implications. Finally, we discuss the range of scientific tools that can be used to map out habitat and species distributions, and thus facilitate effective marine reserve design.

Given these unknowns, it is important to consider our knowledge of marine reserves relative to other management tools for the sea and to look carefully at what we can realistically expect to know. Considering the uncertainties surrounding oceans in general, marine reserves are quite well understood. We know, for example, that marine reserves lead to bigger and more abundant populations of many types of target species within their borders (see chapter 4 for further detail). We also know that for the vast majority of species, some of that production is going to extend to fishing areas (e.g., Roberts et al. 2001). We do not know exactly how much, nor whether these enhancements will fully offset losses to fishing communities from reduced fishing grounds, especially in the short term. However, it is generally accepted and even legally mandated that some regulation is necessary to protect fished populations from overfishing. We regularly use a suite of tools to achieve this objective, including catch quotas, effort quotas, size limits, gear restrictions, closed seasons, temporary closed areas, and even stocking of fish by artificial means. None of these techniques have more scientific validation than marine reserves.

To illustrate this point, let's consider minimum size limits, a common fishery management technique that requires fishers to throw back fish if they are too small. Theory shows they can enhance catches. If fish are allowed to survive until they can reproduce at least once, they are much harder to overfish (Myers and Mertz 1998). Yet the real world provides some challenges to this theory. It is unclear how the size limits will affect fishing behavior, but people are likely to respond by fishing harder for larger fish, which research indicates contribute disproportionately to future production. Nevertheless, theoretical studies have shown that this effort shift will still result in healthier fish populations if the smaller fish are truly protected.

But, size limits might not protect the undersized fish after all nor protect the health of the ecosystems all fish require. Many forms of fishing gear have the potential to kill fish before they are brought to the surface, measured, and thrown back. Trawling—dragging nets behind a boat either in the water column or along the bottom—is one of the most common forms of commercial fishing. Fish that end up at the back of a trawl net are likely to be squeezed to death by the fish that are caught later on. It is tempting to think that this problem could be resolved through regulating the mesh size of trawls, but it is not that easy. Trawls typically catch more than one species of fish at a time. Choosing a mesh size would be tricky because different species will have different ages at maturity. A trawl net designed to avoid all immature fish would probably not catch very many adults. Moreover, once the first layer of fish is caught at the back of a trawl, the effective mesh size becomes quite small even if the initial mesh size was large.

Other gear has similar problems. For example, longlines are another common commercial fishing gear composed of hundreds of fishhooks set out along lines that can be miles long. Longlines are deployed and retrieved many hours later. Fish that bite hooks early on are unlikely to survive until the line is retrieved. Even low impact gear has a high potential to kill certain types of fish. Most deep dwelling fish have swim bladders that inflate as they are brought to the surface. Many are likely to die on the way up whether they are caught by a trawl or a by a hook. Additionally, fish caught at any depth face increased risks of (1) being eaten while they are being reeled in or are swimming back to shelter, (2) infection or other damage from the hook wound, and (3) vulnerability from shock and exhaustion from fighting. These risks are greatest for fish that live a long time and are thus more likely to be caught and released multiple times (A. Bartholomew and J. Bohnsack, personal communication, 11/19/03).

Given this information, do minimum size limits really work? Despite these shortcomings, they are widely accepted and used in management systems and probably do help in many cases. It is intellectually fascinating that many people are unwilling to accept similar sorts of scientific validation for marine reserves. In fact, when compared to other more common fishery management techniques, marine reserves may produce benefits in a more efficient manner—providing greater benefits with fewer short-term costs (Sladek Nowlis 2000).

## LIMITS TO WHAT WE CAN KNOW SCIENTIFICALLY

Marine reserves are often held to a higher standard of scientific proof than other management tools. In fact, some people demand proof of marine reserve effectiveness that may be impossible to provide.

Take the issue of how marine reserves affect the ecosystems within and around them. To assess the impacts within, we could use standard scientific protocol and compare areas inside and outside of the reserve before and after the reserve designation. We would want these areas to be in the same general vicinity so they were affected by the same external events and had similar ecologies prior to the reserve establishment. If there were minimal differences prior to and many more fish in the reserve after establishment, we might conclude that the reserve was responsible for the buildup of fish. But some other phenomenon the experimenter did not notice might have been responsible—say an unusually large new cohort of fish settling in the reserve by chance.

To factor out these concerns, we should ideally use replication—examining multiple independent reserve and fishing areas. Once replicated, statistical tools can help us distinguish random effects from real effects. But replication is difficult to achieve. Reserves are rarely established as well-designed ecological experiments because socioeconomic concerns usually dictate the choice of reserve sites. Consequently, most are not replicated. When they are, a scientist might be lucky to have three or four reserves to work with. Under these circumstances, statistical tools lack the power to distinguish real from random effects unless the differences between reserve and fishing areas are tremendously large (e.g., Paddack and Estes 2000) or unless long-term data sets are available (e.g., Russ and Alcala 1996).

When reserves are replicated, it is usually as part of a reserve network where there is more interest in the effects of reserves on outside areas. Once we view the reserves as interconnected with surrounding areas, we can no longer consider them as independent. This issue throws an even bigger wrench into the scientific paradigm because the statistical tools that distinguish random from real effects require independence of sampling units. So to examine the effects

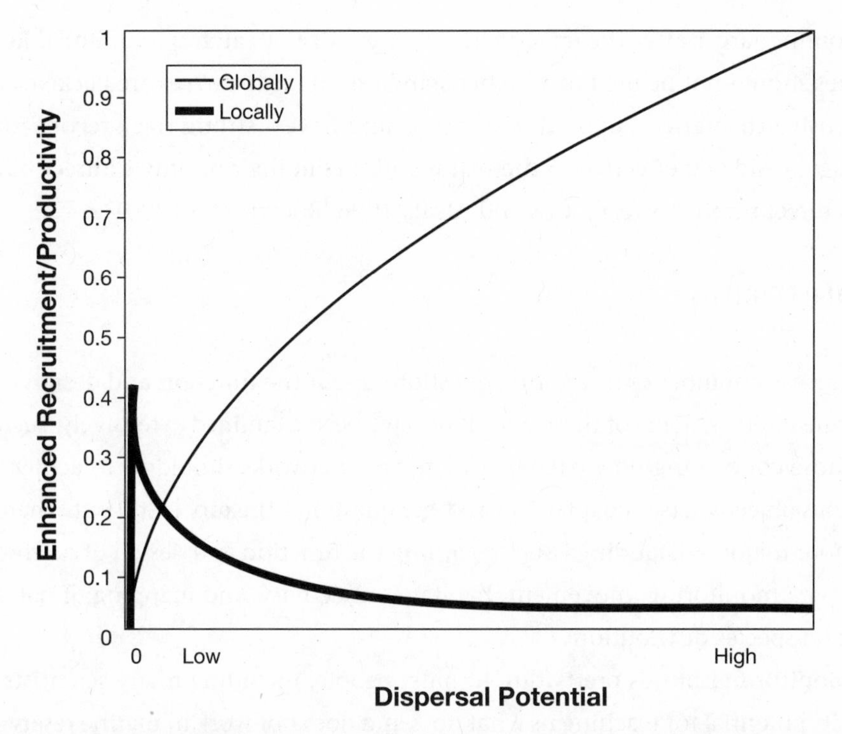


**FIG. 7.1 Limits to What We Can "Prove."** A simple modeling exercise of a population where adults stay in reserves and eggs and larvae vary in their dispersal potential shows a nonintuitive result. Global benefits from the marine reserve, in the form of productivity transported to fishing areas, were greatest with high dispersal potential. However, these same conditions diluted benefits so that they were small in any local area, and thus difficult to detect amidst natural variability and sampling error. Local benefits were most likely to be detected when dispersal potential was low, the conditions least likely to provide global benefits. Source: Data from Sladek Nowlis, unpublished.

of reserve networks we need multiple reserve networks and multiple areas where reserves were not established, all of which should have been similar before the reserve establishment and in close enough proximity to be affected by the same major events afterwards. Considering that a single reserve network is likely to span across an entire region, it is readily apparent that it is virtually impossible to design such an experiment in the real world (see also Fig. 7.1). These problems do not only affect marine reserves. Studies have shown that scientific proof of success may be difficult to achieve for other management tools as well (e.g., size limits, Allen and Pine 2000).

The proof we generally require of other management tools consists of logical consistency (i.e., theoretical support) combined with real-world experience where they seem to work based on anecdotal information and scientific evidence of a positive, but usually indirect, response (e.g., the average size of fish becoming larger after the enactment of a new size or catch quota limit). Reserves should not be held to a higher standard. In fact, reserves are backed by extensive theoretical support (NRC 2001) and have a strong track record for being considered effective in the real world, including not only indirect but also direct responses (e.g., Russ and Alcala 1996; Roberts et al. 2001).

## MONITORING

There are a number of interesting questions about the function and design of marine reserves. Some of these questions have been examined extensively, such as those concerning how large marine reserve networks should be to achieve various objectives (see chapter 5). On other questions, the jury is still out. There are four major outstanding issues regarding the function and design of marine reserves: monitoring, movement, habitat connectivity, and mapping of habitat and species distributions.

Monitoring sounds pretty dull to most people, including many scientists, but its potential for teaching us what does and does not work in marine reserve design should not be overlooked (Carr and Raimondi 1999; Underwood 1995). Monitoring plays a crucial role in evaluating the effectiveness of a marine reserve or reserve network and is an important tool for improving the principles of marine reserve design. The ecosystems both within and around marine reserves should be the subject of monitoring (Carr and Raimondi 1999; Murray et al. 1999), but it is at least as important to monitor social attitudes toward marine reserves (e.g., Roberts et al. 2001; Russ and Alcala 1996).

Monitoring also allows us to learn a great deal about marine ecosystems because marine reserves offer less impacted systems to contrast to fishing areas. Doing so has the potential to teach us a great deal about how to better manage these systems and a better appreciation for how they operate with fewer human impacts. Here, we consider some of the questions one might answer through monitoring, and the tools and techniques available to address them.

Fundamentally, monitoring can answer the question: Is the management system working? The answer to this question will be dependent on both marine reserve performance and management goals, and monitoring systems should be tailored toward the goals of the marine reserve or reserve network. Goals will invariably have biological and social elements, and monitoring should address both. A basic monitoring system should consist of before and after results from direct or indirect counting techniques, basic ecosystem attribute measurements, and surveys of public attitudes regarding the state of the ocean and ocean resources and perceptions about marine reserves. Surveys should at minimum compare populations within the reserve to populations living in similar habitat outside. If effects of reserves on outside areas are of interest, it is preferable to include several sites inside and out that varied in their distance from the reserve center to examine the dilution of the marine reserve effect. More thorough monitoring systems could include tag-recapture studies to identify movement of adults and infer growth and survival rates, detailed study of ecosystem attributes, and examination of how fishing effort is distributed before and after the reserve is established. These additional efforts can provide valuable general information for managers as well as specifics for improving management systems.

## Tools and Techniques

A monitoring program should be properly designed to ensure it is answering relevant questions, and monitoring data should be properly analyzed to ensure that efforts are directed in an efficient manner. It is also important to ensure that monitoring results are widely available to other scientists and the general public because these data are the primary means for assessing the success or failure of existing marine reserves.

### *What to Monitor*

Monitoring programs should be tailored to focus on management issues of particular interest, but a big-picture, long-term view is also important. Obvious monitoring choices would include targeted, rare, and especially vulnerable species. Less obvious but potentially important choices could include abundance of species classified into trophic guilds (e.g., piscivores, herbivores), measures of habitat quality (e.g., structural complexity and abundance of key shelter-providing organisms), and other measures of ecosystem integrity. Finally, the human element should not be overlooked. Monitoring of fishing effort, catches, and attitudes will all help evaluate the success of a marine reserve network and potentially identify ways to improve it. We recommend at minimum a program that examines resident fish species assemblages, physical benthic characteristics (e.g., structural complexity, habitat type), benthic communities, other species of special interest, and attitudes toward reserves and about fishing success. This approach can be used to address a broad range of management issues using a few simple procedures and provides the most essential information on costs and benefits to both the ecosystem and human populations.

After deciding what to monitor comes the question of how to measure it. Regardless of reserve goals, managers will want to see whether populations are recovering from fishing pressure. Indicators of recovery could come in the form of an increase in the density of fish (number per unit area), the average size of fish, or both. Numbers may increase in the absence of fishing because of increased survival and ultimately because of enhanced reproduction. Numbers may not increase if large fish maintain territories, are predators, or otherwise keep smaller individuals out.

Size is even more likely to increase in the absence of fishing, sometimes dramatically so. Fishing pressure tends to target larger fish, which also are generally more aggressive and thus end up getting caught more often. However, even if large fish were not caught preferentially, one would expect the average size of a population to increase in response to the elimination of fishing pressure. Solely by surviving longer, fish will grow larger and average size should increase.

Numbers and sizes can be combined into a useful concept of biomass, which represents the total weight of fish in an area. It can be calculated by multiplying the number of fish in the area by their average weight. Weights can be measured directly if fish are caught, or calculated from length estimates from visual surveys using a standard length–weight conversion of the form $W = aL^b$, where $W$ represents weight, $L$ is length, and $a$ and $b$ are species-specific constants. Length–weight fitting parameters can be found for a number of species in published and Web-based sources (e.g., Fishbase at http://www.fishbase.org). In cases where length–weight information does not exist for a given species, the parameters for similar-bodied relatives can be used.

If reserve goals include the capacity to support fishing in areas outside of them, it may be helpful to monitor recruitment. Techniques for measuring recruitment will be discussed later in the context of movement of adults and their offspring.

*Sampling Designs*

A monitoring program, as with any scientific data collection, requires sampling because even a modest-sized management area is too big to measure in its entirety. Ideally a sampling program should aim to achieve some form of randomness so as to maximize the chance that the sample will be representative of the management area. The key to making sense of data, though, is in identifying patterns. Certain departures from randomness can help to answer key management issues (Green 1979).

Investigators may already know of some patterns prior to the study. If so, stratified random sampling is an effective way to acknowledge the known patterns and maximize the chances of identifying new ones. A stratum is a subset of the population that shares at least one common characteristic. The researcher first identifies the relevant strata (e.g., males versus females, various age classes) and their actual representation in the population. Then samples are drawn randomly from each stratum. Stratified sampling is often used when one or more of the strata in the population have a low incidence relative to the other strata or when a population is naturally divided into categories of particular interest to managers. For example, one might emphasize data collection about mangroves in a tropical reef system even though these ecosystems are often rare. Mangrove lagoons are important ecologically and rare, in part due to past human disturbances. Emphasizing their inclusion in a sampling design will help to ensure there is enough information on which to base any future decision about the management system.

A population or set of populations that spans several habitat types provides a perfect example. If habitat maps are available to guide a monitoring effort, it may be desirable to sample each habitat type multiple times, counting the density of fish and other organisms at each survey site. The manner in which habitat types are categorized, though, is crucial. If the types do not correspond to real ecological differences or are so broad that they miss such distinctions, protections may not be as effective as possible. Fortunately, evidence suggests that experienced scientists can work with their knowledge of ecosystems and existing habitat maps to devise ecologically relevant habitat types to serve as the basis for a zoning plan (Friedlander et al. 2003a). A stratified random sampling program using habitat types as strata would be an appropriate way to monitor under these circumstances.

For example, coral reef fish assemblages and biomass vary significantly among reef habitat types around Hawaii (Friedlander et al. 2003b). Within a single reef or bay, estimates of fish biomass can vary by several orders of magnitude depending on the habitat type sampled. Habitat stratified fish surveys acknowledge these major differences, and in doing so they have greater power to detect differences within habitat types based on other factors such as the presence or absence of marine reserves.

Ideally, sampling would take place across a wide range of areas inside and at varying distances outside of replicated reserves and would span a long time-series starting before and continuing well after the reserve designation. It is extremely unlikely that this ideal will be met. Politics typically constrain our ability to examine an area for much time before reserve designation and to effectively replicate reserves or comparable fishing areas. Moreover, resources often limit the frequency and extent of sampling efforts after reserve designation. As a result, important choices need to be made regarding the timing and location of sampling efforts.

An excellent example of effective sampling design was employed to survey the Dry Tortugas, Florida, prior to the creation of marine reserves in the area (Ault et al. 2002). Scientists classified habitats based on their understanding of coral reef ecology, mapped these habitats using a variety of methods, and then sampled randomly among areas of the same habitat type but ensured each habitat type was adequately sampled. If pursued on a regular basis, this sampling protocol is sure to provide insight into the changes that result from the creation of marine reserves in the Dry Tortugas.

*Timing.* One way to detect changes due to establishing a marine reserve is to examine the same site before and after. Although this simple analysis cannot rule out other causes of observed changes, data may be convincing if they demonstrate that changes happened in conjunction with marine reserve establishment, elimination, or both (e.g., Russ and Alcala 1999). It is preferable to know habitat qualities and fish abundances before reserve establishment as a way of showing that areas being compared were similar to start. Even if data are not available beforehand, time series can be convincing if they show a consistent trend after reserve establishment (e.g., Russ and Alcala 1996).

If data are only available from paired inside and outside areas without an adequate time span, a space-for-time substitution may be used (Pickett 1989). By looking at a number of different areas, it is possible to use space in the form of fishing and no-fishing areas as a substitute for time. However, this sort of comparison lacks statistical power unless the number of sites is quite large, so it is not necessarily recommended when other comparisons are possible.

*Location.* Most commonly, marine reserve monitoring programs compare areas inside to those outside the reserve (Halpern 2003). These sorts of comparisons can provide useful information about how fishing is affecting the outside areas with the reserves serving as a control. In this manner, a marine reserve monitoring program can follow a standard ecological before-after-control-impact-pairs design. These samples are paired in the sense that the control (reserve) and impact (fishing) sites are examined more or less concurrently. Replication comes from collecting such paired samples at a number of times both before and after the reserve designation. The approach is to test whether the differences between the control and impact sites changed after the reserve was established. A variety of statistical tests are available to analyze these data (e.g., randomized intervention analysis, Welch-Satterwaite-Aspin modified *t*-test, Box and Jenkins intervention analysis) and the appropriate method will depend on the behavior of the data set.

This approach can only give evidence of how reserves compare to outside areas, as opposed to how the reserve itself might be affecting areas outside. The inside versus outside approach should be used carefully. If the outside area is too close to a reserve, it is more likely to be influenced by adults and larvae spilling over and thus underrepresent the effects of fishing. If it is too far away, though, it may be influenced by different ecological processes and events, and as such be less valuable for comparison. A useful way to get around this difficulty is to look at multiple sites at varying distances away from the reserve center (e.g., Rakitin and Kramer 1996). Using this approach, one can actually look for where the effect of the reserve tails off and, in doing so, gain some insight into how organisms are moving in and out of the reserve.

### *Sampling Analysis*

In addition to the challenges associated with the general design of a monitoring program, there are important considerations regarding analysis of sampling data. Some of these considerations are important in the design phase, whereas others are key to detecting patterns in the data after it is collected.

*Sample Size and Sampling Effort.* Monitoring programs are typically designed to examine more than one attribute or to look for change related to more than one factor. It is important to understand how the data will be analyzed so that the sampling effort is expended where it can achieve the best results. At the core of this decision is the issue of statistical power.

The power of a statistical test is its ability to identify a difference when the difference really exists. The trick is separating real differences from random variation. A powerful statistical test will allow an investigator to identify a real but small difference in, say, the abundance of fish inside and outside a marine reserve, whereas a low-power test might record the same difference but chalk it up to random chance.

Statisticians distinguish between two types of error. Type I errors occur when a perceived difference is identified as real when it was really just a random happenstance. Type II errors, on the other hand, occur when a perceived difference is rejected as a random effect when it was in fact real. Current management focuses on reducing the type I error because this kind of error results in catching fewer fish and can result in short-term economic loss (Dayton 1998). Ignoring type II errors, however, can result in serious long-term damage such as the collapse of fisheries or environmental damage. Dayton (1998) advocated reversing the burden of proof to require fisheries to demonstrate that their activities do not cause significant long-term ecological changes (a type II error). Consequently, increasing statistical power is important for moving management in a precautionary direction.

Three factors shape a statistical test's power to identify a difference between two populations: the number of samples taken, the actual difference between two populations, and the variance, a measure of the expected difference between any two samples within a population. The chance of detecting a difference is greater with more samples, a larger difference, and a smaller variance. Researchers can reduce the variance by pooling similar individuals through stratified sampling. Researchers can also effectively increase sample sizes by putting the monitoring effort where it can do the most good. Finally, researchers can calculate the statistical power that a test has so they can decide how likely it was that a difference existed but was simply not detectable.

Monitoring effort should be focused carefully. Within a site, it is valuable to apply enough effort so that the data collected are likely to represent the state of that site. However, extensive surveying within a sampling site at a given time is counterproductive. In the end, statistical tests will pool all of this effort into a single data point. Therefore, one is better off sampling just enough within a site to give an accurate picture of its current state and focusing the effort in increasing the number of sites and times sampled.

*A Bayesian Approach.* Fishery scientists have shown increasing interest in a Bayesian approach to data analysis (Thompson 1992; Walters 1986). Unlike conventional approaches to management exercises such as stock assessments, which tend to produce single best answers, a Bayesian approach explicitly incorporates probabilities into the mix. Bayesian data analysis starts with estimates of the probability of various outcomes and then modifies those probabilities based on experience (Gelman et al. 1995). These modifications represent an explicit effort to learn and improve knowledge, whereas more conventional approaches often rely on the same unreliable estimates year after year. Most importantly, the Bayesian approach provides a range of possible outcomes for any management action, with associated probabilities. These probabilities can help guide managers by highlighting the potential for bad outcomes as well as indicating which management actions might have the best chances of achieving good ones. Monitoring data can be fed back into a Bayesian adaptive management approach to marine reserve design.

### *Monitoring Plants, Invertebrates, and Habitat Attributes*

Monitoring plants, invertebrates, and habitat attributes is often quite straightforward. With the exception of a few types of invertebrates, these things tend to stay put and are relatively easy to count and measure. The key questions with respect to monitoring them have to do with the scale of interest and the degree of physical and fiscal limitations.

When conditions permit, most plants, invertebrates, and habitat attributes can be censused visually with the use of transects—swaths of habitat of a fixed width and length. These transects may each follow a different depth contour, may be straight lines following a compass heading, or may follow a narrow patch of habitat if one is using habitat stratified sampling. Often a transect is subsampled using quadrats, a square area of habitat that is measured in its entirety, randomly chosen along the length of the transect (Davis et al. 1997; Rogers et al. 1994). Quadrats can be quantified in the field or photographs can be taken and analyzed later on land, the latter being especially useful when depth or current or other physical factors limit dive times. Alternatively, the entire transect can be sampled using a fixed distance on either side of the midline. This approach can be valuable for rare but important species (Davis et al. 1997). Quadrats are two-dimensional views and may not adequately represent the three-dimensional structure of some habitats. To measure the peaks and valleys of an underwater terrain, a chain allowed to mold to the bottom contour works well. Each link in the chain then represents a sampling unit, rather than a horizontal distance of measure. Chain link surveys can be tediously slow, though, and have the potential to cause damage to the physical structure of the habitat (Rogers et al. 1994). Recent advances in computer-aided design software have now made three-dimensional analysis possible through underwater photography (Bythell et al. 2001), but these techniques are time consuming and better suited for monitoring individual organisms rather than whole habitats.

Transects can be permanent or randomly selected. Permanent transects have the advantage that one observes the same organisms over time, allowing for detailed tracking of events. They suffer, though, from their vulnerability to chance disturbances, which could give an impression of widespread disaster when only the transect itself was impacted. Random transects cannot track the response of individuals or local biological communities to specific events but are better at putting disturbances into a broader framework.

Quadrats and transects can be performed by trained divers in the field or can be captured on film or video and analyzed later on land. Field collection has the advantage that data are available immediately without risk of loss through equipment failure. It is also easier to clearly identify species in the field than from photo or video images, although even a trained diver will not be able to replicate the measurement precision of images analyzed on a computer. Photo and video work is especially valuable in habitats that are difficult to access because of depth or currents. Images can be collected by divers in less time than would be required to count quadrats or transects, or can be collected remotely by towing a camera behind a ship or on a submersible.

There are a few invertebrate species capable of long-distance movement, for example spiny lobsters (Davis and Dodrill 1989). Lobsters are also nocturnal, making visual surveys unproductive. For invertebrates that are mobile and nocturnal, trapping can serve as an acceptable substitute to direct counting. With trapping, one loses a direct measure of density since the animals trapped represent only a portion of the population over an unknown area. Nevertheless, trapping may represent a reasonable alternative in situations where visual surveys are of limited value.

*Monitoring Fish Abundance*

There are a number of highly effective methods for counting fish that require disturbing them and in some cases their habitat, such as through the use of the fish poison rotenone or bottom trawling. Counting fish underwater is a valuable way to gather information on the patterns and changes of fish populations with little disruption. Underwater fish counts are known to produce underestimates of nocturnally active fish, fish that reside in crevices, and those that flee approaching divers (Brock 1982). Visual fish counts for diurnally active fish have compared well to numbers achieved using complete, destructive sampling methods (Brock 1982).

*Area-Based Fish Counts.* Belt transects are the oldest and most frequently used method of visually surveying fishes and invertebrates (Brock 1954). Divers swim along a transect, recording species abundance and size in a three-dimensional corridor (Bortone and Kimmel 1991). As with surveys of plants, invertebrates, and habitats, these surveys can be conducted along permanently marked or randomly selected transects. The length and width of a belt transect may vary according to the species targeted for a census but should use the same dimensions for all transects sampled. Most standard transects are 25 to 50 meters long and 3 to 5 meters wide. A narrow transect (1–2 m) may be good for small, cryptic species or newly settled fishes. Longer (100 m) and wider (5–10 m) transects may be useful for examining rare or highly mobile resource species. Any transect involves counting fish within a well-defined area and therefore provides an estimate of species density.

In the stationary point count method, a diver remains at a randomly selected fixed point and counts species within a prescribed area or volume (Bohnsack and Bannerot 1986; Bortone and Kimmel 1991). All fish species observed are listed within a 7.5 m radius cylinder for five minutes. Numbers and sizes (in separate size classes) are added following the five-minute listing period.

In interval counts, also referred to as timed swims, a diver swims in a haphazard pattern along a reef tract for a prescribed period of time, typically fifty minutes (Jones and Thompson 1978; Kimmel 1985). The time period is broken up into intervals (normally ten minutes) during which species are recorded as present or absent. Rank abundance scores are based on the number of time intervals in which the species appears.

Mark–recapture techniques can also be used to assess abundance. In this technique, a known number of marked individuals are released into a population. The population is later surveyed, and its size inferred from the proportion of marked individuals. The total population size is approximated by taking the number of marked fish and dividing it by the proportion of marked fish in the population. This technique is more time consuming than visual counting but may produce more accurate results (Davis and Anderson 1989), although inaccuracies can occur if fish move outside of the counting area or if they have discreet territories.

A comparison of sampling methods (belt, random point, and interval counts) showed that belt transects and point counts were preferable to interval counts because these methods recorded the number of species and individuals as values relative to area (Bortone et al. 1989). Because interval counts rank abundances based on frequency of encounter and disregard variations in spatial distribution, they overemphasize the importance of widespread but locally sparse fishes, while underrepresenting patchy but locally abundant species (DeMartini and Roberts 1982). Belt transects and point counts each have their inherent biases but both methods yield abundance estimates that are sufficiently robust to examine questions related to marine protected area (MPA) design and effectiveness. The applicability and limitations of various techniques for estimating reef fish abundance have been reviewed by a number of authors (e.g., Bortone and Kimmel 1991; Brock 1982). See Table 7.1 for a summary comparison.

Table 7.1 Visual Survey Methods

| Method | Description | Advantages | Disadvantages | Use |
|---|---|---|---|---|
| Belt transects | Fish counted along transect line. Transect widths normally 1–5 m Transect lengths normally 25–100 m | Quantitative, well-defined transect boundaries, easily learned, most used method so large number of studies for comparison, can derive density estimates that are comparable among transects of different dimensions | Larger transects can span multiple microhabitats leading to inaccurate variance estimates, following behavior of some fishes causes overestimates of these species, poor at assessing cryptic species, changes in swimming speed affect estimates. | General method can sample entire diurnal noncryptic fish assemblage, good for patchily distributed species, interested in assessing large area |
| Variable distance transects | Variation of belt transect method. The distance and sometimes angle between the fish and the transect line is estimated | Quantitative, high precision density estimates information on behavior and vagility | Difficult to learn, high observer variability, data analysis more difficult and difficult to compare with other methods, normally restricted species list | Used for wary species, highly vagile species, and those that range widely |
| Point counts | Stationary diver counts within a fixed cylinder for a fixed period of time<br>Standard method Bohnsack and Bannerot used 7.5 m radius with a 5 min count | Quantitative, ease of deployment, fast so a large number of replicates possible during a single dive, fixed time so time/space relationship unambiguous, somewhat better at detecting cryptic species particularly closer to observer, stationary observers less likely to cause distortion of abundance estimates | Small area surveyed, probability of detection decreases with distance from observer, more difficult to compare with belt transects or point counts that use different times and dimensions | Can sample entire diurnal noncryptic fish assemblage |

Table 7.1 Visual Survey Methods (continued)

| Method | Description | Advantages | Disadvantages | Use |
|---|---|---|---|---|
| Instantaneous point count | Variation of previous method but instantaneous counts made at random locations | Fast, simple, large sample size | Limited number of species can be assessed | Used for highly mobile, large species |
| Timed swims | Occurrence of a species is noted during a set time period, normally 5–10 min time intervals | Cover large area, high number of species detected, free moving diver can sample that are difficult to access using other survey methods, sample in currents | Qualitative, a dive usually constitutes a single sample | Interest in species richness, relative abundance estimates over large areas |

*Tow Boards.* Tow boards allow trained divers to census organisms and classify habitats over extensive reef areas quickly. The manta tow technique is used to provide a general description of large areas of reef and to gauge broad changes in abundance and distribution of organisms on coral reefs. The technique involves towing a snorkel diver at a constant speed behind a boat (English et al. 1997). The advantage of manta tows over other survey techniques is that they enable large areas of reefs to be surveyed quickly and with minimal equipment. Studies have shown the manta tow technique to be a relatively accurate and cost effective way of determining the abundance of noncryptic crown-of-thorns starfish and corals over large areas (Fernandes 1990; Fernandes et al. 1990; Moran and De'ath 1992). This technique has, in fact, been used to assess the distribution and abundance of crown-of-thorns starfish and corals on the Great Barrier Reef for over fifteen years (Sweatman et al. 2001). Tow boards can also be used to assess the presence and abundance of large mobile species such as sharks and jacks that are often too sparse to sample with conventional diver surveys.

*Deep Ocean Assessments.* Scientists have recently begun using video cameras in underwater housings to monitor fish and evaluate habitat at depths inaccessible by scuba (e.g., Auster et al. 1991; Parrish et al. 1997; Uzmann et al. 1977). Video cameras can either touch the bottom or be suspended from the surface and can be used singly, arranged in arrays, or mounted on rotating platforms (similar to stationary diver census). Video cameras can conveniently take pictures at preselected intervals, allowing an extended census time, and mounted lasers can help estimate transect width as well as fish length. They can also be used alone or with bait to attract and concentrate individuals of species of interest. Video cameras towed above the bottom from surface vessels can be used to conduct belt transects. In all video censuses, low image resolution can be a problem for identifying organisms to the species or sometimes even the genus level.

Other mechanisms for surveying deep ocean environments include the use of remotely operated vehicles (ROVs) or manned submersibles. These techniques allow direct counts of organisms and habitat assessments at depths much greater than divers can reach, for longer, and in strong currents. Scientists in submersibles, using either direct observations or video, often have a much greater field of view than an ROV and are therefore better able to accurately survey the sample area. New single operator submersibles like the DeepWorker 2000, which can dive to a depth of 600 meters and allow dives of twelve hours or more, will expand current efforts to survey deeper habitats (e.g., Coleman and Williams 2002).

The development of tethered sonar mapping and imaging vehicles and autonomous underwater vehicles (AUVs) has greatly expanded our ability to observe, map, and sample the deep ocean and seafloor beyond the capabilities of ROVs and submersibles and will continue to provide new and exciting information.

Technology has also made it possible for divers to reach deep ocean environments. Divers can now breathe controlled mixtures of gases that give them greater working time at greater depths. Additionally, divers can utilize rebreathing technology, which vastly increases the supply of air they can carry with them (Pyle 2000).

*Acoustic Surveys.* Fish can be detected underwater using an acoustic signal because various body parts (e.g., swim bladder and muscle) have densities different from the surrounding water (Brandt 1996). A transducer mounted on a vessel sends out an acoustic signal and receives the returning echoes created when the sound reflects off the fish back to the surface (Gunderson 1993). This method should really only be used for single-species aggregations of similar size fish (Hedgepeth and Condiotty 1995), which greatly limits its utility.

*Experimental Fishing.* When other techniques are impractical, experimental fishing can provide useful information on fish abundance. These techniques are by nature destructive and thus at odds with the mandate of a marine reserve. They also only produce a measure of relative abundance because of differences in catchability and the lack of a defined sampling area. To the extent possible, it is better to rely on nondestructive techniques. But there are a number of environments in which visual fish counts are impractical because of great depth, poor visibility, or wave energy. Because of its ability to sample large tracts fairly quickly, experimental fishing will be especially valuable for sampling nocturnally or widely dispersed animals and for sampling large regions in a short time frame.

Many types of fishing gear can be used to sample fish, including hook-and-line gear, nets, traps, and trawls. Traps and trawls serve as a good example of the advantages and disadvantages of these techniques. Traps are mesh cages with funnels that are designed to keep individuals from getting out once they have entered. They are used around the world to capture fish and invertebrates. Trap design, volume, and mesh size can greatly influence the catch (e.g., Munro et al. 1971) and catch rates and effective area fished vary greatly depending on habitat type and reef complexity (Friedlander et al. 2002; Wolff et al. 1999).

Trawls drag a net attached to a frame either in the water column or along the bottom. Because of their efficiency, they can be a valuable tool in assessing finfish, shellfish, and other invertebrates, particularly those species exploited by similar commercial gear. They sample a discrete area or volume over a specific time and provide quantitative indices of population abundance (Hayes et al. 1996). There are a variety of trawls and dredges (bottom trawls that actually dig into the sediment) that can be used in assessing populations of marine organisms with bottom composition and target species often determining the type of gear used. Biomass and population size are the main parameters of interest during trawl or dredge surveys (Gunderson 1993). Standardized dredge surveys for Atlantic sea scallops conducted by the National Marine Fisheries Service (NMFS) off Georges Banks closely tracked landings data and have shown the effectiveness of large-scale closed areas as a fishery management tool (Murawski et al. 2000). Traps and trawls can be used as a fisheries stock assessment device but should be considered with caution due to the variations in fishing efficiency among species and gear configurations (Recksiek et al. 1991). Similar concerns surround every other form of experimental fishing.

### *Monitoring Eggs and Larvae*

Surveys of eggs and larvae can provide a better understanding of fluctuations in the production and survival of offspring, the biotic and abiotic processes that determine recruitment, the size of the adult population, and potential sources or sinks of offspring (Gunderson 1993). Spawning and nursery areas can be identified by looking for areas with high concentrations of early eggs or larvae (Heath and Walker 1987). Spatial and temporal differences in spawning characteristics of exploited populations can be inferred from larval fish studies (Graham et al. 1984). Sample designs should take into consideration any known locations and timing of spawning and settlement.

Marine organisms have a diverse array of reproductive modes, including differential behavior and growth of larvae, spawning behavior, and habitat preferences, and these attributes make studying larvae a challenge. Studying larvae is also hampered by the fact that their concentrations are usually very low for any given species. To sample a sparse group of larvae that may vary tremendously in size, behavior, and location requires a wide variety of egg- and larvae-collecting devices.

The low densities of reef fish larvae make the study of ocean dynamics of

Table 7.2 Collection Techniques for Eggs and Larvae

*Active*

Plankton nets—Plankton nets are the most common gear type used and are typically towed at speeds of less than 2 m/s for periods of 30 s to an hour.

Benthic plankton samplers—Benthic plankton sleds are used to sample eggs and larvae on or just above the bottom.

Pelagic trawls—Low to moderate speed (0.5–3 m/s) midwater trawls can sample a large volume of water and are normally used to sample large larvae and small juveniles in pelagic areas. However, pelagic Tucker trawls were shown to be ineffective in estimating the density and size composition of pelagic reef fish larvae, particularly small individuals (Choat et al. 1993).

Neuston nets—These nets sample the water surface by towing the top edge of the net above the water and are ideal for surveying organisms that reside in this habitat.

High-speed samplers—Plankton nets can be mounted inside rigid cylinders and towed at speeds of up to 9 m/s. These devices can sample a large volume of water and reduce net avoidance by mobile larvae but can clog easily and damage larvae and eggs.

Pumps—Most systems involve pumping a target volume of water from an intake hose to a filter or net. Pumps provide discrete quantitative sampling that can be operated from a stationary or moving platform. Disadvantages of pumps include larvae avoidance, damage to samples, and limited volume of water sampled.

*Passive*

Drift samplers—Stationary sets of standard plankton nets. Useful in shallow or confined bodies of water.

Emergence traps—Demersal eggs, from fish such as salmonids, can be sampled after emergence using fixed nets with collecting bags.

Activity traps—These traps, developed for free swimming larvae and juveniles, are often used in shallow or confined bodies of water.

Light traps—Nighttime traps with artificial light sources are effective in collecting larvae that are positively phototactic (Choat et al. 1993)

the pelagic stage problematic, with different gear types selectively capturing different taxa and sizes within a taxon (Choat et al. 1993). Table 7.2 briefly describes some of the methods used to collect eggs and larvae.

### *Monitoring People*

The monitoring of the human dimension of marine reserves is an important tool for understanding the full impact of reserves on both people and ecosystems. There are a number of different attributes that are worth consideration when designing a monitoring program.

*Recreational/Subsistence Creel Surveys.* The recreational fishing community is very large, particularly in developed countries. These anglers can have a significant impact on fish stocks and the economy of coastal communities. On coral reefs, the recreational and subsistence catch is often equal to or greater than the catch by commercial fishers, and these fishers also catch a wider variety of species using a broader range of fishing gear than do their commercial counterparts (Friedlander and Parrish 1997). Recreational fishing, or creel, surveys are one of the few opportunities for management agency personnel to interact with the fishing community on a personal basis (Malvestuto 1996). This allows for the collection of fisheries data as well as the occasion to gain support and educate the public on management actions such as marine reserves.

The two basic steps in developing a recreational fishery survey are (1) selecting the statistical survey design that provides the best quantitative estimates of the fishery and (2) finding the most effective method of carrying out this design with the human and financial resources available (Malvestuto 1983). Stratified random creel surveys of fishing effort provided a useful sample of almost all types of fishing activity in a small bay in Hawaii (Friedlander and Parrish 1997). Within each sampling period, census of nearly all fishing effort and more than 70 percent of all catches was possible because of favorable local geography and the small size of the bay and the fisheries associated with it. A thorough treatment of recreational creel surveys can be found in Malvestuto (1996), Guthrie et al. (1991), and Pollock et al. (1994).

*Landings and Logbooks.* Catches are another important measure of the human dimension of marine reserves. Fishing men and women are often most concerned about how marine reserves will affect their ability to catch fish. Commercial fishers often keep log books of what they caught by time and location, and these can be an invaluable resource. Catch data can also be collected by monitoring the amount of fish that are landed or brought back to port.

Recreational fishers also keep records in the form of record-size fish caught in different locations using various tests of fishing line. An examination of world record catch data can identify an important recreational benefit from large adult fish moving out of reserves and into fishing areas (Roberts et al. 2001).

*Interviews.* Whether or not scientific tools can prove that reserves enhance overall fish catches, fishers who experience reserves are bound to develop their own opinion. Interviews of this sector of society can provide very useful information about whether reserves are working from a socioeconomic perspective. Interview questions generally focus on attitudes about fish catches and efforts expended. In combination, these questions can indicate how livelihoods have changed both in terms of product gained and the time required to do so. These techniques have been used successfully even in closed, small, island communities (Roberts et al. 2001; Russ and Alcala 1996). Of course, it is crucial to supplement this information with independent scientific sampling, in part because all of us are prone to distorting our memories over time and because nonscientists rarely sample in as random or extensive a manner as scientists doing a survey.

## MOVEMENT

The fluidity of marine systems via adult movement and reproductive dispersal is a major factor in marine reserve design (see chapter 5). The importance of this phenomenon to the performance and design of marine reserves warrants a fuller discussion of what we presently do and do not know on the subject, and what questions are useful to address in the future.

### Adult Movement

Many fished species are incapable of adult movement. These sessile invertebrates and plants cannot cross reserve boundaries, and those in reserves will only be vulnerable to fishing in the form of poaching. On the opposite end of the spectrum, some fish are built for swimming long distances. Bluefin tuna (*Thunnus thynnus*) are huge swimming machines that cover distances measured in thousands of kilometers on a regular basis (NRC 1994) creating a management challenge requiring international cooperation.

The rare studies of species that fall in between have sometimes yielded surprising results. A Hawaiian study of the blue trevally (*Caranx melampygus*)—a fish capable of swimming tens of kilometers—used multiple methods to track movements (Holland et al. 1996). These data indicated that nearly 90 percent of fish moved fewer than 2 kilometers, although a small proportion (less than 5 percent) of fish dispersed tens of kilometers away. Similar results have been found for the galjoen (pronounced galleon, *Coracinus capensis*) in South Africa (Attwood and Bennett 1994). Over 80 percent of tagged and recaptured galjoen were caught within 5 kilometers of their release site, while others had moved as far as 1450 kilometers away. Other fish move even less. A Hawaiian study of the white goatfish (*Mulloides flavolineatus*) showed that most fish moved no more than a few hundred meters (Holland et al. 1993). Seven percent of fish were "recaptured" at a commercial fish market. Since the study area was closed to fishing, these may have migrated several kilometers out of the reserve. However, it is also possible they were poached from within the reserve.

Movements often follow patterns. A number of fish show regular movement patterns on a daily basis between resting and feeding areas. Examples include grunts (Ogden and Ehrlich 1977) and goatfish (Holland et al. 1993). Fish also move as they develop, utilizing different habitats as they grow larger (Roberts 1996). Some species also move seasonally, including moving potentially hundreds of kilometers to aggregate at specific places to reproduce (Harding et al. 1978; Johannes et al. 1999).

However, the movement of relatively few species has been studied. As we learn more about how far fish move, particularly as movements relate to habitats and seasons, we will gain valuable tools for creating better marine reserves. Until then, reserves are likely to benefit homebodies more than vagrants (Bohnsack 1996).

### Reproductive Dispersal

Eggs and larvae can persist for as short as minutes or as long as months, and travel distances ranging from decimeters to thousands of kilometers (Shanks et al. 2003). The duration of the development phase may play an important role in potential dispersal distance, as may the strength and direction of prevailing currents. Eggs and larvae are generally incapable of swimming large distances themselves, at least early on in their development, so currents are the means by which they move from place to place. However, larvae may often disperse shorter distances than surface currents would suggest. Larval behavior can reduce dispersal distance by utilizing current changes that occur with depth, especially turbulence and drag that occur near the bottom (e.g., Breitburg et al. 1995). Nearshore turbulence and drag also can serve to retain larvae (Wolanski and Sarsenski 1997). And, even those larvae that do disperse face slim odds between diffusion, which may take them far from a desired habitat, and the poor survival that results (Cowen et al. 2000). These factors are probably responsible for the fact that studies of animal populations often find significant amounts of retention of eggs and larvae near the site where they were spawned (e.g., Brogan 1994; Jones et al. 1999; Swearer et al. 1999). However, local dispersal is not always the rule (e.g., the intertidal acorn barnacle, Bertness and Gaines 1993) and studies have identified what are apparently consistent patterns of larval dispersal and retention, with some sites producing great numbers of offspring and others receiving numerous new settlers (Lipcius et al. 1997).

Collectively, the body of work on larval dispersal suggests that many coastal marine populations retain some larvae locally while allowing others to disperse long distances. This strategy parallels dispersal and dormancy strategies exhibited by terrestrial and freshwater animals and plants, and in those contexts has been shown to have profoundly positive ecological and evolutionary benefits (Cohen and Levin 1987).

There are still large gaps in our knowledge about the dispersal of ocean offspring, in particular a clear understanding of what biological and physical factors influence the level of local retention of eggs and larvae (Sponaugle et al. 2002). The preceding conclusions are compatible with the data gathered in the few studies to really examine reproductive dispersal, but many more studies will be invaluable toward furthering our understanding of reproductive dispersal. Fortunately, a number of exciting new techniques now enable us to study this topic more effectively than ever before.

## Key Questions

Marine reserves themselves provide a chance to study movement patterns. Study of the distribution of fish across a reserve border can give an indication of how much adults move (e.g., Kramer and Chapman 1999; Rakitin and Kramer 1996). Moreover, there is much to learn by tagging individuals within reserves and observing the degree to which they move to outside areas (Attwood and Bennett 1994; Johnson et al. 1999). In addition to learning about general patterns, reserves provide an opportunity to look at whether and how fish move differently when they are densely versus sparsely populated (Sánchez Lizaso et al. 2000).

Reserves can also be used to study larval dispersal. Tegner (1992) transplanted adult green abalone (*Haliotis fulgens*) into an area where the species had been previously extirpated but were now protected. She then studied patterns of recruitment at varying distances from the transplant site, much in the same way one could look for enhanced recruitment at varying distances away from a marine reserve.

Studies of fluidity need not rely on reserves. The same techniques to look at movement from reserves can be used more generally to understand how fish and their offspring disperse. There are a number of key questions that remain to be answered about these phenomena. In addition to filling in many gaps in our understanding of basic movement tendencies, we would benefit from having substantially more information about how movement patterns change throughout development and respond to the local density of fish in an area. We would also benefit from having a clearer picture of the degree to which offspring are retained locally versus dispersed broadly and whether current patterns and behavior result in patterns of settlement. Since these phenomena will affect how many fish and their offspring leave a reserve, it will be difficult to predict the response of populations within and around marine reserves with any real precision until they are better understood.

## Tools and Techniques

Information on the movement of adult and larval marine organisms is crucial to evaluating the effectiveness of existing MPAs and determining the optimal location for establishing new ones. Local fisheries yields can be improved through export of adult biomass from protected areas to adjacent areas (DeMartini 1993; Johnson et al. 1999; Polacheck 1990; Russ and Alcala 1996), although those improvements are very small compared to the catch enhancements possible if adults stay within reserves but provide offspring to fishing areas (Sladek Nowlis and Roberts 1999). As a result, it is important to understand the dispersal patterns of both adults and their offspring.

### *Traditional Knowledge*

Fishermen are vastly more numerous than biologists and have lifetimes of experience pursuing and capturing fish and other marine resources (Johannes 1997). This large body of knowledge is often passed on from generation to generation and can be extremely valuable in understanding the dynamics of marine ecosystems and changes that have occurred over time. Traditional knowledge, particularly the timing and location of fish movement patterns, is highly relevant to the management of these resources (Friedlander et al. 2002b; Johannes and Yeeting 2001).

An interview process can be helpful in acquiring this information for incorporation into management decisions. Maps of habitat distributions and important landmarks can engage groups of fishers and others who have spent substantial time on the water, and help them to contribute their knowledge to the process. The resultant maps can also be invaluable in guiding management decisions (Friedlander et al. 2003a; Fig. 7.2), but are only one piece of a comprehensive assessment that should also include data collection in the field.

### *Long-Term Tagging and Recapturing*

Marking fish has already been discussed in the context of estimating abundance. The rate at which fish disperse from marine reserves can be determined

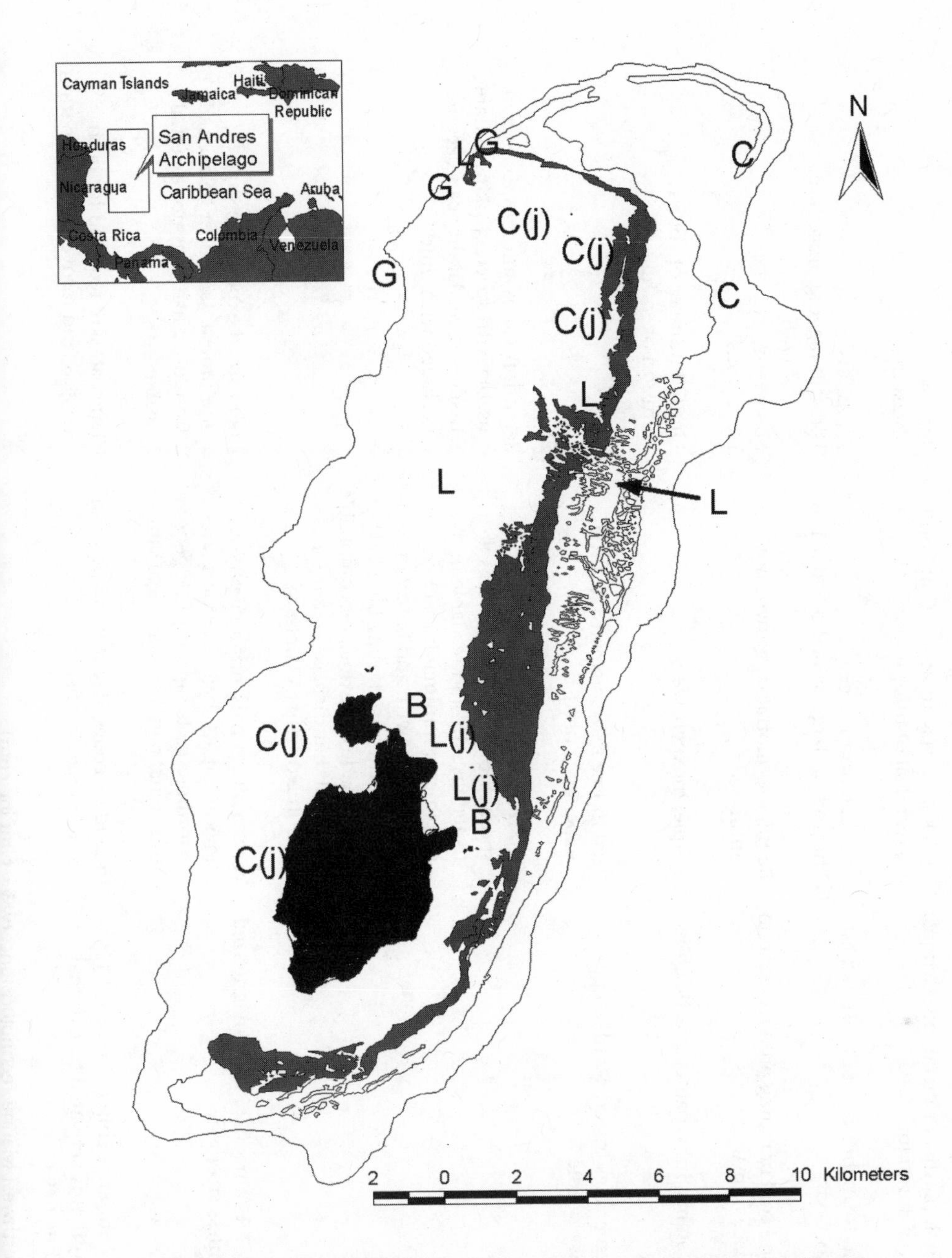


**FIG. 7.2 Traditional Knowledge around Old Providence Island, Seaflower Biosphere Reserve, Colombia.** Important ecological locations were identified by local fishermen on Old Providence Island, San Andrés Archipelago, Colombia. These data were collected through interviews as part of a process to zone portions of the Seaflower Biosphere Reserve. They provided useful biological information to the scientists making recommendations on zoning. Symbols: B—baitfish, C—conch, C(j)—conch juvenile, G—grouper spawning site, L—lobster, L(j)—lobster juveniles (see Friedlander et al. 2003a for more detail).

Table 7.3 Tagging Methods

| Tag type | Description | Advantages | Disadvantages |
|---|---|---|---|
| External tags | | | |
| Dart and t-bar tags | External, plastic, or metal, inserted just below dorsal fin | Quick and easy to apply. Can contain great deal of information | Tag loss<br>Abrasions |
| Internal anchor tags | Similar to anchor tags but inserted into body cavity | High retention rate<br>Effective on large, wide-bodied fishes | Abrasions<br>Difficult tagging procedure |
| Branding | Hot and cold branding produces scar on surface of fish | Rapid, low mortality, growth not inhibited | Short term<br>Dangerous |
| Pigment marks | Dyes, stains, inks, paints, plastic chips | Simple, inexpensive | Limited number of colors<br>Diffusion over time |
| Internal tags | | | |
| Visible implant tags | Elastomer tags injected with small-gauge needle | Read in live fish, inexpensive, little experience needed | |
| Visible implant alphanumeric tags | | High retention in suitable tissue or species; low capital costs; tags detected visually and readable in live specimens; minimal impact on survival, growth, and behavior; visibility is enhanced using blue LED light; made of flexible, biocompatible material that increases retention over the original VIalpha | Retention varies among species, not usable with species lacking suitable tissue, tag readability can become occluded with time |
| Coded wire tags | Small (1.1 mm × 0.25 mm) magnetized stainless steel wire | Very small animals, minimal physiological impact, high retention rates, enormous code capacity, inexpensive individual tags, can scan large sample | High capital expense for tagging equipment, tags must be excised from dead fish, tags not externally visible |
| Passive integrated transponder tags (PIT) | Small computer chip (12 mm × 2.1 mm) that is activated by external reading device | Individual codes, live fish, no performance impact on fish | High cost, skill required for both tagging and recovering information |

Sources: Guy et al. 1996 and Northwest Marine Technology, Inc. (www.Nmt.inc.com)

using "tag and release" methodology, with fish being recaptured either through fishing activity or underwater visual surveys. Resighting of marked fish underwater can also be used as a method to obtain information on the movement patterns of these marked individuals. This is particularity true when the mark possesses unique features (e.g., color, location on the fish's body, unique individual tag code) that can identify individual fish or provide information on release locations (Guy et al. 1996).

A wide range of tags are available, each of which has its own set of advantages and disadvantages (Table 7.3).

*Sonic Monitoring*

Acoustic telemetry is an effective way to track the daily movement patterns of individual fish. In this technique, one or more fish are fitted with a cylinder that gives off a unique frequency sound and released back into the study area (Fig. 7.3). Multiple fish can be monitored simultaneously if their transmitters use different frequencies. Scientists can track fish from boats using a hydrophone and global positioning system, or remotely using one or more stationery underwater hydrophones. Endemic Hawaiian goatfish (*Parupeneus porphyreus*) were continually monitored using an omnidirectional hydrophone for up to one month (Meyer et al. 2000). Information gained from tracking fishes can help to explain questions of immigration/emigration, residence time, habitat preference, site fidelity, and many other important life history traits, all of which have important implications for the design of effective marine reserves. Electronic tagging and tracking technologies have progressed rapidly in recent years and the tools and techniques used will vary depending upon the application and environment. Results from a recent symposium on electronic tagging and tracking of marine organisms document the current state of the art of this rapidly expanding field (Sibert and Nielsen 2001), which includes tags that record data on depth and position and then transmit this data via satellite, in some cases after the tag has deliberately popped off and floated to the surface.

Tracking and movement data can be analyzed in a number of ways, including maximum area covered (using minimum convex polygon analysis) (e.g., Klimley and Nelson 1984) and frequency of visits (using grid-square analysis). "Animal Movement" is an ArcView extension (see http://www.esri.com/) that is designed to implement a wide variety of animal movement functions in an integrated geographic information system (GIS) environment.

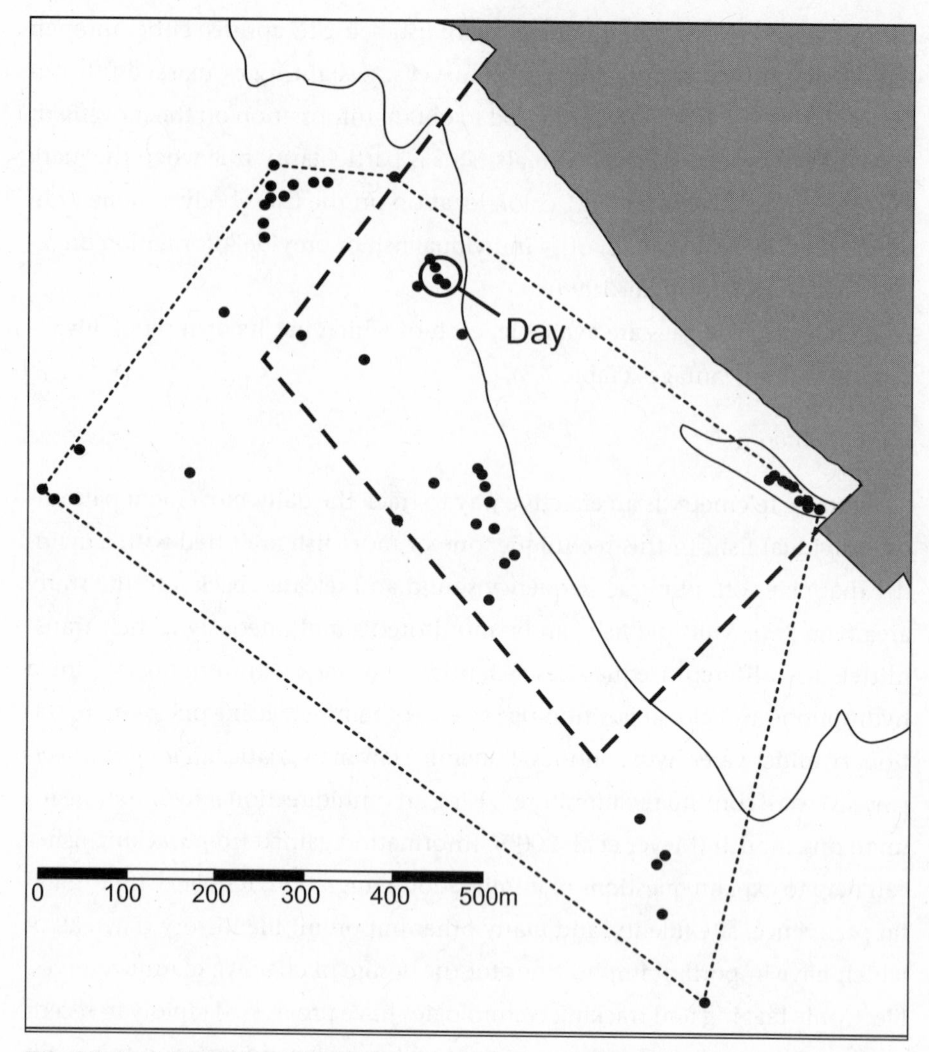


**FIG. 7.3 Sonic Tag Track.** Minimum Convex Polygon (MCP) home range of a 32.3 cm (Fork Length) yellowstripe goatfish, *Mulloidichthys flavolineatus,* tracked continuously for 46 h in the Waikiki marine reserve, Hawaii. Shaded area = land, solid line = 1 m depth contour, dashed bold line = reserve boundary, dotted line = MCP home range boundary, solid points = position fixes for the *M. flavolineatus*. Source: Adapted from Meyer 2003.

*Inferring Larval Dispersal*

By studying current patterns and larval behavior, we can get a sense of how larvae might disperse over time. Drifters are expendable systems launched from ships or aircraft into specific ocean areas. As they drift in response to ocean currents and winds, they make measurements of the atmospheric pressure, air and sea temperature, wind speed, and wind direction. Data from drifting buoys are relayed to ground stations via National Oceanic and Atmospheric Administration (NOAA) polar orbiting environmental satellites (POES). Because the great majority of tropical reef fishes have planktonic larvae, drifters are helpful in determining the settlement patterns of juvenile fish. Current patterns of the types drifters can provide have been proposed as a maximum limit on the dispersal capability of fishes from various Caribbean islands (Roberts 1997).

As we learn more about actual dispersal patterns, though, there is an increased focus on the influences of turbulent coastal current patterns (e.g., Wolanski and Sarsenski 1997) and larval behavior (e.g., Breitburg et al. 1995).

*Surveying Settlement of Larvae*

Monitoring of juvenile fishes will help to assess the future health and population dynamics of the assemblage. Larval fishes, which settle to local reefs, spend weeks in the plankton and are transported large distances from adult spawning sites. Therefore, local abundance of adults may not mean numerous juveniles on a specific reef. If changes in adult fish species richness or abundances are detected, settlement levels from preceding years should be examined to determine if the changes are most likely a result of settlement variation or from some other cause, like fishing level. Since most reef fishes have high site fidelity during their juvenile and adult stages, local reef fish populations must be replenished from the planktonic pool of larvae. Local settlement is highly variable over space and time and profoundly influences the population dynamics of the adult assemblage structure. Areas of locally retained larvae have obvious implications for MPA site designations. Visual transects can be used to survey juvenile fish that have recently settled out of the plankton. Typically a 50- by 2-meter strip is used (Fowler et al. 1992).

*Inferring Larval Dispersal from Fish Ear Bones*

Fish otoliths (crystalline structures in the inner ear) deposit new material daily and incorporate trace elements from the surrounding sea water into their structure (Swearer et al. 1999). Because productivity and trace-element concentrations differ between coastal and oceanic waters, otoliths can be used to reconstruct the dispersal history of larvae. Higher concentrations of trace-elements in otoliths should indicate faster growth rates because coastal waters are generally more productive. Trace-element concentrations in otoliths can be measured under laboratory conditions using mass spectrometers, which allows for reconstruction of past environmental conditions experienced by individual fish. The technique is proving invaluable for examining population connectivity, spawning behavior, and stock associations in a number of fish species (Thorrold et al. 1998, 2001).

*Inferring Dispersal and Movement by Tracking Introduced Species*

Though not a natural phenomenon, there have been and continue to be introductions of nonnative species into new ecosystems through ballast water of ships, the aquarium trade, and intentional releases to provide new recreational fishing opportunities. Some of these introduced species establish themselves and even flourish in their new environment. Tracking their spread offers the opportunity to see firsthand how far and quickly a population disperses. For example, in the 1950s and 1960s, eleven species of snapper and grouper were intentionally introduced in the waters surrounding the main Hawaiian Islands. Four of these species became established (three snappers and one grouper, Randall 1987). One snapper—*Lutjanus kasmira*—adapted especially well to its new location, and its spread along the greater Hawaiian archipelago is reasonably well documented (Fig. 7.4). Introduced in 1958 (Randall 1987), it had spread throughout the archipelago by the early 1990s (Randall et al. 1993), averaging 33 to 130 km per year. This rate of spread supports the idea that seed areas, such as marine reserves, can sustain fisheries outside their borders. However, it should be noted that such rates do not necessarily contradict the idea that much reproductive output is retained in a relatively small local area.

## HABITAT CONNECTIVITY

Habitats may be connected through the flow of nutrients, through animal movements, or both. Understanding these connections is a crucial step toward designing better marine reserves. To the extent that marine reserves can encompass complementary habitat types in a connected manner, they will be more effective at retaining productive populations within their borders. Advancing our understanding of habitat connectivity will require habitat mapping exercises, habitat categorizations of ecological relevance, and direct and indirect studies of fish themselves (Christensen et al. 2003).

In a few cases, we understand the linkages among habitat types. Shallow, nearshore, tropical ocean environments, for example, are characterized by the connectivity among coral reefs, mangrove lagoons, and sea grass beds (Ogden 1988). These connections can result in different fish assemblages on a coral reef depending on whether there are mangroves and sea grass beds nearby (Appel-

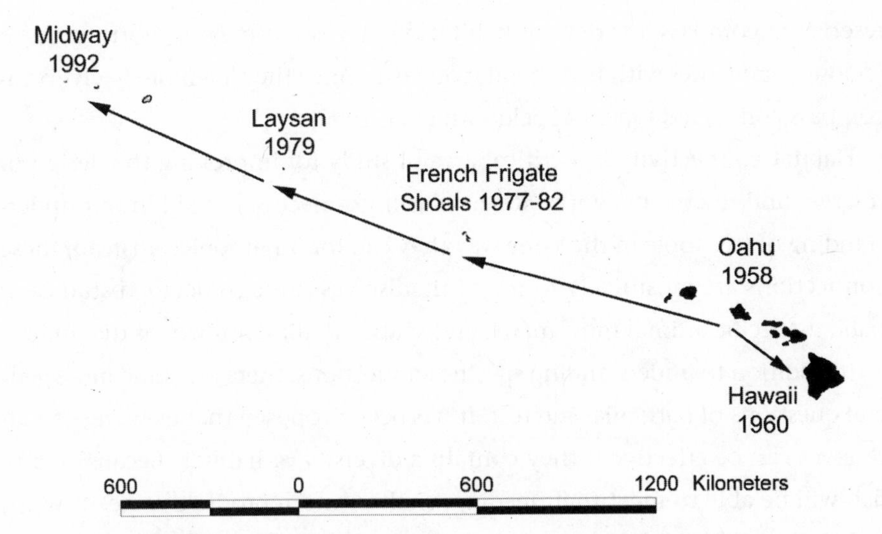


**FIG. 7.4 Spread of Ta'ape, *Lutjanus kasmira,* throughout the Hawaiian Archipelago.** Introduced in the 1950s, this species of tropical snapper spread through the main Hawaiian Islands and then the Northwestern Hawaiian Islands. Tracking its appearances along this island chain gives us the opportunity to measure its rate of spread and infer the degree to which the population disperses.

doorn et al. 2003; Christensen et al. 2003). Understanding these connections plays two important roles. First, it is valuable to ensure that any marine reserve or reserve network in a shallow nearshore tropical ocean environment contains reef, mangrove, and sea grass habitats because many fish species move among these habitat types as they grow, feed, and seek shelter and mates. Second, in cases where the management area includes coral reefs that are near these other coastal habitats as well as reefs that are some distance away, it may be useful to distinguish these types of reefs and ensure that both are represented in a reserve or reserve network because these reef types will have different living communities associated with them, all of which should receive some degree of protection (Friedlander et al. 2003a).

### Key Questions

It is likely there are many important ecological linkages between habitat types that have not yet been recognized. At present we have only a primitive understanding of the movement patterns of ocean animals and the distribution and ecological functions of various habitat types. Because linkages may well exist that we do not yet recognize, it is important to ensure that marine reserves encompass a variety of habitat types. Also, it may be worthwhile to choose reserve sites with high habitat diversity since they are more likely to encompass connected types (Appeldoorn et al. 1997).

Habitat connectivity is a critical area of study for improving the design of reserves and reserve networks. Reserves will not necessarily aid in our understanding of this topic in the same way they can for other topics. Instead, these connections are most likely to reveal themselves through detailed studies of habitat-specific animal movements and spatial studies of nutrient dynamics.

In addition to understanding specific connections, there are some more general questions of particular interest. It has been proposed that even very small reserves can be effective if they contain a diversity of habitats because many fish will be able to meet their feeding, shelter, and reproductive needs without leaving. This hypothesis is still largely untested and would benefit from further study. We would also benefit from a better understanding of the influence of human impacts on habitat connectivity. Identifying the connections is one step in this understanding. The next step would be to examine how different human impacts are likely to affect these connections. A more thorough understanding of these connections will help to design management systems that conserve whole functioning natural systems. Finally, there has been great attention paid to the value of reserve networks, or a series of marine reserves that are interconnected through the dispersal of adults and larvae.

### Tools and Techniques

In addition to monitoring and adult and larvae movement patterns, an understanding of habitat connectivity is critical in the design of marine reserves. If reserves are to maintain system integrity and structure, all components necessary to ensure system function must be included within the reserve design, either by networking smaller areas or by creating large, all-encompassing areas (Appeldoorn et al. 1997, 2003).

Mobile animals play a key role in connections among habitat types. In moving from one habitat to another, adult animals make direct ecological connections and can even shuttle nutrients from one habitat to the next (Parrish 1989). Movements throughout development also contribute to habitat connections. Finally, eggs and larvae are a means by which adult habitats connect back to nurseries via a waterborne stage. Thus we can learn a lot about habitat connectivity through the study of movement patterns, as already discussed.

Movement patterns can also be inferred by looking at changes in habitat usage seasonally or through development by studying the distribution of different size classes or at different times of year. For example, small grunts and snappers are consistently found in shallow habitat types, primarily sea grasses and mangroves, whereas larger individuals are most often found offshore on coral reefs. These patterns indicate movement tendencies from nearshore nursery habitats to offshore adult habitats as these fishes grow and mature (Appeldoorn et al. 2003; Christensen et al. 2003). Such indirect studies can provide a valuable broad-scale perspective when studying the way that habitats are interconnected ecologically. Such studies, when combined with habitat mapping exercises (discussion follows) can teach us a great deal about these connections.

## HABITAT AND SPECIES DISTRIBUTIONS

Having discussed issues relevant to the performance attributes of marine reserves as a management tool, we will now describe some data requirements for designing specific marine reserves in practice and how best to collect these data. In particular, it is valuable to collect and compile data about the distribution of habitats and species in a management area. Marine reserve design can be greatly enhanced by an understanding of habitat and species distributions and interactions.

Habitat-oriented marine reserve design can achieve a wide range of potential goals (see chapter 5). Sometimes it is necessary, though, to study species distributions as a way of categorizing habitats in an ecologically relevant manner. For example, Friedlander and colleagues (2003a) used species distributions to confirm a system of habitat types they used to advise an ocean zoning process in Colombia. They recategorized habitats from existing habitat maps based on their understanding of coral reef ecosystems. They then used surveys of species distributions and abundance to confirm that their habitat type definitions were ecologically based.

Good habitat maps are rare, and even when they do exist the habitats may not be categorized in a manner conducive to reserve design. In many cases, we may need to rely on the simplest of physical features, such as separating sheltered bays from open coastlines, and nearshore from offshore areas (e.g., Ballantine 1997). Alternatively, it may be desirable to map habitats if resources are available to do so quickly. Where habitat maps already exist, they should guide marine reserve design (e.g., Friedlander et al. 2003a).

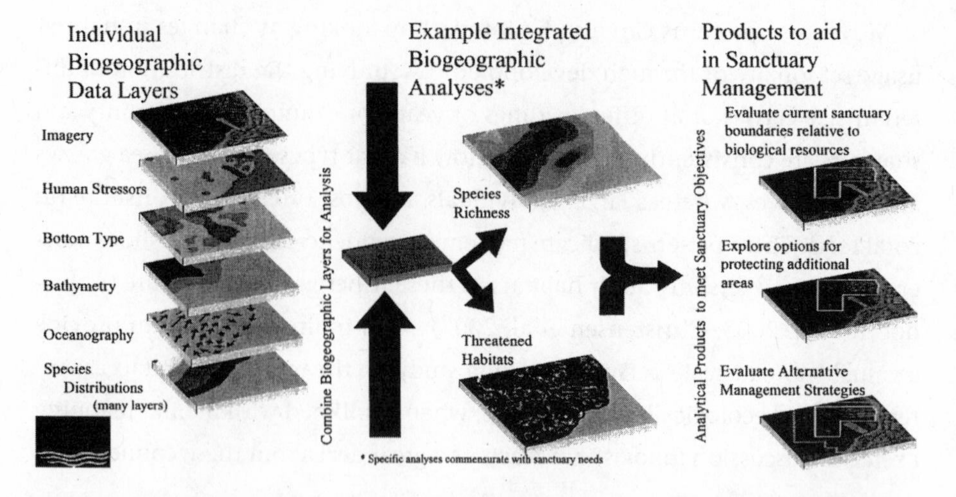


**FIG. 7.5 Biogeographic Assessment for Marine Protected Area Design.** A range of geographic data are collected and can be combined to gain insight into threatened habitats and other priority areas for protection. These insights can be communicated to managers through various illustrative products. Source: From Kendall and Monaco 2003 with permission.

## Tools and Techniques

The quality, quantity, and distribution of habitats are some of the most important factors influencing fish and invertebrate distributions and population dynamics and thus should be considered when designing effective marine reserves.

It's important to know the exact location of habitats within a reserve because of the key role habitats play in sustaining ocean fisheries and ecosystems. Geophysical techniques, such as sideways looking sonar (side-scan) and seismic reflection profiling, which uses sound waves to distinguish among habitat types, can help to define large-scale benthic features and to map protected areas. NOAA's Biogeography Program recently produced a series of highly accurate GIS-based maps of coral reef habitats using digital georeferencing and orthorectified aerial photography (Monaco et al. 2001). These images came from photographs taken from directly above, with reference points and other measures used to remove any distortion and to link adjacent photos. The resulting habitat classifications and underlying hierarchical spatial structure were used to look at resource distribution, abundance, and habitat utilization (Fig. 7.5).

Knowing the layout of habitat types goes hand in hand with knowing the fish assemblages that use them. There is a strong link between coral reef fish diversity and physical habitat, as demonstrated by a number of authors (e.g., Luckhurst and Luckhurst 1978; Friedlander and Parrish 1998). For example, habitats with low spatial relief and limited shelter in Hawaii harbored a lower biomass of reef fishes than highly complex habitats (Friedlander and Parrish 1998). This strong link means that we must know the location, distribution, and extent of habitats necessary for successful recruitment, growth, feeding, and reproduction (Christensen et al. 2003; Friedlander and Parrish 1998). Habitats and species can be mapped using techniques already discussed in the context of monitoring, or with more sophisticated remote sensing technologies.

### *Remote Sensing*

Different technologies allow the measurement of habitat attributes at a wide range of spatial resolutions and varying costs (Table 7.4). Habitats can be identified and mapped remotely using reflected light and sound. Light reflectance only works for relatively shallow waters, whereas sound reflectance can paint a picture of deeper areas.

High-resolution digital imagery gathered from either an aircraft or a satellite can be used to create habitat maps. Maps can be generated by either human observers or computerized image analysis software. Imagery from satellites typically has spatial resolution on the order of tens of meters of spatial resolution and hundreds of nanometers of spectral resolution (Hochberg and Atkinson 2000). The recent advent of high spatial (1-m) and high spectral (1- to 10-nm) resolution digital images should provide fine-scale identification and mapping of coral reefs and other benthic communities (Mumby et al. 1997). Aerial surveys can also be conducted at lower altitudes using conventional light aircraft, kites and balloons, remote-controlled aircraft, or ultralights (McManus et al. 1996). LIDAR (light detection and ranging) integrates light and sound signals to determine the properties and ranges of target objects. LIDAR can be mounted on aircraft to quantify bathymetry and habitat complexity over tens and hundreds of kilometers of area.

Techniques using sound reflectance are usually performed in the water using ship-based equipment. They send sound signals out and analyze the returning echoes. Sound surveys can be useful in fisheries for stock assessment (Karp 1990), for remotely sensing characteristics of bottom habitat (e.g., Able et al. 1987; Yoklavich et al. 1997), and for assessing impacts to fisheries habitat (Collie et al. 1997). Acoustic remote sensing also presents a promising independent approach to evaluate fishing effort on a spatial scale consistent with commercial fishing activities. For example, this technology can identify marks left by bottom trawls as a method of determining the frequency of impact of trawl gear over a range of depths and habitat types (Friedlander et al. 1999).

Table 7.4. Remote-Sensing Methods with Associated Resolutions and Costs.

| Technology | Spatial Resolution | Spectral Resolution | Cost to Acquire |
|---|---|---|---|
| Space-based | | | |
| IKONOS | 16 m$^2$ | 4 bands | \$27–62 km$^2$ |
| Quickbird II | 6.25 m$^2$ | 4 bands | \$30 km$^2$ |
| LandSat | 900 m$^2$ | 6 bands | \$0.0015 km$^2$ |
| Hyperion | 900 m$^2$ | 220 bands | NA |
| Aircraft | | | |
| Hyperspectral | 9 m$^2$ | 74 bands | \$325 km$^2$ |
| Color Aerial Photography | 1 m$^2$ | 3 bands | \$175 km$^2$ |
| Ship-board | | | |
| Multibeam | 0.01–10 m$^2$ | Bathymetry + backscatter | Shallow Water –\$27,000 km Deep Water –\$300 km |
| Sidescan | 0.01–10 m$^2$ | Backscatter | Shallow Water –\$24,000 km Deep Water –\$200 km |

In addition to providing the basis for quantifiable, accurate determination of the coral reef ecosystem, digital maps are easily incorporated into a computerized GIS for analysis along with other map information. They can be used to support the designation and conservation of MPAs and essential habitats. They also have an important role in shaping public awareness and education (Monaco et al. 1998, 2001).

The coupling of ecology, remote sensing, and GIS technology has recently been used to map and monitor U.S. coral reefs in Florida, the Caribbean, and Hawaii (Monaco et al. 2001). Digital benthic habitat maps derived from high-resolution aerial photography have been used to help define spatial and temporal distributions by life stage of fishes and invertebrates and to determine species habitat affinities (Monaco et al. 1998). Coupling the distribution of habitats and species habitat affinities using GIS technology enables the elucidation of species habitat utilization patterns for a single species or for assemblages of animals (Kendall et al. 2003). This integrated approach is useful in quantitatively defining essential fish habitat (Clark et al. 2003) and defining biologically relevant boundaries of marine protected areas (Christensen et al. 2003). Such approaches are also critical to understanding ecological connections among habitats as already discussed here.

## CONCLUSIONS

Marine reserves present an unparalleled opportunity to study marine ecosystems in the absence of fishing pressure and other major human impacts. The ocean is an inherently unpredictable place. This unpredictability makes it difficult to scientifically prove that marine reserves work, especially when it comes to export of benefits to surrounding fishing grounds. However, we do not demand such proof of other management tools we regularly enact, wisely so because it is nearly impossible to provide. Compared to other management tools, extensive scientific support backs marine reserves as an effective conservation and fishery management tool. Better still, extensive marine reserve networks can serve as an effective insurance policy against the unpredictable nature of the ocean.

In spite of their promise as a management tool, there are a number of areas where more study would enhance our ability to design marine reserves. Monitoring of existing and new marine reserves will be an invaluable tool for improving our ability to design them effectively while providing broader insight into management issues and natural systems in general. Movement of adults and their offspring is another major topic of interest related to reserve design. As we learn more about how fish move, we will gain valuable insight into creating better marine reserves. Fish movement is one crucial way in which habitats are connected, and additional study on this topic will aid greatly in designing reserves that are effective at protecting fish throughout their life cycle. One of the most important reasons that reserves are important is to protect against ecosystemwide crashes. Our lack of knowledge warrants such protection, and reserves themselves can teach us how to avoid crashes. Design of marine reserves would not be complete without habitat maps and species distributions, so these exercises will be valuable as part of a process of reserve designation in an area.

We know everything necessary to improve on current use of marine reserves. Nevertheless, a number of tools and techniques are available to enhance our ability to design and implement effective reserve networks. Reserves themselves provide an invaluable research tool and give us the opportunity to learn a great deal about areas relatively unaffected by humans. Populations of important species in many marine ecosystems are now so sparse that they cannot exert their former ecological role (Dayton 1998), and the indirect effects of the reductions of these species are unknown because no baseline data exist for comparison (Dayton et al. 1998). Modern studies of marine ecosystems began long after enormous changes in these systems had occurred (Jackson et al. 2001). Moreover, the "shifting baseline syndrome" causes people to identify ecosystems as near pristine when they first experience them, even if they may have already been severely degraded (Pauly 1995; Sheppard 1995). These problems make it difficult to determine what constitutes a natural ecosystem and how to manage these ecosystems accordingly. One irreparable consequence of overfishing and habitat destruction is the loss of opportunity to study and understand intact communities (Dayton 1998). Marine reserves can provide that baseline if the ecosystems are not already too degraded to recover.

Other tools, ranging from simple plastic tags to sophisticated satellites, can help to advance our understanding of how to design marine reserves better. In the short run, most designs will represent an improvement over present management practices. In the long run, the techniques discussed in this chapter will be vital to improving our ability to design effective marine reserves for conservation and fisheries.

## REFERENCES

Able, K. W., D. C. Twichell, C. B. Grimes, and R. S. Jones. 1987. Sidescan sonar as a tool for detection of demersal fish habitats. *Fishery Bulletin, U.S.* 85(4):725–736.

Allen, M. S., W. E. Pine III. 2000. Detecting fish population responses to a minimum length limit: Effects of variable recruitment and duration of evaluation. *North American Journal of Fisheries Management* 20:672–682.

Appeldoorn, R. S., C. W. Recksiek, R. L. Hill, F. E. Pagan, and G. D. Dennis. 1997. Marine protected areas and reef fish movements: The role of habitat in controlling ontogenetic migration. *Proceedings of the 8th International Coral Reef Symposium* 2:1917–1922.

Appeldoorn, R. S., A. Friedlander, J. Sladek Nowlis, P. Usseglio, and A. Mitchell-Chui. 2003. Habitat connectivity in reef fish communities and marine reserve design in Old Providence–Santa Catalina, Colombia. *Gulf and Caribbean Research* 14(2):61–78.

Attwood, C. G., and B. A. Bennett. 1994. Variation in dispersal of galjoen (*Coracinus capensis*) (Teleostei: Coracinidae) from a marine reserve. *Canadian Journal of Fisheries and Aquatic Sciences* 51(6):1247–1257.

Ault, J. S., S. G. Smith, G. A. Meester, J. Luo, J. A. Bohnsack, and S. L. Miller. 2002. *Baseline Multispecies Coral Reef Fish Stock Assessment for Dry Tortugas*. NOAA Technical Memorandum NMFS-SEFSC-487. Silver Spring, MD: U.S. Department of Commerce.

Auster, P. J., R. J. Malatesta, S. C. LaRosa, R. A. Cooper, and L. L. Stewart. 1991. Microhabitat utilization by the megafaunal assemblage at a low relief outer continental shelf site—Middle Atlantic Bight, USA. *Journal of Northwest Atlantic Fishery Science* 11:59–69.

Bakun, A. 1993. The California Current, Benguela Current, and Southwestern Atlantic Self ecosystems: A comparative approach to identifying factors regulating biomass yields. In Sherman, K., L. M. Alexander, and B. D. Gold, eds. *Large Marine Ecosystems: Stress, Mitigations, and Sustainability,* 199–221. Washington, DC: AAAS Press.

Ballantine, W. J. 1997. Design principles for systems of "no-take" marine reserves. *Fisheries Centre Research Reprints* 5(1):4–5. Vancouver, BC: UBC Fishery Centre.

Bertness, M., and S. D. Gaines. 1993. Larval dispersal and local adaptation in acorn barnacles. *Evolution* 47:316–320.

Bohnsack, J. A. 1996. Maintenance and recovery of reef fishery productivity. In Polunin, N. V. C., and C. M. Roberts, eds. *Reef Fisheries,* 283–313. London: Chapman and Hall.

Bohnsack, J. A., and S. P. Bannerot. 1986. A stationary visual census technique for quantitatively assessing community structure of coral reef fishes. NOAA Technical Report NMFS-41. Silver Spring, MD: U.S. Department of Commerce.

Bortone, S. A., and J. J. Kimmel. 1991. Environmental assessment and monitoring of artificial habitats. In Seaman, W. Jr., and L. M. Sprague, eds. *Artificial Habitats for Marine and Freshwater Fisheries,* 177–236. New York: Academic.

Bortone, S. A., J. J. Kimmel, and C. M. Bundrick. 1989. A comparison of three methods for visually assessing reef fish communities: Time and area compensated. *Northeast ASIF Science* 10(2):85–96.

Brandt, A. B. 1996. Acoustic assessment of fish abundance and distribution. In Murphy, B. R., and D.W. Willis, eds. *Fisheries Techniques,* 2nd ed., 385–432. Bethesda, MD: American Fisheries Society.

Breitburg, D. L., M. A. Palmer, and T. Loher. 1995. Effects of flow, structure and larval schooling behavior on settlement behavior of oyster reef fish. *Marine Ecology Progress Series* 125:45–60.

Brock, R. E. 1982. A critique of the visual census method for assessing coral reef fish populations. *Bulletin of Marine Science* 32:269–276.

Brock, V. E. 1954. A preliminary report on a method of estimating reef fish populations. *Journal of Wildlife Management* 18:297–308.

Brogan, M. W. 1994. Distribution and retention of larval fishes near reefs in the Gulf of California. *Marine Ecology Progress Series* 115:1–13.

Bythell, J. C., P. Pan, and J. Lee. 2001. Three-dimensional morphometric measurements of reef corals using underwater photogrammetry techniques. *Coral Reefs* 20:193–199.

Carr, M. H., and P. T. Raimondi. 1999. Marine protected areas as a precautionary approach to management. *California Cooperative Oceanic Fisheries Investigations Reports* 40:71–76.

Choat, J. H., P. J. Doherty, B. A. Kerrigan, and J. M. Leis. 1993. Sampling of larvae and pelagic stages of coral reef fishes: A comparison of towed nets, purse seine and light-aggregation devices. *Fishery Bulletin* 91:195–201.

Christensen, J. D., C. F. G. Jeffrey, C. Caldow, M. E. Monaco, M. S. Kendall, and R. S. Appeldoorn. 2003. Cross-shelf habitat utilization patterns of reef fishes in southwestern Puerto Rico. *Gulf and Caribbean Research* 14(2):9–28.

Clark, R. D., W. Morrison, J. D. Christensen, M. E. Monaco, and M. S. Coyne. 2003. Modeling the distribution and abundance of spotted seatrout: Integration of ecology and GIS technology to support management needs. In Bortone, S.A., ed. *Biology of the Spotted Seatrout.* Boca Raton, FL: CRC Press.

Clark, W. G. 1993. The effect of recruitment variability on the choice of a target level of spawning biomass per recruit. In Kruse, G., D. M. Eggers, R. J. Marasco, C. Pautzke, and T. J. Quinn II, eds., *Proceedings of the International Symposium on Management Strategies for Exploited Fish Populations,* 233–246. Alaska Sea Grant College Program, AK-SG-93-02.

Cohen, D., and S. A. Levin. 1987. The interaction between dispersal and dormancy strategies in varying and heterogeneous environments. *Lecture Notes in Biomathematics* 71:110–122.

Coleman, F. C., and S. Williams. 2002. Overexploiting marine ecosystem engineers: potential consequences for biodiversity. *Trends in Ecology and Evolution* 17:40–44.

Collie, J. S., G. A. Escanero, and P. C. Valentine. 1997. Effects of bottom fishing on the benthic magafauna of Georges Bank. *Marine Ecology Progress Series* 155:159–172.

Cowen, R. K., K. M. M. Lwiza, S. Sponaugle, C. B. Paris, and D. B. Olson 2000. Connectivity of marine populations: Open or closed? *Science* 287:857–859.

Davis, G. E., and T. W. Anderson. 1989. Population estimates of four kelp forest fishes and an evaluation of three in situ assessment techniques. *Bulletin of Marine Science* 44:1138–1151.

Davis G. E., and J. W. Dodrill. 1989. Recreational fishery and population dynamics of spiny lobster, *Panulirus argus,* in Florida Bay, Everglades National Park, 1977–1980. *Bulletin of Marine Science* 44:78–88.

Davis, G. E., D. J. Kushner, J. M. Mondragon, J. E. Mondragon, D. Lerma, and D. V. Richards. 1997. *Kelp Forest Montoring Handbook,* vol. 1, *Sampling Protocol.* Ventura, CA: Channel Islands National Park.

Dayton, P. K. 1998. Reversal of the burden of proof in fisheries management. *Science* 279:821–822.

Dayton, P. K., M. J. Tegner, P. B. Edwards, and K. L. Riser. 1998. Ghost communities and the problem of reduced expectation in kelp forests. *Ecological Applications* 8:309–322.

DeMartini E. D. 1993. Modeling the potential of fishery reserves for managing Pacific coral reef fishes. *Fishery Bulletin* 91:414–427.

DeMartini, E. E., and D. Roberts. 1982. An empirical test of biases in the rapid visual technique for species-time censuses of reef fish assemblages. *Marine Biology* 70:129–134.

Earle, S. A. 1995. *Sea Change: A Message of the Oceans*. New York: G. P. Putnam's Sons.

English, S., C. Wilkinson, and V. Baker, eds. 1997. *Survey Manual for Tropical Marine Resources*. Townsville: Australian Institute of Marine Science.

Fernandes, L. 1990. Effect of the distribution and density of benthic target organisms on manta tow estimates of their abundance. *Coral Reefs* 9:161–165.

Fernandes, L., H. Marsh, P. J. Moran, and D. Sinclair. 1990. Bias in manta tow surveys of *Acanthaster planci*. *Coral Reefs* 9:155–160.

Fowler, A. J., P. J. Doherty, and D. McB. Williams. 1992. Multiscale analysis of recruitment of a coral reef fish on the Great Barrier Reef. *Marine Ecology Progress Series* 82:131–141.

Friedlander, A. M., and J. D. Parrish. 1997. Fisheries harvest and standing stock in a Hawaiian Bay. *Fish. Res* 32(1):33–50.

———. 1998. Habitat characteristics affecting fish assemblages on a Hawaiian coral reef. *J. Exp. Mar. Biol. Ecol* 224(1):1–30.

Friedlander, A. M., G. W. Boehlert, M. E. Field, J. E. Mason, J. V. Gardner, and P. Dartnell. 1999. Sidescan sonar mapping of benthic trawl tracks on the shelf and slope off Eureka, California. *Fishery Bulletin* 97(4):786–801.

Friedlander, A. M, J. D. Parrish, and R. C. DeFelice. 2002a. Ecology of the introduced snapper *Lutjanus kasmira* in the reef fish assemblage of a Hawaiian bay. *Journal of Fish Biology* 60:28–48.

Friedlander A., K. Poepoe, K. Helm, P. Bartram, J. Maragos, and I. Abbott. 2002b. Application of Hawaiian traditions to community-based fishery management. *Proceedings of the 9th International Coral Reef Symposium* 2:813–818.

Friedlander A., J. Sladek Nowlis, J. A. Sanchez, R. Appeldoorn, P. Usseglio, C. McCormick,

S. Bejarano, and A. Mitchell-Chui. 2003a. Designing effective marine protected areas in Seaflower Biosphere Reserve, Colombia, based on biological and sociological information. *Conservation Biology* 17:1–16.

Friedlander, A. M., E. K. Brown, P. L. Jokiel, W. R. Smith, and K. S. Rodgers. 2003b. Effects of habitat, wave exposure, and marine protected area status on coral reef fish assemblages in the Hawaiian archipelago. *Coral Reefs* 22:291–305.

Gelman, A., J. B. Carlin, H. S. Stern, and D. B. Rubin. 1995. *Bayesian Data Analysis*. London: Chapman and Hall.

Graham, J. J., B. J. Joule, C. L. Crosby, and D. W. Townsend. 1984. Characteristics of Atlantic herring (*Clupea harengus* L.) spawning population along the Maine coast, inferred from larval studies. *Journal of Northwest Atlantic Fisheries Science* 5(2):131–142.

Green, R. H. 1979. *Sampling Design and Statistical Methods for Environmental Biologists*. New York: John Wiley and Sons.

Gunderson, D. R. 1993. *Surveys of Fisheries Resources*. New York: John Wiley and Sons.

Guthrie, D., J. M. Hoenig, M. Holliday, C. M. Jones, M. J. Mills, S. A. Moberly, K. H. Pollock, and D. R. Talhelm, eds. 1991. *Creel and Angler Surveys in Fisheries Management*. Bethesda, MD: American Fisheries Society.

Guy, C. S., H. L. Blankenship, and L. A. Nielsen. 1996. Tagging and marking. In Murphy, B. R., and D. W. Willis, eds. *Fisheries Techniques*, 2nd ed., 353–383. Bethesda, MD: American Fisheries Society.

Halpern, B. 2003. The impact of marine reserves: Do reserves work and does reserve size matter? *Ecological Applications* 13(1, suppl):S117–S137.

Harding, D., J. H. Nicholas, and D. S. Tungate. 1978. The spawning of plaice (*Pleuronectes platessa* L.) in the Southern North Sea and English Channel. *Rapp. Proces Verb. Reun. Cons. Int. Explor. Mer.* 172:102–113.

Hayes, D. B., C. Paola Ferreri, and W. W. Taylor. 1996. Active fish capture methods. In Murphy, B. R., and D. W. Willis, eds. *Fisheries Techniques*, 2nd ed., 193–220. Bethesda, MD: American Fisheries Society.

Heath, M. R., and J. Walker. 1987. A preliminary study of the drift of larval herring (*Clupea harengus* L.) using gene-frequency data. *J. Cons. Ciem.* 43:139–145.

Hedgepeth, J. B., and J. Condiotty. 1995. Radar tracking principles applied for acoustic fish detection. ICES International Symposium on Fisheries and Plankton Acoustics (poster), Aberdeen (UK).

Hochberg, E. J., and M. J. Atkinson. 2000. Spectral discrimination of coral reef benthic communities. *Coral Reefs* 19:164–171.

Holland, K. N., J. D. Peterson, C. G. Lowe, and B. M. Wetherbee. 1993. Movements, distribution and growth rates of the white goatfish *Mulloides flavolineatus* in a fisheries conservation zone. *Bulletin of Marine Science* 52(3):982–992.

Holland, K. N., C. G. Lowe, and B. M. Wetherbee. 1996. Movements and dispersal patterns of blue trevally (*Cranx melampygus*) in a fisheries conservation zone. *Fisheries Research* 25:279–292.

Jackson J. B. C., M. X. Kirby, W. H. Berger, K. A. Bjorndal, L. W. Botsford, B. J. Bourque, R. H. Bradbury, R. Coke, J. Erlandson, J. A. Estes, T. P. Hughes, S. Kidwell, C. B. Lange, H. S. Lenihan, J. M. Pandolfi, C. H. Peterson, R. S. Steneck, M. J. Tegner, and R. R. Warner. 2001. Historical overfishing and the recent collapse of coastal ecosystems. *Science* 293:629–638.

Johannes, R. E. 1997. Traditional coral-reef fisheries management. In Birkeland, C., ed. *Life and Death of Coral Reefs,* 380–385. New York: Chapman and Hall.

Johannes, R. E., and Yeeting B. 2001. I-Kiribati knowledge and management of Tarawa's lagoon resources. *Atoll Research Bulletin* Vol. 489.

Johannes, R. E., L. Squire, T. Graham, Y. Sadovy, and H. Renguul. 1999. *Spawning Aggregations of Groupers (Serranidae) in Palau.* The Nature Conservancy Marine Conservation Research Series 1. Arlington, VA: The Nature Conservancy.

Johnson, D. R., N. A. Funicelli, and J. A. Bohnsack. 1999. Effectiveness of an existing estuarine no-take fish sanctuary within the Kennedy Space Center, Florida. *North American Journal of Fisheries Management* 19:436–453.

Jones, R. S., and M. J. Thompson. 1978. Comparison of Florida reef fish assemblages using a rapid visual survey technique. *Bulletin of Marine Science* 28:159–172.

Jones, G. P., M. J. Milicich, M. J. Emslie, and C. Lunow. 1999. Self-recruitment in a coral reef fish population. *Nature* 402:802–804.

Karp, W. A. 1990. Results of echo integration midwater-trawl surveys for walleye pollock in the Gulf of Alaska in 1990. In *Stock Assessment and Fishery Evaluation Report for the 1991 Gulf of Alaska Groundfish Fishery.* Anchorage, AK: North Pacific Fishery Management Council.

Kelso, W. W., and D. A. Rutherford. 1996. Collection, preservation, and identification of fish eggs and larvae. In Murphy, B. R., and D. W. Willis, eds. *Fisheries Techniques,* 2nd ed., 255–302. Bethesda, MD: American Fisheries Society.

Kendall, M. S., and M. E. Monaco. 2003. *Biogeography of the National Marine Sanctuaries.* NOAA/NOS/NCCOS/Coastal Center for Monitoring and Assessment Technical Report.

Kendall, M. S., J. D. Christensen, and Z. Hillis-Starr. 2003. Multi-scale data used to analyze the spatial distribution of French grunts, *Haemulon flavolineatum,* relative to hard and soft bottom in a benthic landscape. *Environmental Biology of Fishes* 66:19–26.

Kimmel, J. 1985. A new species-time method for visual assessment of fishes and its comparison with established methods. *Environmental Biology of Fishes* 12:23–32.

Klimley, A. P., and D. R. Nelson. 1984. Diel movement patterns of the scalloped hammerhead shark (*Sphyrna lewini*), in relation to El Bajo Espiritu Santo: A refuging central-position social system. *Behav. Ecol. Sociobiol.* 15:45–54.

Kramer, D. L., and M. R. Chapman. 1999. Implications of fish home range size and relocation for marine reserve function. *Environmental Biology of Fishes* 55:65–79.

Lipcius R. N., W. T. Stockhausen, D. B. Eggleston, L. S. Marshall, and B. Hickey. 1997. Hydrodynamic decoupling of recruitment, habitat quality and adult abundance in the Caribbean spiny lobster: Source-sink dynamics? *Australian Journal of Marine and Freshwater Research* 48:807–815.

Luckhurst, B. E., and K. Luckhurst. 1978. Analysis of the influence of substrate variables on coral reef fish communities. Marine Biolology 49:317–323.

Malvestuto, S. P. 1983. Sampling the recreational fishery. In: Nielsen, L. A., and D. L. Johnson eds. *Fisheries Techniques,* 397–419. Bethesda, MD: American Fisheries Society.

———. 1996. Sampling the recreational creel. In: Murphy, B. R., and D. W. Willis, eds. *Fisheries Techniques,* 591–624. Bethesda, MD: American Fisheries Society.

McManus, J. W., C. L. Nanola, A. G. C. del Norte, R. B. Reyes, Jr., J. N. P. Pasamonte, N. P. Armada, E. D. Gomez, and P. M. Alino. 1996. Coral reef sampling methods. In Gallucci, V. F., S. B. Saila, D. J. Gustafson, and B. J. Rothschild, eds. *Stock Assessment: Quan-*

*titative Methods and Applications for Small-Scale Fisheries,* 226–270. Boca Raton, FL: CRC Press.

Meyer, C. G. 2003. *An Empirical Evaluation of the Design and Function of a Small Marine Reserve (Waikiki Marine Life Conservation District).* Ph.D. diss., Manoa, HI: University of Hawaii.

Meyer, C. G, K. N. Holland, B. M. Wetherbee, and C. G. Lowe. 2000. Movement patterns, habitat utilization, home range and site fidelity of whitesaddle goatfish, *Parupeneus poryphyreus,* in a marine reserve. *Environmental Biology of Fishes* 59:235–242.

Monaco, M. E., S. B. Weisberg, and T. A. Lowery. 1998. Summer habitat affinities of estuarine fish in U.S. mid-Atlantic coastal systems. *Fisheries Management and Ecology* 5:161–171.

Monaco, M. E., J. D. Christensen, and S. O. Rohmann. 2001. Mapping and monitoring of U.S. coral reef ecosystems: The coupling of ecology, remote sensing, and GIS technology. *Earth System Monitor* 12(1).

Moran, P. J., and G. De'ath. 1992. Suitability of the manta tow technique for estimating relative and absolute abundances of crown-of-thorns starfish (*Acanthaster planci* L.) and corals. *Australian Journal of Marine and Freshwater Research* 43:357–78.

Mumby P. J., E. P. Green, A. J. Edwards, and C. D. Clark. 1997. Coral reef habitat mapping: How much detail can remote sensing provide? *Marine Biology* 130:193–202.

Munro, J. L., P. H. Reeson, and V. C. Gaut. 1971. Dynamic factors affecting the performance of the Antillean fish trap. *Proc. Gulf Caribbean Fish Inst.* 23:184–194.

Murawski, S. A., R. Brown, H.-L. Lai, P. J. Rago, L. Hendrickson. 2000. Large-scale closed areas as a fishery-management tool in temperate marine systems: The Georges Bank experience. *Bulletin of Marine Science* 66(3):775–798.

Murray, S. N., R. F. Ambrose, J. A. Bohnsack, L. W. Botsford, M. H. Carr, G. E. Davis, P. K. Dayton, D. Gotshall, D. R. Gunderson, M. A. Hixon, J. Lubchenco, M. Mangel, A. MacCall, D. A. McArdle, J. C. Ogden, J. Roughgarden, R. M. Starr, M. J. Tegner, and M. M. Yoklavich. 1999. No-take reserve networks: protection for fishery populations and marine ecosystems. *Fisheries* 24(11):11–25.

Myers, R. A., and G. Mertz. 1998. The limits of exploitation: a precautionary approach. *Ecological Applications* 8(1, suppl):S165–169.

National Research Council (NRC). 1994. *An Assessment of Atlantic Bluefin Tuna.* Washington, DC: National Academy.

———. 2001. *Marine Protected Areas: Tools for Sustaining Ocean Ecosystems.* Washington, DC: National Academy.

Nielsen, L. A. 1992. *Methods of Marking Fish and Shellfish.* American Fisheries Society Special Publication 23. Bethesda, MD: American Fisheries Society. Table.

Ogden, J. C. 1988. The influence of adjacent systems on the structure and function of coral reefs. *Proceedings of the 6th International Coral Reef Symposium* 1:123–129.

Ogden, J. C., and P. R. Ehrlich. 1977. The behavior of heterotypic resting schools of juvenile grunts (Pomadasyidae). *Marine Biology* 42:273–280.

Paddack, M. J., and J. A. Estes. 2000. Kelp forest fish populations in marine reserves and adjacent exploited areas of central California. *Ecological Applications* 10:855–870.

Parrish, J. D. 1989. Fish communities of interacting shallow-water habitats in tropical oceanic regions. *Marine Ecology Progress Series* 58(1–2):143–160.

Parrish, F. A., E. E. DeMartini, and D. E. Ellis. 1997. Nursery habitat in relation to production of juvenile pink snapper, *Pristipomoides filamentosus,* in the Hawaiian Archipelago. *Fishery Bulletin* 95:137–148.

Pauly D. 1995. Anecdotes and the shifting baseline syndrome of fisheries. *Trends in Ecology and Evolution* 10:430.

Pickett, S. T. A. 1989. Space-for-time substitution as an alternative to long-term studies. In Likens, G.E., ed. *Long-Term Studies in Ecology: Approaches and Alternatives,* 110–135. New York: Springer-Verlag.

Polacheck, R. 1990. Year around closed areas as a management tool. *Natural Resource Modeling* 4:327–353.

Pollock, K. H., C. M. Jones, and T. L. Brown. 1994. *Angler Survey Methods and Their Application in Fisheries Management.* American Fisheries Society Special Publication 25. Bethesda, MD: American Fisheries Society.

Pyle, R. L. 2000. Use of advanced self-contained diving technology for asessing udiscovered fish biodiversity on deep coral reefs. *Marine Technology Society Journal* 33(4).

Rakitin, A., and D. L. Kramer. 1996. Effect of a marine reserve on the distribution of coral reef fishes in Barbados. *Marine Ecology Progress Series* 131:97–113.

Randall, J. E. 1987. Introductions of marine fishes to the Hwaiian Islands. *Bulletin of Marine Science* 41:490–502.

Randall, J. E., J. L. Earle, R. L. Pyle, J. D. Parrish, and T. Hayes. 1993. Annotated checklist of the fishes of Midway Atoll, northwestern Hawaiian Islands. *Pacific Science* 47:356–400.

Recksiek, C. W., R. S. Appeldoorn, and R. G. Turingan. 1991. Studies of fish traps as stock assessment devices on a shallow reef in south-western Puerto Rico. *Fisheries Research* 10:177–197.

Roberts, C. M. 1996. Settlement and beyond: Population regulation and community structure of reef fishes. In Polunin, N. V. C., and C. M. Roberts. *Reef Fisheries,* 85–112. London: Chapman and Hall.

Roberts C. M. 1997. Connectivity and management of Caribbean coral reefs. *Science* 278:1454–1457.

Roberts, C. M., J. A. Bohnsack, F. Gell, J. P. Hawkins, and R. Goodridge. 2001. Effects of marine reserves on adjacent fisheries. *Science* 294:1920–1923.

Rogers, C. S, G. Garrison, R. Grober, Z. Hillis, and M. A. Franke. 1994. *Coral Reef Monitoring Manual for the Caribbean and Western Atlantic.* St. John, Virgin Islands: National Park Service.

Russ, G. R., and A. C. Alcala. 1996. Do marine reserves export adult fish biomass? Evidence from Apo Island, central Philippines. *Marine Ecology Progress Series* 132:1–9.

———. 1999. Management histories of Sumilon and Apo marine reserves, Philippines, and their influence on national marine resource policy. *Coral Reefs* 18:307–319.

Sánchez Lizaso, J. L., R. Goñi, O. Reñones, J. A. García Charton, R. Galzin, J. T. Bayle, P. Sánchez Jerez, A. Pérez Ruzafa, and A. A. Ramos. 2000. Density dependence in marine protected populations: A review. *Environmental Conservation* 27:144–158.

Shanks, A. L., B. Grantham, and M. Carr. 2003. Propagule dispersal distance and the size and spacing of marine reserves. *Ecological Applications* 13(1, suppl):S159–S169.

Sheppard C. 1995. The shifting baseline syndrome. *Mar Poll Bull* 30:766–767.

Sibert, J. R., and J. L. Nielsen. 2001. *Electronic Tagging and Tracking in Marine Fisheries.* Dordrecht, the Netherlands: Kluwer Academic Publishers.

Sladek Nowlis, J. 2000. Short- and long-term effects of three fishery-management tools on depleted fisheries. *Bulletin of Marine Science* 66:651–662.

Sladek Nowlis, J., and C. M. Roberts. 1999. Fisheries benefits and optimal design of marine reserves. *Fishery Bulletin* 97:604–616.

Sponaugle, S., R. K. Cowen, A. Shanks, S. G. Morgan, J. M. Leis, J. Pineda, G. W. Boehlert, M. J. Kingsford, K. C. Lindeman, C. Grimes, and J. L. Munro. 2002. Predicting self-recruitment in marine populations: biophysical correlates and mechanisms. *Bulletin of Marine Science* 70(1, suppl):341–375.

Swearer, S. E., J. E. Caselle, D. W. Lea, and R. R. Warner. 1999. Larval retention and recruitment in an island population of a coral-reef fish. *Nature* 402:799–802.

Sweatman, H., A. Cheal, G. Coleman, S. Delean, B. Fitzpatrick, I. Miller, R. Ninio, K. Osborne, C. Page, and A.Thompson. 2001. *Long-Term Monitoring of the Great Barrier Reef: Status Report Number 5.* Townsville: Australian Institute of Marine Science.

Tegner, M. J. 1992. Brood stock transplants as an approach to abalone stock enhancement. In Shepherd, S. A., M. J. Tegner, and S. A. Guzman del Proo. *Abalone of the World: Biology, Fisheries and Culture,* 461–473. Oxford: Blackwell Scientific.

Thompson, G. G. 1992. A Bayesian approach to management advice when stock-recruitment parameters are uncertain. *Fishery Bulletin* 90:561–573.

Thorrold, S. R., C. M. Jones, P. K. Swart, and T. E. Targett. 1998. Accurate classification of nursery areas of juvenile weakfish (*Cynoscion regalis*) based on chemical signatures in otoliths. *Mar. Ecol. Prog. Ser.* 173:253–265.

Thorrold, S. R., C. Latkoczy, P. K. Swart, and C. M. Jones. 2001. Natal homing in a marine fish metapopulation. *Science* 291:297–299.

Underwood, A. J. 1995. Ecological research and (and research into) environmental management. *Ecological Applications* 5:232–247.

Uzmann, J. R., R. A. Cooper, R. B. Theroux, and R. L. Wigley. 1977. Synoptic comparison of three sampling techniques for estimating abundance and distribution of selected megafauna: Submersible vs. camera sled vs. otter trawl. *Mar. Fish. Rev.* 39:11–19.

Walters, C. J. 1986. *Adaptive Management of Renewable Resources.* New York: Macmillan.

Wolanski E., and J. Sarsenski. 1997. Larvae dispersion in coral reefs and mangroves. *American Scientist* 85:236–243.

Wolff N., R. Grober-Dunsmore, C. S. Rogers, and J. Beets. 1999. Management implications of fish trap effectiveness in adjacent coral reef and gorgonian habitats. *Environmental Biology of Fishes* 55:81–90.

Yoklavich, M. M. 1997. Applications of side-scan sonar and in situ submersible survey techniques to marine fisheries habitat research. In Boehlert, G. W., and J. D. Schumacher, eds. *Changing Oceans and Changing Fisheries: Environmental Data for Fisheries Research and Management,* 140–141. Silver Spring, MD: Department of Commerce.NOAA Tech. Memo. NOAA-TM-NMFS-SWFSC-239.

# Global Experience and Case Studies

EIGHT

# California's Channel Islands and the U.S. West Coast

JOSHUA SLADEK NOWLIS

Marine reserves have been promoted as a means to address a wide variety of management challenges, from conservation of natural systems to enhancement of fisheries production. There is a substantial and growing body of evidence and experience to inform marine reserve efforts, some significant parts of which were gained along the West Coast of North America. These results demonstrate that reserves maintain and restore exploited species and consequently maintain more natural balances within their borders. Despite strong evidence of their value, marine reserves are extremely rare and small along the West Coast, with the exception of a recently designated and extensive network of marine reserves surrounding California's Channel Islands, which greatly expanded the region's total marine reserve area. Several ongoing processes also have the potential to dramatically increase the region's collection of marine reserves, but these processes face big political challenges.

Several different pieces of state and national legislation have created a mosaic of marine protected areas (MPAs)—ocean areas with site-specific protections—along the West Coast. Few of these areas provide the comprehensive protections required of a marine reserve. In California, for example, of 113 MPA designations, 20 prohibit all commercial and recreational fishing while only 4 prohibit all forms of extraction, including scientific take (McArdle 1997; McArdle et al. 2003). Half of California's marine reserves and seven-eighths of their area were added in 2003 through designation of a marine reserve network in the Channel Islands. This designation makes up a substantial proportion of all ocean areas conserved along the U.S. West Coast and represents one of the most comprehensive and scientifically driven marine reserve or reserve networks in the world. As such, it offers a unique opportunity to gain experience with this new form of ocean management and to learn more about the design and function of marine reserves.

This chapter first reviews the history of MPAs along the West Coast, from California to Washington, with a focus on the areas that receive the greatest protection. Next, it discusses some of the key insight scientists have gained from reserves along the West Coast of North America. It then highlights the recently completed Channel Islands process that created one of the world's first extensive networks of marine reserves. This process has set a new standard for conservation while providing a great opportunity to learn more about how and how well marine reserves work. Finally, it discusses the future of marine reserves in the region and several ongoing processes that might build on the successful experience to date and create additional highly protected ocean areas.

## WHAT HAS BEEN ESTABLISHED

MPAs have existed on the West Coast for nearly a century, dating back to 1913. California has designated the first and the most MPAs and has also led in the designation of marine reserves (Fig. 8.1a–c). In all three states, these areas make up fewer than 1.5 percent of state[1] ocean waters along the West Coast (see Table 8.1). As a fraction of the United States' West Coast Exclusive Economic Zone, these marine reserves are almost imperceptible at less than 0.04 percent, or less than 1/2,500 even after the creation of a substantial network of reserves in the Channel Islands.

### California

The oldest MPA along the West Coast is the Cabrillo National Monument, created in 1913 near San Diego, California. The area has been expanded since its creation and allows only fishing for finfish by hook and line (McArdle 1997). Two marine life refuges were added in the first half of the twentieth century: San Diego Marine Life Refuge established in 1929 and Hopkins Marine Life Refuge established in 1931. Both refuges were primarily intended to protect intertidal areas. Thus they prohibited the collection of invertebrate animals and marine plants but allowed fishing for fish (Joseph Wible, Hopkins Marine Station Librarian, personal communication, 10/30/01). The Pacific Grove Marine Gardens Fish Refuge was also established in 1931 and limited fishing to hook

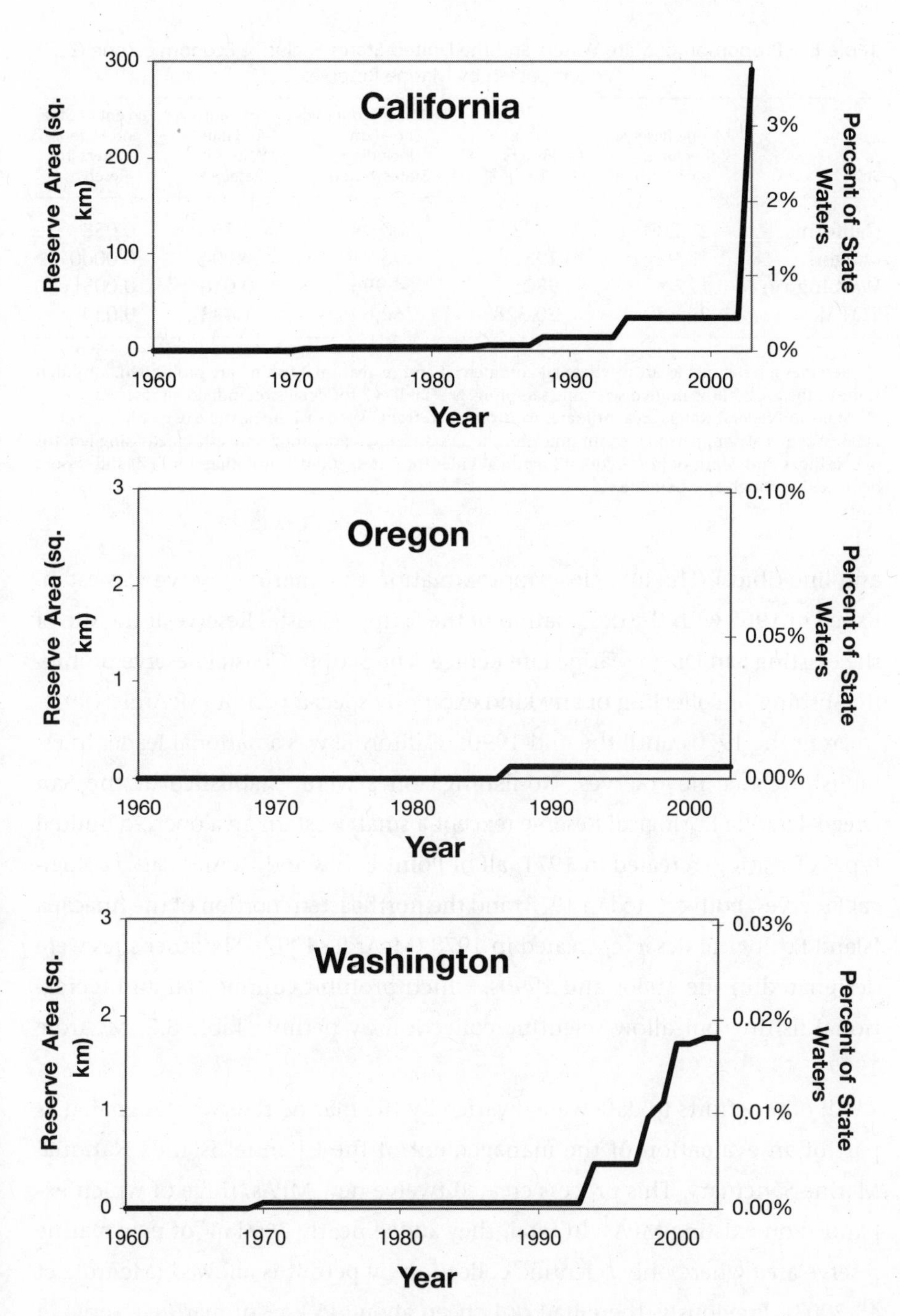


**FIG. 8.1 West Coast Marine Reserve Area.** Chronology indicates the build-up of marine reserves along the Pacific Coast. Note that state waters of this region consist of over 20,000 sq. kilometers. Also note that the Oregon and Washington's marine reserve areas are too small to be visible at the same scale as California's. Individual reserves are identified in Table 8.2.

Table 8.1 Proportion of State Waters and the United States Exclusive Economic Zone (EEZ) Encompassed by Marine Reserves

| State | Marine Reserve (hectares = 0.01 km$^2$) | State Waters (km$^2$) | Exclusive Economic Zone (km$^2$), Including States Waters | Percent of State Waters in Reserves | Percent of State and Federal Waters in Reserves |
|---|---|---|---|---|---|
| California | 29,230 | 7772 | 500,987 | 3.76 | 0.058 |
| Oregon | 12.9 | 2951 | 196,713 | 0.004 | 0.00007 |
| Washington | 177 | 9805 | 64,996 | 0.018 | 0.003 |
| TOTAL | 29,420 | 20,528 | 762,696 | 1.433 | 0.039 |

Reserves are defined as areas where all commercial and recreational fishing are prohibited, although some of the areas allow limited scientific sampling. See Table 8.2 for details on individual reserves.

State and federal waters area estimates were obtained from experts where possible (e.g., McArdle et al. 2003; Wayne Palsson, personal communication, 6/30/03). For Oregon, they were estimated using lengths of coastlines and width of jurisdiction (3 nautical miles for states, 200 nautical miles for EEZ) and should be treated as rough approximations.

and line (ibid.). The first close approximation to a marine reserve was established in 1965 with the designation of the Scripps Coastal Reserve at the site of the existing San Diego Marine Life Refuge. The Scripps Coastal Reserve prohibited fishing or collecting of any kind except by special permit (McArdle 1997).

From the 1970s until the mid-1990s, California was a national leader in establishing marine reserves. No-fishing zones were established at the San Diego–La Jolla Ecological Reserve (except a small western area open to limited types of fishing), created in 1971; all of Point Lobos and Heisler Park Ecological Reserves, both created in 1973; and the northeastern portion of the Anacapa Island Ecological Reserve, created in 1978 (McArdle 1997). Six other sites were designated in the 1980s and 1990s, which prohibit commercial and recreational fishing but allow scientific collection by permit (Table 8.2; McArdle 1997).

All other efforts to date were dwarfed by the marine reserves designated as part of an evaluation of the management of the Channel Islands National Marine Sanctuary. This process created twelve new MPAs, three of which expanded on existing MPAs. In total, they added nearly 260 km$^2$ of new marine reserve area where only scientific collection by permit is allowed (McArdle et al. 2003). Previously, there had only been about 35 km$^2$ of marine reserve in waters off of California and only 37 km$^2$ along the entire West Coast (see Fig. 8.1). With the addition of the Channel Islands network, reserves now encompass almost 4 percent of California state waters but still do not put an appreciable dent in the larger federal plus state area (see Table 8.1). Nevertheless, the magnitude and scientific underpinnings of the marine reserves in the Channel

Islands make them both a new standard for conservation and an unparalleled opportunity to examine how species and ecosystems change as a result of protection in a large network of marine reserves.

## Oregon

Only a handful of MPAs exist in the state of Oregon. Most areas with potential to protect marine life in Oregon stop at the tide line and do not include subtidal habitats. Most are also open to some form of commercial or recreational fishing. Only the Whale Cove Intertidal Research Reserve includes subtidal habitats and prohibits the taking of marine fish, shellfish, and invertebrates (Didier 1998), but it does not prohibit collection of intertidal algae. The Whale Cove Reserve was created in the late 1960s and includes 12.9 hectares (0.129 $km^2$) of subtidal ocean habitat (see Table 8.2). Beyond this one small area, Oregon has nothing approximating a marine reserve at present (David Fox, Oregon Department of Fish and Wildlife, personal communication, 10/31/01), and the one area makes up 0.004 percent of state waters and an even less significant portion of state plus federal waters (see Fig. 8.1; Table 8.1).

## Washington

Washington State created its first MPA in 1923, when the San Juan County/Cypress Island area was established. Within this area kelp may be collected as well as any type of seafood, but no other biological material can be taken without permission (Murray 1998). Even weaker regulations accompanied the creation of the Olympic National Park in 1938 and the San Juan National Historical Park in 1966, and these regulations remain weak today (Murray 1998).

It wasn't until the creation of the Edmonds Underwater Park in 1970 that Washington had something akin to a marine reserve. Edmonds prohibited the take of food fish or shellfish. However, it is situated around an artificial structure created by a dry dock sunk in 1935 (Palsson and Pacunski 1995). This park was expanded as part of the Brackett's Landing Shoreline Sanctuary Conservation Area in 1999. Marine reserves have also been designated in the past decade through several Conservation Areas: Octopus Hole and Sund Rock established in 1994; Orchard Rocks, City of Des Moines Park, and S 239th Street Park established in 1998; Saltars Point Beach and Waketickeh Creek established in 2000; and Keystone Harbor created in 2002 (Wayne Palsson, Washington

Table 8.2 Marine Reserves of the U.S. West Coast

| Name | Location | Established | Size (hectares) | Notes |
|---|---|---|---|---|
| Scripps Coastal Reserve | CA | 1965 | 35.2 | |
| San Diego–La Jolla Ecological Reserve | CA | 1971 | 145 | Bait fishing for squid using a handheld scoop net is allowed in a narrow band on the western edge, otherwise no exceptions. |
| Point Lobos Ecological Reserve | CA | 1973 | 278.9 | No exceptions. |
| Heisler Park Ecological Reserve | CA | 1973 | 12.8 | No exceptions. |
| Anacapa Island Ecological Reserve Natural Area | CA | 1978 | 12 | No exceptions. |
| Hopkins Marine Life Refuge | CA | 1984 | 32.5 | |
| Catalina Marine Science Center Marine Life Refuge | CA | 1988 | 844.9 | |
| King Range (Punta Gorda) MRPA* Ecological Reserve | CA | 1994 | 611. 5 | |
| Vandenberg MRPA* Ecological Reserve | CA | 1994 | 617 | |
| Big Sycamore Canyon MRPA* Ecological Reserve | CA | 1994 | 517.7 | |
| Big Creek MRPA* Ecological Reserve | CA | 1994 | 378.6 | |
| Richardson Rock State Marine Reserve | CA | 2003 | 6297 | |
| Judith Rock State Marine Reserve | CA | 2003 | 997 | |
| Harris Point State Marine Reserve | CA | 2003 | 3441 | Small harbor exempted, estimated at 1/30 of the total area of 0.3559 ha. |
| Skunk Point State Marine Reserve | CA | 2003 | 274 | |
| Carrington Point State Marine Reserve | CA | 2003 | 2601 | |
| South Point State Marine Reserve | CA | 2003 | 2112 | |
| Gull Island State Marine Reserve | CA | 2003 | 3149 | |
| Scorpion State Marine Reserve | CA | 2003 | 2014 | |
| Santa Barbara State Marine Reserve | CA | 2003 | 2581 | |
| Anacapa Island State Marine Reserve | CA | 2003 | 2288 | Expanded on existing 12 ha Anacapa Island Ecological Reserve Natural Area. Total area expressed. |
| Whale Cove Intertidal Research Reserve | OR | Late 1960s | 12.9 | Also includes intertidal area where algae but no animals can be collected. |
| Edmonds Underwater Park | WA | 1970 | 6.8 | No exceptions. |
| Octopus Hole Conservation Area | WA | 1994 | 11 | No exceptions. |
| Sund Rock Conservation Area | WA | 1994 | 28.8 | No exceptions. |
| Orchard Rocks Conservation Area | WA | 1998 | 41.9 | No exceptions. |
| City of Des Moines Park Conservation Area | WA | 1998 | 3.7 | No exceptions. |
| S 239th Street Park Conservation Area | WA | 1998 | 2.1 | No exceptions. |
| Bracketts Landing Shoreline Sanctuary Conservation Area | WA | 1999 | 17 | No exceptions. Additional area added to Edmonds Underwater Park. |

Table 8.2 (continued)

| Name | Location | Established | Size (hectares) | Notes |
|---|---|---|---|---|
| Saltars Point Beach Conservation Area | WA | 2000 | 1.6 | No exceptions. |
| Waketickeh Creek Conservation Area | WA | 2000 | 59.2 | No exceptions. |
| Keystone Harbor | WA | 2002 | 4.6 | No exceptions. |

These 32 areas are closed to all forms of fishing, with an exception only by permit for scientific purposes unless otherwise noted. The earliest site was designated in 1965, whereas the 10 most recent marine reserves were created in 2003.

* MRPA = Marine Resources Protection Act

Sources: McArdle (1997); Murray (1998); Didier (1998); McArdle et al. (2003); Gary Davis, Channel Islands National Park, personal communication, 10/30/01; Wayne Palsson, Washington Department of Fish and Wildlife, personal communication, 10/30/01; David Fox, Oregon Department of Fish and Wildlife, personal communication, 10/31/01.

Department of Fish and Wildlife, personal communication, 6/30/03). These new areas have made Washington State a national leader in establishing marine reserves, though they still encompass less than 0.02 percent of state waters.

Some additional areas are closed to all forms of fishing in Washington State but have not yet been adequately described or cataloged because of their nonconventional designations. The Zella P. Schultz/Protection Island area is leased by the United States Fish and Wildlife Service, which keeps all boats 200 yards away from shore and prohibits shore access from land. There are also other areas where a no-approach zone is maintained for security reasons. These areas include McNeill Island, the site of a federal penitentiary, and the Bangor and Bemerton Navy bases (ibid.).

## HAVE THE RESERVES WORKED?

Despite the small size, paucity, nonscientific design, and uncertain enforcement of most marine reserves along the West Coast, we have learned a surprising amount from them and related studies. A limited number of studies have been conducted on marine reserves of the West Coast, but they have produced some important results. In addition to studies of actual reserves, there are some relevant studies of areas closed to particular types of fishing and studies examining the potential effects of large-scale closures. Although any closure can provide evidence for how individual populations or species may respond to a marine reserve, only actual marine reserves can provide insight into how ecosystems will respond to protection within a reserve while fishing takes place outside it (Tegner and Dayton 2000).

It is also relevant to look at the performance of conventional ocean management tools along the West Coast. The growth and spectacular crash of the sardine fishery are widely known. This pattern of boom and bust has continued despite modern management techniques and institutions, including traditional California staple fisheries for abalone and rockfish and, more recently, sea urchins (Fig. 8.2a). This pattern has led to a reduction of total landings and an increasing reliance on the squid and kelp fisheries (Fig. 8.2b), which target important sources of food and shelter for other members of those ecosystems. Given the booming nature of the squid fishery (see Fig. 8.2a), there is reason to believe that it, too, may go bust.

In total, the research conducted in this region provides strong evidence that marine reserves provide a variety of benefits, highlighting the promise of reserves as a valuable ocean conservation and management tool. Specific questions are addressed following here.

### Do Reserves Increase the Size and Abundance of Target Species?

Increases in the size and abundance of target species are a crucial test of marine reserve function. If successful in this endeavor, and reserve design promotes some larval dispersal, reserves are likely to provide substantial fisheries benefits, particularly to overfished species (Sladek Nowlis and Roberts 1999). These benefits are especially promising for long-lived species because older, larger individuals typically produce far more offspring than their younger, smaller counterparts (Bohnsack 1996).

*Rocky Reef–Associated Fish in Puget Sound, Washington.* Palsson and Pacunski (1995) compared Edmonds Underwater Park—primarily used by recreational divers and part of the Bracketts Landing Shoreline Sanctuary Conservation Area, a marine reserve created in 1970—to four other nearby sites open to fishing. Three of these fishing sites had consistently fewer and smaller lingcod, copper rockfish, and quillback rockfish than Edmonds, but a fourth fishing area, Boeing Creek, had comparable numbers and sizes of quillback rockfish. However, there were more and larger copper rockfish and larger lingcod in Edmonds. The authors also observed more lingcod at Edmonds but this difference was not large enough to distinguish it from differences due to natural variability.

The authors also compared Shady Cove, an area closed in 1990 to all fishing except herring and commercial salmon, to a nearby area open to fishing. Their results here were less dramatic but still promising. Lingcod were 75 percent more abundant inside the closed area with two to three times the num-

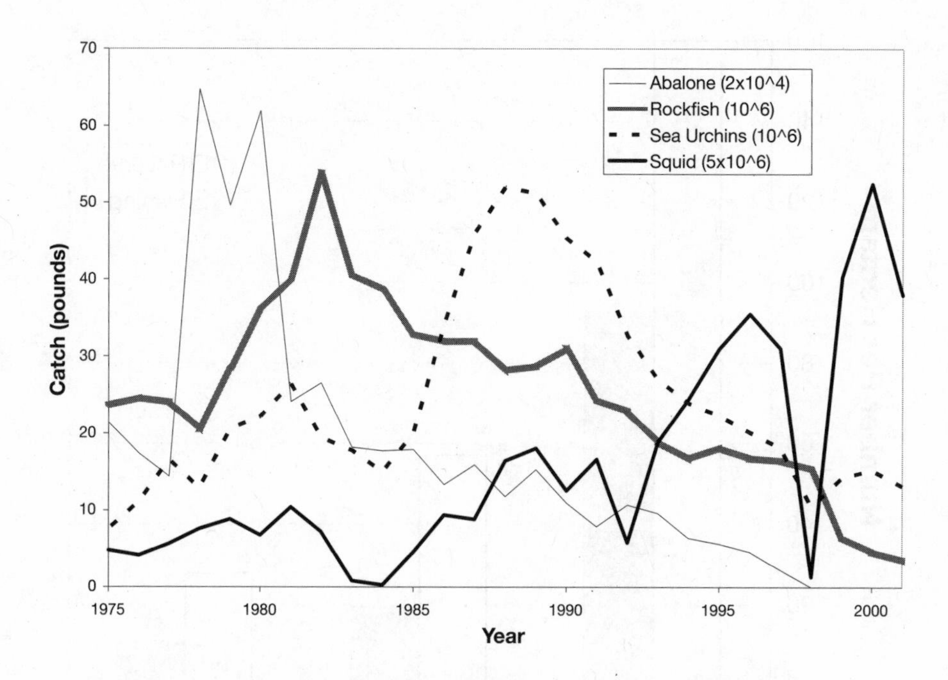


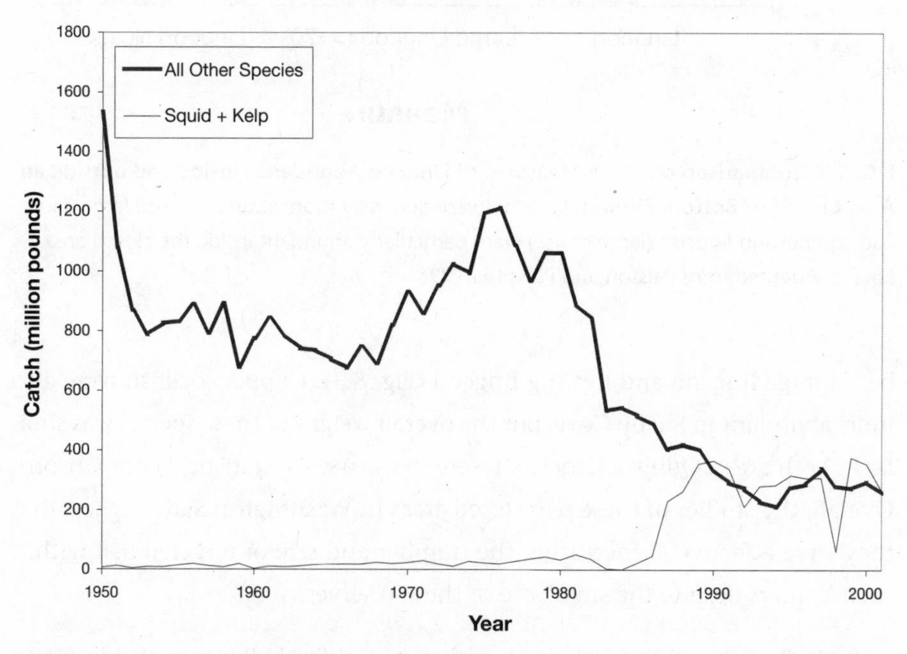


**FIGS. 8.2a and 8.2b. Status of California Stocks.** Traditional staple fisheries in California, like rockfish and abalone, have collapsed. Overall landings are also down sharply, with an increasing reliance on the squid and kelp fisheries. From the National Marine Fisheries Commercial Landings Data (http://www.st.nmfs.gov/st1/commercial/).

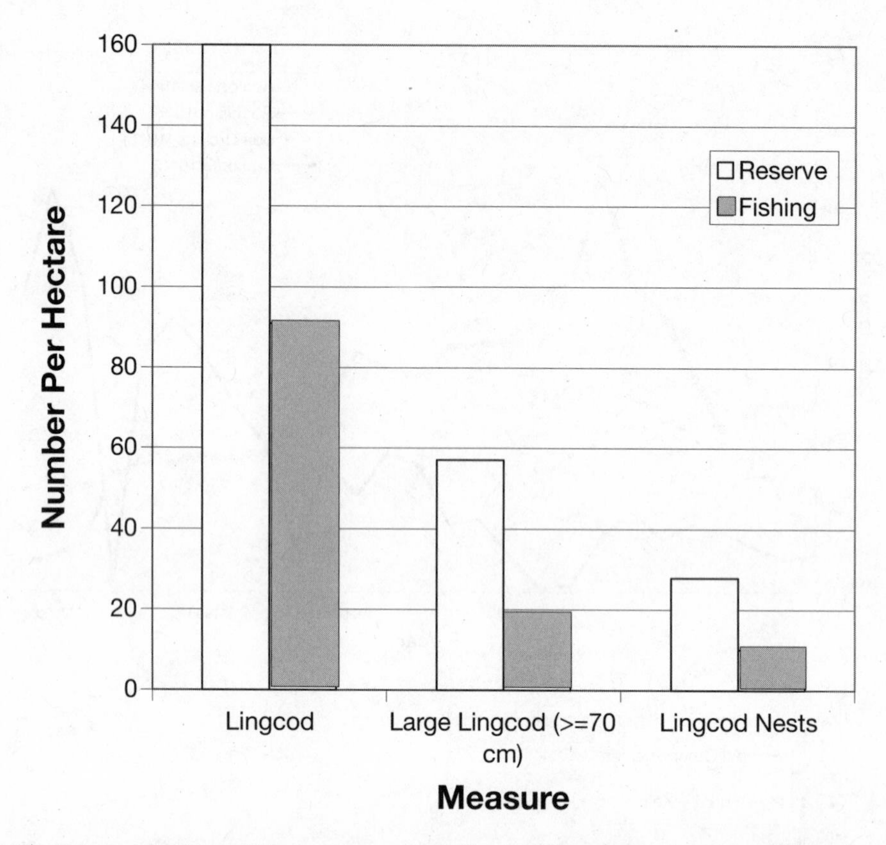


**FIG. 8.3 Comparison of Three Measures of Lingcod Abundance inside and outside an Area Closed to Bottom Fishing.** Lingcod were generally more abundant, and large lingcod and nesting lingcod (large males) were particularly abundant inside the closed area. Source: Adapted from Palsson and Pacunski 1995.

ber of large lingcod and nesting lingcod (Fig. 8.3). Copper rockfish were also more abundant in Shady Cove, but the overall weight of these species was similar in both areas. Quillback rockfish were too sparse for statistical comparisons. Overall, the studies of these two closed areas in Washington State suggest that they were effective at increasing the number and size of targeted fish within their borders despite the small size of these reserves.

*Rocky Reef–Associated Fish at Three Central California Reserves.* Paddack and Estes (2000) examined three areas closed to commercial and recreational fishing and compared them to adjacent fishing areas along California's central coast. Point Lobos Ecological Reserve was created in 1973 and bans all forms of fishing or collection. Hopkins Marine Life Refuge was created in 1931, though it did not prohibit all commercial and recreational fishing until it was expanded in 1984. The Big Creek Ecological Reserve was established in 1994.

Like Hopkins, it prohibits all commercial and recreational fishing but allows collection for scientific purposes. Point Lobos and Hopkins had larger fish than adjacent sites open to fishing, but a similar difference was not detected at Big Creek, the most recently designated site. And, while all three reserves showed indications of more abundant fish than adjacent fishing areas—54 percent more at Hopkins, 17 percent more at Point Lobos, and 25 percent more at Big Creek—these differences were not confirmed by the statistical method chosen by the authors. The differences are nevertheless worth considering because the statistical test lacked the power to confirm even fairly large differences between reserve and fishing areas.

*Northern Abalone around Southern Vancouver Island, British Columbia.* Wallace (1999) examined the abundance of northern abalone (*Haliotis kamtschatkana*) at eight sites in southern British Columbia, an area that is ecologically interconnected with Puget Sound, Washington. Due to a provincewide closure on the collection of northern abalone, abalone should have been recovering in most sites comparably, although the prison site had potential for greater abalone because it has been closed sine 1958.

Instead, the abundance of abalone was strongly correlated with the level of enforcement at each site. The military site, with the greatest enforcement, had more than twenty-four times as many abalone as the five undesignated and relatively poorly enforced sites (Fig. 8.4). The prison reserve and ecological reserve, with intermediate levels of enforcement, had fewer abalone than the military site but still fourteen to fifteen times as many abalone as the five undesignated sites. This study provides one of the best demonstrations that enforcement is crucial for the performance of marine reserves and other fishery closures. Moreover, it provides evidence that marine reserves are better and possibly more easily enforced than areas where some types of fishing are allowed.

### Do Reserves Export Production to Fishing Areas?

Though many studies indicate a greater reproductive potential inside than outside reserves (e.g., Paddack and Estes 2000; Palsson and Pacunski 1995; Wallace 1999), few studies have been able to demonstrate the link between the buildup of fish in a reserve and increased production in nearby fishing areas. This demonstration can be tricky because of the biology of marine organisms and the small size of most existing reserves. It should be noted, though, that ocean managers routinely enact other regulations—such as quotas and size

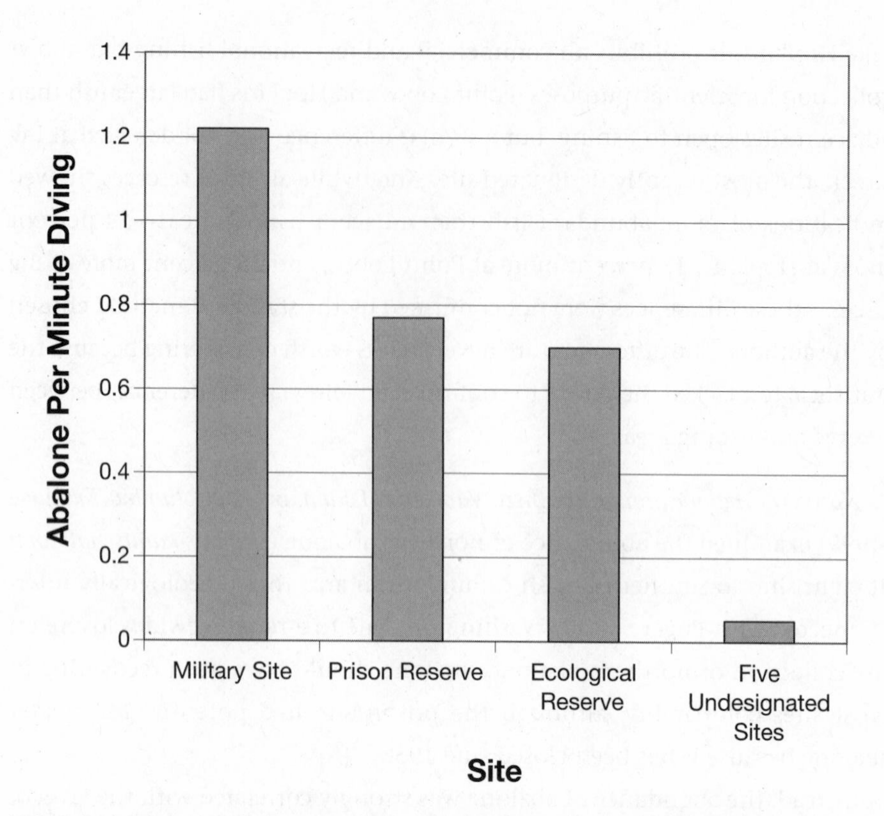


**FIG. 8.4 Northern Abalone Abundance at Eight Sites in Southern British Columbia.** Abalone were encountered with greatest frequency near a military base where enforcement of a provincewide abalone closure was best enforced. Abalone were moderately abundant at two no-fishing zones, one by a prison and another an ecological reserve, and virtually absent from five other sites where fishing for other species was permitted. Source: Adapted from Wallace 1999.

limits for fisheries—that also remain unproven in their ability to increase production of the fished population (see chapter 5).

Most marine organisms go through a waterborne larval stage during which they may drift or swim long distances. Because of the small size and high mobility of larvae, it is very difficult to actually study them in the ocean. Nevertheless, one of the best studies to examine the potential dispersal of larvae from a reserve was conducted along the West Coast.

*Green Abalone, Southern California.* Tegner (1992) studied the dispersal of green abalone (*Haliotis fulgens*) from a concentration of individuals created by transplanting adults to areas devoid of this species but within its historical range. This arrangement provided a reservelike situation where a concentration of adults might seed surrounding areas. The author found that after three years, abalone had spread a few kilometers from the transplant sites and that most of the abalone were new recruits, presumably coming from the transplanted adults. Although recruitment patterns could not rule out the possibility that some larvae traveled to sites 45 to 100 kilometers away, the levels of recruitment at these distant locales was only 4 to 23 percent of the level of recruitment near the transplant sites. Abalone have a shorter larval life than most marine organisms (Prince et al. 1987). Consequently, there is a higher likelihood that abalone from a closed area will provide a measurable increase in recruitment to nearby areas but that this recruitment enhancement will tail off quickly farther away from the reserve.

### Do Reserves Promote Ecosystem Function?

*Anacapa Island Ecological Reserve, Channel Islands, California.* The National Park Service, in cooperation with the State of California and the U.S. Department of Commerce, is responsible for monitoring the health of park ecosystems and has maintained a monitoring program of kelp forests around the Channel Islands for two decades. This monitoring protocol includes two sites within a marine reserve. The Natural Area of Anacapa Island Ecological Reserve lies along the north shore of East Anacapa Island, where all fishing was prohibited in 1978. The protocol also monitors fourteen fishable sites, including three near the reserve, allowing for comparison of fished and unfished ecosystems over a twenty-year period.

The results of this long-term monitoring study provide strong evidence that the ecosystem within the reserve has maintained a more natural ecological balance while fishing areas have shifted to an unnatural state. Fishing has reduced the abundance of several targeted populations, including California spiny lobster, which was 10 times more abundant inside the reserve, and a fish called the California sheephead, the large, colorful, and visible males of which were 1.5 times as abundant inside the reserve (Fig. 8.5a). Spiny lobsters and sheephead are key urchin predators in Southern California kelp bed ecosystems (Cowen 1983; Tegner and Levin 1983). Red sea urchins are also collected by people, but purple and white sea urchins are rarely fished and their populations have grown dramatically in fishing areas in the absence of predators or competitors. In fact, while nontargeted urchin densities have grown by only a factor of four in the reserve, they have increased by a factor of over fifteen in nearby fished survey sites since 1983 (Fig. 8.5b).

Giant kelp has suffered from heavy grazing by these nontargeted urchins in fishing areas, all but disappearing from fished survey sites near the reserve (Fig. 8.5c). Kelp is especially vulnerable to purple urchin outbreaks because these urchins cause disproportionate damage to holdfasts (Tegner et al. 1995), the part of kelp plants that anchors them to the bottom. In contrast, kelp has actually increased by more than 10 percent inside the reserve since 1983. Giant kelp provides food and shelter for many species in the kelp bed ecosystem (Dayton et al. 1998). Therefore, the protections afforded by the marine reserve have maintained a healthier and more natural kelp forest ecosystem than nearby fishing areas where kelp has virtually disappeared or at least been slow to recover from the most recent El Niño event. These findings bolster claims that overfishing has wrought havoc on kelp forest ecosystems over timescales of decades or more (Tegner and Dayton 2000) and highlight that even small reserves have the potential to protect ecosystems. These ecosystem benefits are more likely from a reserve than a less restrictive MPA where some types of fishing are allowed because fishing, even if restricted to only a few species, has the potential to disrupt natural balances and cause negative effects to ecosystems.

A study of warty sea cucumbers from the same area provided additional insight into the value of marine reserves as a control area with which to understand the effect of human activities. Schroeter et al. (2001) examined long-term monitoring data on the populations of this relative of starfish and sea urchins but more wormlike in appearance, which is the target of a new and growing fishery. These data showed declines of sea cucumbers in fishing areas, whereas populations stayed relatively stable within the Anacapa reserve. The scientists concluded that fishing was responsible for population declines of 33 to 83 percent. These results were especially dramatic considering that a more traditional assessment of this stock based on the ease of catching sea cucumbers had predicted that stocks were stable or increasing.

**FIG. 8.5 Ecosystem Protection for a Small Channel Islands Reserve.**
(A) The two dominant urchin predators—spiny lobster and male California sheephead—were more abundant inside the Natural Area of Anacapa Island Ecological Reserve, a marine reserve, than at three survey sites open to fishing nearby. Fishing heavily targets both of these species. (B) Consequently, urchin species that are not fished have increased markedly in fishing areas. (C) The nontargeted urchins have caused a dramatic reduction in giant kelp, the species that provides the foundation for the kelp forest ecosystem. Data from long-term kelp forest monitoring program of the Channel Island National Park.

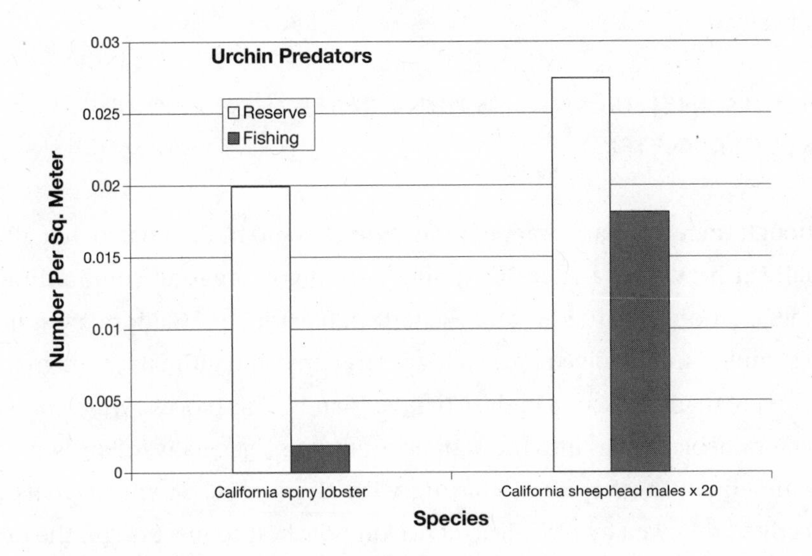
Urchin Predators
Reserve
Fishing
Number Per Sq. Meter
0.03
0.025
0.02
0.015
0.01
0.005
0
California spiny lobster
California sheephead males x 20
Species

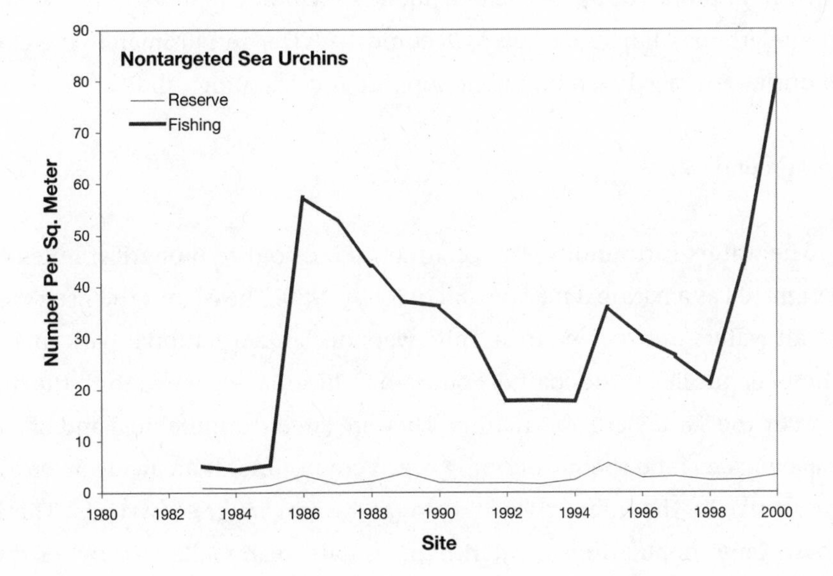
Nontargeted Sea Urchins
Reserve
Fishing
Number Per Sq. Meter
90
80
70
60
50
40
30
20
10
0
1980
1982
1984
1986
1988
1990
1992
1994
1996
1998
2000
Site

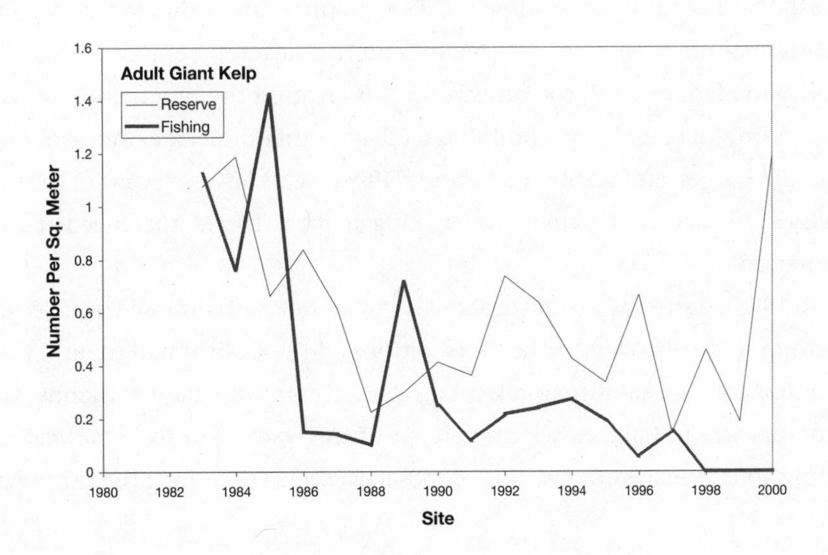
Adult Giant Kelp
Reserve
Fishing
Number Per Sq. Meter
1.6
1.4
1.2
1
0.8
0.6
0.4
0.2
0
1980
1982
1984
1986
1988
1990
1992
1994
1996
1998
2000
Site

## THE CHANNEL ISLANDS MARINE RESERVE NETWORK—A NEW FRONTIER

Although there are many ways marine reserves could be designated, not all fare equally at providing a scientific foundation, involving local communities in the designation, or garnering and wielding public support. California was home to a complex and involved process that considered and ultimately designated a network of marine reserves in the Channel Islands. This process, which provides several valuable lessons into the designation process, generally followed the one recommended in chapter 5, starting with the establishment of goals and objectives, followed by technical advice on how best to meet them, the development of proposals by the general public—including representatives of stakeholder groups—but supported by technical advice, and ultimately the choice of one alternative based on public support and scientific validity.

### Background

The waters surrounding the Channel Islands out to 6 nautical miles were designated as a national marine sanctuary in 1980. The islands themselves plus ocean waters out to 1 nautical mile were declared a national park in 1986. These islands lie off the coast of Southern California, separated from the mainland by the Santa Barbara Channel. They are largely uninhabited and of a natural character, and the surrounding ocean environments are made diverse and productive by the intersection of cold- and warm-water ecosystems. The area boasts large populations of marine mammals, seabirds, and a rich array of coastal and oceanic environments. The channel has also been the target of substantial offshore oil development, and this development and a major oil spill provided much of the motivation for creating the sanctuary. The sanctuary's original regulations prohibited oil and other mineral extraction, pollutant discharges, and seabed alterations. Fifteen years later, a review of the sanctuary's management plan found fishing to be a threat that needed to be addressed.

In 1998, spurred by a concern about declines they had witnessed over decades, a group of sportfishing enthusiasts petitioned the California Fish and Game Commission to establish no-take marine reserves in the Channel Islands. Their proposal would have closed 20 percent of the waters out to 1 nautical mile from shore. Needless to say, this proposal caused a stir among the fishing and conservation communities, but it also touched a nerve in the Commission, which had just been granted greater authority to manage California's oceans. The Commission did not accept the proposal but instead charged the California Fish and Game Department to develop a process in collaboration with the national marine sanctuary. The sanctuary staff was also motivated to take on such a task because they were beginning the first ever review of their management plan.

## Process

Together, the sanctuary and the Fish and Game Department developed a science-informed stakeholder-driven process, whereby the responsibility of developing management alternatives was given to the Marine Reserves Working Group, a group representative of the various interests in the Channel Islands. The working group was supported by a scientific advisory panel, who provided general scientific criteria, developed analytical tools, and ultimately evaluated proposals on the degree to which they followed the scientific design criteria. In selecting scientists, the working group was solicited for nominations and also given a chance to veto scientists they felt were biased. Fishing interests utilized the veto power to oppose the selection of any scientist who had previously published on MPAs. They did so out of fear that some scientists had been co-opted on this issue by the conservation community, but this also eliminated from consideration the most knowledgeable scientists. The scientific panel was still of high quality, mainly because the rapidly growing interest in the subject among marine scientists meant that many of them were eager to work on this project and had already done background research and thinking about marine reserves. Had the scientific interest not been as dramatic, the broad veto of scientists who had published on the subject could have limited the value of scientific advice fed into this process.

One of the first steps taken by the working group was to adopt goals and objectives, or reasons why marine reserves should be considered in the Channel Islands. They developed five broad goals, two of which had the greatest influence on the process: ecosystem biodiversity—to protect representative and unique marine habitats, ecological processes, and populations of interest; and sustainable fisheries—to achieve sustainable fisheries by integrating marine reserves into fisheries management. These goals and objectives were forwarded to the scientific advisers, who were asked to provide relevant design criteria to achieve them.

## Scientific Analyses and Recommendations

The scientists performed a broad literature review to develop general criteria, the most controversial of which was a recommendation that 30 to 50 percent of each habitat type in the sanctuary should be designated as marine reserves in a networked design (Airamé et al. 2003). Scientists on the advisory board had different rationales for this recommendation. Some believed that reserves needed to encompass this range of area to provide an adequate insurance factor given the uncertainty and poor health of many populations around the Channel Islands. Others took a simpler and more controversial view, eventually dubbed the "scorched earth" hypothesis. According to this line of thinking, 30 to 50 percent of each fishery population had to be protected to ensure that fisheries would be sustainable. These scientists did not believe other management techniques could guarantee this level of protection, and they concluded that reserves of this size would be an adequate and appropriate way to do so.

The scientists also focused on distributing the reserve areas so that they covered each of three regions within the Channel Islands—a cooler water region more characteristic of ecosystems to the north, a warmer water region more characteristic of ecosystems to the south, and an intermediate zone with some characteristics of each of the other two regions (Airamé et al. 2003). They also recommended that each habitat type within each region be well represented in the reserve network. They based habitat types on a variety of different measures and made habitats the focus because of their ability to serve as a surrogate for biodiversity (see chapter 5 for more details).

In addition, the scientific advisory panel developed a decision-making tool to guide stakeholders as they looked for favorable reserve siting alternatives (Fig. 8.6) that might also meet the scientific design criteria (Airamé et al. 2003). The decision tool was based on habitat and fishing effort data and was designed to give anyone feedback about the design of a marine reserve network. The tool was made available to members of the working group and the general public. Interested people could draw lines designating a potential marine reserve network, and the decision tool would provide feedback as to how well the network encompassed representative habitat types and on the amount of fishing it would displace. The concept was that the habitat inclusiveness gave an indication of the ecological value of a reserve proposal while the displaced effort gave an idea of how much short-term socioeconomic cost it would create. The designer could then make small modifications to improve their design or scrap the idea and try something else. The scientists also did their own analysis using

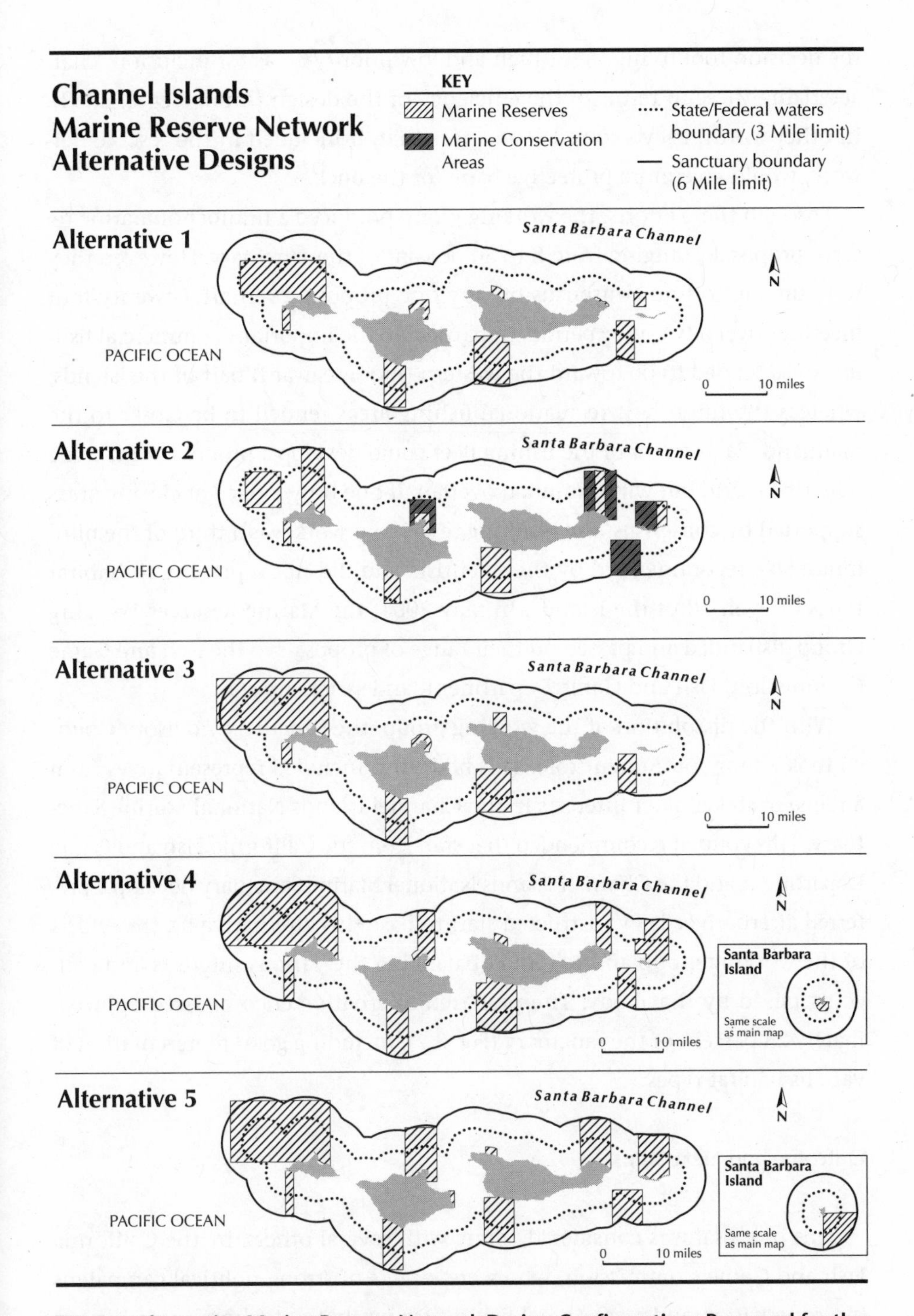


**FIG. 8.6 Alternative Marine Reserve Network Design Configurations Proposed for the Channel Islands during the Designation Process.** The SITES computer decision-making tools provided an ability to both create and evaluate alternative designs with respect to a variety of criteria.

the decision tool to highlight high and low priority areas for inclusion, characterizing these in terms of the efficiency of the design (Airamé et al. 2003). In other words, they showed some areas that, if included in the reserve network, would give more protective bang for the buck.

Through these efforts, the working group produced a number of marine reserve proposals, ranging from 8 to 50 percent of the sanctuary. However, they were unable to find consensus on any one proposal even after twenty-four meetings over a two-year period. In particular, the important commercial fishing areas tended to be toward the western (more seaward) part of the islands, whereas the important recreational fishing areas tended to be closer to the mainland. Each sector of the fishing fleet could develop a proposal with some scientific merit, but without much overlap. If one only looked at closing areas supported by consensus, the resulting reserve network was a third of the minimum size recommended by the scientists and did not represent all habitat types or even all of the islands. In May 2001, the Marine Reserves Working Group disbanded and passed on their range of proposals to the Fish and Game Commission, Fish and Game Department, and sanctuary staff.

With the dissolution of the working group, the Sanctuary Advisory Council took a more prominent role. This body also included representatives from a range of stakeholder interests in the Channel Islands National Marine Sanctuary. This council recommended that staff from the California Fish and Game Department and the Channel Islands National Marine Sanctuary develop a preferred alternative. They instructed staff to base this alternative on the efforts of the working group and to aim at balancing the various interests and concerns raised by that body. The preferred alternative encompassed approximately 25 percent of the sanctuary (Fig. 8.7), including good representation of various habitat types.

### Outcome and Next Steps

This proposal was considered, along with several others, by the California Fish and Game Commission, who were targets of strong political campaigns by conservation and recreational fishing groups, both of whom turned out in large numbers at public meetings on the subject. Ultimately the Commission chose the preferred alternative in late 2002. Amidst strong political pressure from both sides, the Commission most likely made their decision mindful of the quality and integrity of the multiyear process that had carefully and systematically developed goals and objectives, sought scientific advice on how

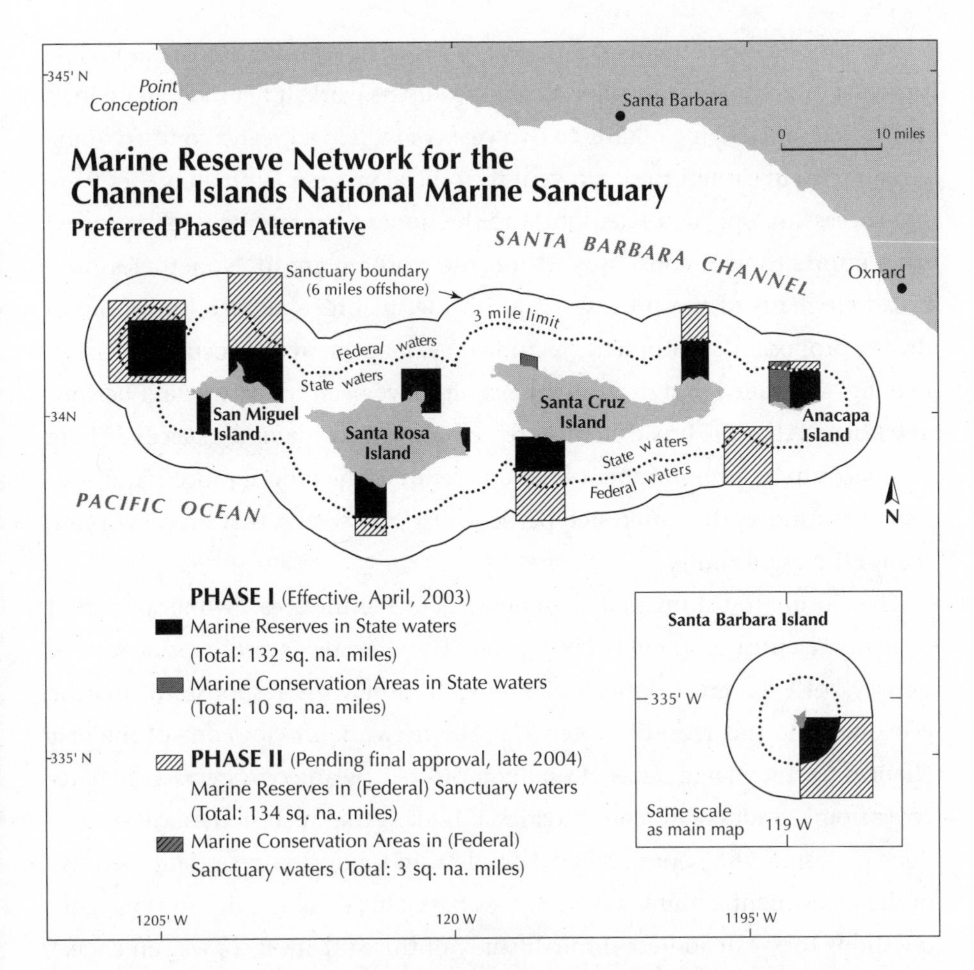


**FIG. 8.7 The Preferred Alternative for the Channel Islands Marine Reserve Network As It Appeared in the California Environmental Impact Review (EIR) and As Approved by the State of California.** This phase approach called for approving and implementing the portion of the network in state water first and then federal waters. The state portion is now final and the federal portion is currently pending approval.

to satisfy them, and strove to develop consensus proposals that met the scientific advice. This decision went into effect in 2003, creating ten no-take marine reserves and two additional MPAs where certain fishing activities were prohibited (see Table 8.2). However, the Commission's jurisdiction only extends 3 nautical miles offshore. Consequently, they were unable to establish marine reserves in the offshore federal waters of the sanctuary (which extend from 3 to 6 nautical miles). The National Oceanic and Atmospheric Administration supported the preferred alternative and thus is expected to designate the federal portions of these marine reserves in late 2004.

The process was certainly not perfect or easy. The working group was unable to achieve consensus, and when it wrapped up its work, it felt more like a melt down than a satisfying ending to two years of work. Of greater concern, some groups tried in earnest to circumvent the official process. Ultimately they were not successful, but such attempts (and the hopes they represented) detracted from efforts to gain consensus within the working group. Nonetheless, the broad elements of the process—goal setting, technical advice, stakeholder-driven proposal development, technical evaluation, and informed decision making—prevailed and the outcome could have been much worse. I personally learned that the broad steps were not enough to guarantee success. There also needs to be a commitment from all groups in the official process and consistent reminders that interested parties must play within that process or lose out on the opportunity.

The closure to fishing of a large network of marine areas, while supported by many scientists, has rarely taken place. The Channel Islands marine reserve network offers a great opportunity to learn about the responses of marine ecosystems to this level of protection. The network provides one of the first chances to study large tracts of highly protected ocean ecosystems as they recover from decades of human activities. It is likely that such studies will be conducted, despite the poor fiscal environment in California at present, because of the investment a number of scientists have already made and the engaging questions they can address through study of this system. If so, we can expect to gain a far greater understanding of how marine reserves operate not only individually but also as a network in sustaining ocean ecosystems, populations, and fisheries.

## WHAT IS NEXT FOR THE WEST COAST?

A number of processes offer the potential to create additional MPAs and marine reserves along the West Coast. These include state and federal efforts to examine and coordinate MPAs and to manage fisheries.

### California State Legislation

The California Legislature passed two key pieces of legislation with great potential to expand the use of MPAs and marine reserves in state waters: the Marine Life Management Act (MLMA) in 1998 and the Marine Life Protection Act (MLPA) in 1999. The MLMA transferred fisheries management authority from

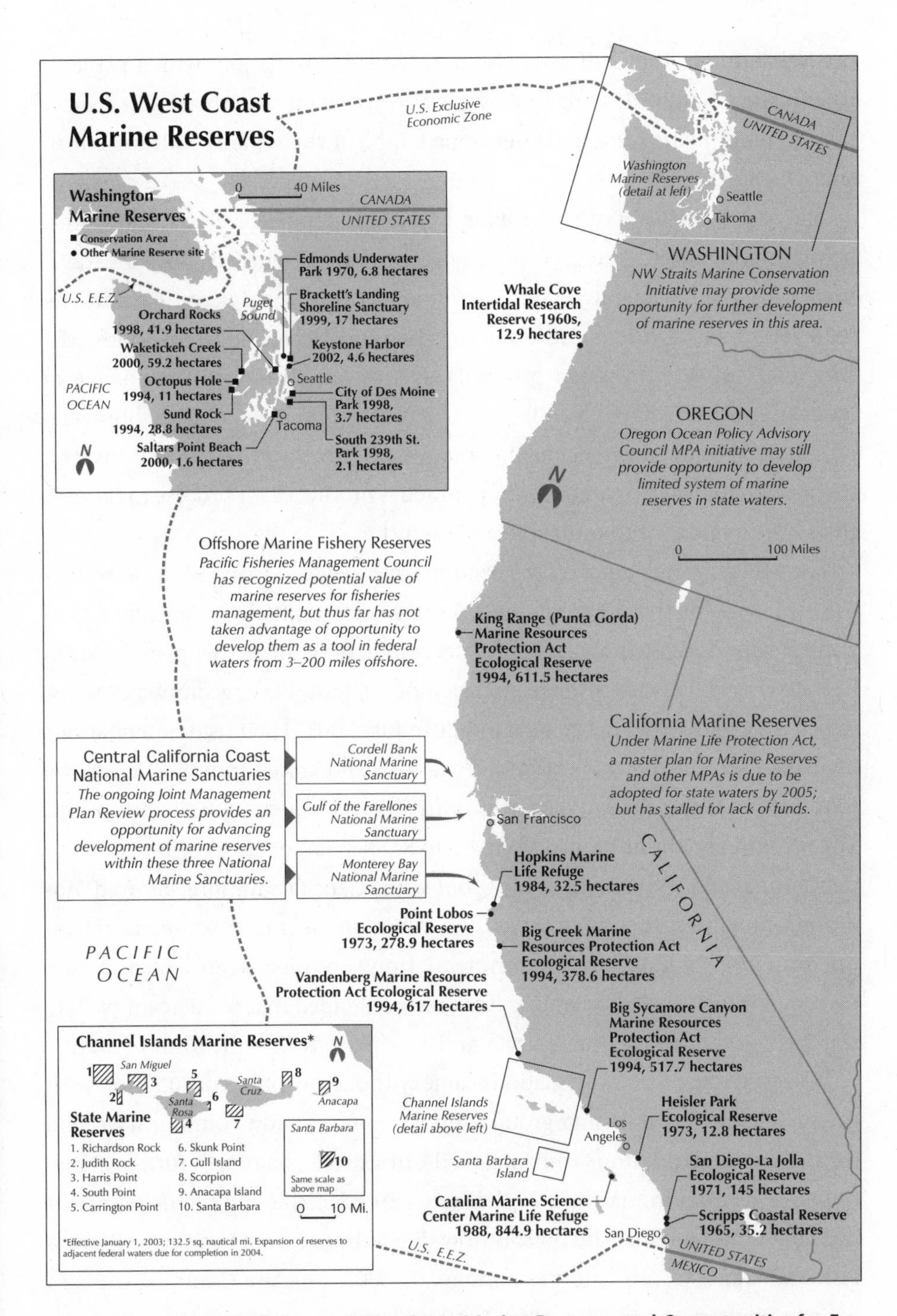


**FIG. 8.8 Locations of All Current West Coast Marine Reserves and Opportunities for Future Marine Reserve Development.**

the legislature to the California Fish and Game Commission, which played a role in the Channel Islands process. It also specified that three new fishery management plans were to be developed. One of these plans addresses nearshore fisheries.

The Nearshore Fishery Management Plan was adopted in 2002 and is now being implemented. The plan development process relied heavily on stakeholder input and generated draft recommendations for 10 percent marine reserves along the northern California coastline and 15 percent to the south. Marine reserves were proposed primarily as a means to reduce the chance of inadvertent overfishing given large uncertainties about the biology of several nearshore species. However, in the final version they were removed. Instead, the nearshore fishery management plan relies on the MLPA process to provide the MPA component to protect nearshore fish.

The MLPA created a process to examine and revise existing MPAs as well as to site new MPAs that might include marine reserves. It also established a scientific team to advise the state on the development of the master plan for MPAs in state waters. This scientific team developed a draft set of recommendations, which they released to the general public in July 2001. Their recommendations identified specific areas for MPA designations and varied in scope across the four regions they delineated (north, north central, south central, and south). The recommendations ranged from 4.5 to 8.9 percent of each region closed to recreational and commercial fishing, but with scientific sampling allowed. Additional areas were identified for closure to bottom fishing or commercial fishing, bringing the total MPA recommendations up to between 14.6 and 24.1 percent. On average, the scientific team recommended that 6.9 percent of state waters should be in marine reserves and 17.7 percent in some form of MPA.

However, these recommendations came without any prior consultation with the public, and both fishing groups and the conservation community called for the recommendations to be delayed until there could be a formal stakeholder consultation. As a result, the state extended the original deadline for the MLPA master plan. The draft is now due to be presented to the California Fish and Game Commission by January 1, 2005, and the Commission is to adopt a final version by December 1, 2005. These recommendations are to be informed by seven formal regional stakeholder panels with representation from a wide range of public interests, including sport and commercial fishing groups, conservation organizations, and representatives from education, tourism, and other interested parties. However, lack of funding has stalled this process and it is unclear how it will proceed during a state budget crisis.

### State of Oregon's Ocean Policy Advisory Council (OPAC)

In July 2000, the governor of Oregon requested that the Ocean Policy Advisory Council (OPAC) review and make recommendations on MPAs in Oregon in consultation with stakeholders. That August, OPAC recommended that the state begin a public process to create marine reserves in state waters. OPAC's recommendations focused on the establishment of a limited system of marine reserves to evaluate their effectiveness, and it identified neither individual sites nor conservation or fisheries goals. The report recommended the use of a locally oriented public process, utilizing a reserve planning committee of stakeholders. In opposition to these steps, some factions have pressed a piece of legislation that would have effectively gutted the entire state ocean program. It is not clear at present whether the bill will pass, although even if it does amendments have taken out some of the most extreme provisions (Robert Bailey, personal communication, 6/20/03). This has not officially ended the process of adopting a series of small experimental marine reserves in Oregon, but the future remains uncertain.

### Northwest Straits Marine Conservation Initiative

In 1998, Congress established the Northwest Straits Marine Conservation Initiative in Washington State. The initiative is a locally based, grassroots approach to improving the health of the marine ecosystem in the Northwest Straits, an area at the mouth of Puget Sound. In addition to addressing issues ranging from water pollution to derelict fishing gear, it is mandated to create a scientifically based system of MPAs. The initiative established seven marine resources committees (MRCs)—one for each county in the region—to discuss and recommend MPAs and marine reserves to a regionwide Northwest Straits Commission. The recommendations are to be implemented using existing local, state, and federal authorities. Several voluntary MPAs have been established by the MRCs and these groups continue to be a key source of advice to the State's Department of Fish and Wildlife as they consider establishing MPAs.

### Federal Fisheries Management

The Pacific Fishery Management Council drafts recommendations to the secretary of commerce regarding federal fisheries management along the West

Coast. After a troubled history of declining fish populations, the Council established a two-phase process to consider marine reserves. In the first phase, the Council and its advisory bodies considered whether marine reserves would be a useful tool and, if so, for what specific purposes. The first phase ran from the spring of 1999 through September 2000. It culminated in February 2001 with the release of a report (Parrish et al. 2001) that asserted marine reserves had potential value as a management tool to address some of the challenges in the region. The Council chose to pursue marine reserves as a means to rebuild overfished fisheries. However, the Council has stalled on taking up the second phase, in which reserves would actually be implemented for this purpose, and has to date not established a single permanent marine reserve. Nevertheless, the Council has closed some large areas to some forms of fishing as a rebuilding measure for several species of rockfish.

### Central California National Marine Sanctuaries

Central California is home to three national marine sanctuaries—Cordell Bank, Gulf of the Farallones, and Monterey Bay—which extend for hundreds of miles from Bodega Bay, north of San Francisco nearly to Morro Bay, near San Luis Obispo. These sanctuaries are undergoing a review of their management plan, one of the factors that spurred the creation of the Channel Islands marine reserve network. The review process does not require the creation of marine reserves but is intended to ascertain whether existing management measures are adequately protecting the marine environments within the sanctuary. Given the troubled state of many fish species in California (see Fig. 8.2a), it would not be out of the question for marine reserves to be expanded or wholly redesigned as part of the management plan review. At present, sanctuary staff are participating in other regional processes to consider and possibly enact marine reserves and other MPAs, and have established a working group, which has identified goals and objectives and various other background information. However, they are going to recognize the authority of the State of California in inshore waters, where marine reserves are most likely to be desired and effective.

## DISCUSSION

The West Coast experience has taught us a great deal about the design, function, and designation of marine reserves. Though still a small fraction of the state and federal waters in this region, marine reserves are becoming more common, larger, and better designed. Studies of these reserves have shown that they increase the size and abundance of many target species, have strong potential to export production to fishing areas, and promote the natural and healthy functioning of marine ecosystems. The West Coast experience has also taught us how to better conduct reserve designation processes.

The West Coast experience with marine reserves and MPAs parallels experiences from other parts of the world (see chapters 4 and 11). As is true in the West Coast studies, the creation of marine reserves has had dramatic effects on biological communities inside them worldwide despite the small size of most individual reserves and limited extent of reserve networks. Although individual studies are not always capable of demonstrating statistical proof of such changes because of a lack of replication or experimental controls (Rowley 1994), comparisons across multiple studies consistently lead to the conclusion that marine reserves contain more and larger fish within their borders than surrounding areas open to fishing (e.g., Halpern 2003).

The West Coast evidence on dispersal from reserves also parallels evidence from elsewhere. Reserves will not necessarily benefit fisheries for organisms with extremely limited dispersal unless the reserve units are small and numerous. With abalone in Southern California, for example, reserves would only provide fisheries benefits if the reserves were small enough to provide for dispersal to fishing areas only a few kilometers away. On the other hand, organisms that disperse too broadly may prove challenging to protect with reserves (Polacheck 1990). Fortunately, despite the long larval lives of most marine organisms and the resulting potential for broad dispersal (Shanks et al. 2003), studies have shown that a combination of oceanography and larval behavior can lead to significant amounts of local retention of larvae (e.g., Cowen et al. 2000; Swearer et al. 1999). Collectively, the body of work on larval dispersal suggests that many coastal marine populations retain a large proportion of larvae locally while some larvae disperse long distances.

The ecosystem shifts seen outside of the marine reserve off of East Anacapa Island, California, have parallel results from a number of other countries, including Australia, New Zealand, Kenya, Spain, and elsewhere, as discussed in chapters 4 and 11. Such shifts are fairly widespread in well-studied environments, but in many cases external factors such as runoff from land and sewage effluent can also play an important and sometimes synergistic role (Pinnegar et al. 2000). These processes introduce extra nutrients, fertilizing the ocean. Nutrients can play an important role because the ecosystem shifts are often seen in the algal community, which for better or worse is more likely to thrive in the presence of extra nutrients.

In other parts of the world, fisheries benefits have often paled in comparison to nonconsumptive benefits from marine reserves. In Kenya, for example, nearly two-thirds of the fishing fleet quit fishing, most for jobs in the tourism industry, after the creation of a marine reserve network encompassing 65 percent of local fishing grounds (McClanahan and Kaunda-Arara 1996). The fishing community on St. Lucia also showed strong interest in tourism jobs during the designation of the Soufriere Marine Park. Though dive tourism is less significant along the West Coast, especially in the state of Oregon, there is a substantial dive industry in places like Puget Sound, Monterey Bay, Catalina Island, and San Diego, which is likely to benefit from the creation of marine reserves.

The designation of the Channel Islands marine reserve network opens a new frontier for marine protection with the potential to greatly improve the health of the region. The science-based, stakeholder-driven process succeeded in establishing the West Coast's first marine reserve network and changing the scale of marine reserve protection. Much can be learned from this experience. It shows that neither initial opposition nor controversy inherently means a death knell for a marine reserve proposal. Managers and supporters of the reserve network neither ran away from nor eliminated controversy, but they did overcome it. Strong science and an open public process were instrumental to this. Defining goals and objectives for the marine reserve network early in the process and developing scientific criteria to meet them were also key to this success. The scientific advisory panel and the decision-making tools provided credibility, flexibility, and support at key times. Although the Marine Reserves Working Group process did not achieve absolute consensus, it did provide a foundation for a set of alternatives that represented a range of views, allowed for a vigorous and public debate over these alternatives, and ultimately led to the development of a preferred alternative with enough public support to succeed.

In sum, there is strong evidence from the West Coast and elsewhere that marine reserves are effective at protecting populations from overfishing, exporting enhanced production to nearby fishing areas, maintaining healthy and balanced ecosystems, and providing nonconsumptive recreational and economic opportunities. Despite this evidence, marine reserves are still rare. The greatest potential for improving both fisheries and conservation on the West Coast lies in the creation of effective networks of marine reserves. There is now evidence that such marine reserves can be successfully established on the West Coast when supported by strong science and sound public process. Whether additional networks will be created remains to be seen and hinges on uncertain processes under way in state and federal governments. Even the Channel Islands network is not yet complete, with the federal portion still needing to be finalized. The successful establishment of the region's first marine reserve network raises the bar for marine conservation and provides an opportunity for raising it further. Many people will be watching to see the impact of this reserve network. A strong research and monitoring program will likely be critical to evaluating this impact. Such a program is important to evaluate the success of the Channel Islands network for its own sake, but the extent to which it can demonstrate positive impact from the reserve network may also go a long way toward determining the success of other reserve efforts.

## ACKNOWLEDGMENTS

I am especially grateful to Gary Davis, David Kushner, and the Channel Islands National Park for the foresight to collect a fantastic dataset on kelp forest ecosystems in the Channel Islands and for permission to have access to this data. I am also indebted to Wayne Palsson, Bob Bailey, Dave Fox, Joe Wible, Alan Baldridge, Warner Chabot, Greg Helms, Doug Obegi, and Karen Reyna for providing valuable information on the history of MPAs and ongoing processes. Finally, I wish to thank Rebecca Sladek Nowlis for her valuable comments on a draft of this manuscript.

## ENDNOTES

1. State waters extend 3 nautical miles along the West Coast, with federal jurisdiction beginning at the end of state waters and ending 200 nautical miles out, at the end of the United States' Exclusive Economic Zone.

## REFERENCES

Airamé, S., J. E. Dugan, K. D. Lafferty, H. Leslie, D. A. McArdle, and R. R. Warner. 2003. Applying ecological criteria to marine reserve design: A case study from the California Channel Islands. *Ecological Applications* 13(1, suppl):S170–S184.

Bohnsack, JA. 1996. Maintenance and recovery of reef fish productivity. In Polunin, N. V. C., and C. M. Roberts, eds. *Reef Fisheries,* 283–313. London: Chapman and Hall.

Cowen, R. K. 1983. The effect of sheephead (*Semicossyphus pulcher*) predation on red sea urchins (*Strongylocentrotus franciscanus*) populations: an experimental analysis. *Oecologia* 58:249–255.

Cowen, R. K., K. M. M. Lwiza, S. Sponaugle, C. B. Paris, and D. B. Olson. 2000. Connectivity of marine populations: Open or closed? *Science* 287:857–859.

Dayton, P. K., M. J. Tegner, P. B. Edwards, and K. L. Riser. 1998. Sliding baselines, ghosts, and reduced expectations in kelp forest communities. *Ecological Applications* 8:309–322.

Didier, A. J., Jr. 1998. *Marine Protected Areas of Washington, Oregon, and California*. Portland, OR: Pacific Fishery Management Council.

Halpern, B. 2003. The impact of marine reserves: Do reserves work and does reserve size matter? *Ecological Applications* 13(1, suppl):S117–S137.

McArdle, D. A., ed. 1997. *California Marine Protected Areas*. La Jolla: California Sea Grant College System, University of California.

McArdle, D., S. Hastings, and J. Ugoretz. 2003. *California Marine Protected Area Update*. La Jolla: California Sea Grant College Program.

McClanahan, T. R., and B. Kaunda-Arara. 1996. Fishery recovery in a coral-reef marine park and its effect on the adjacent fishery. *Conservation Biology* 10:1187–1199.

Murray, M. 1998. *The Status of Marine Protected Areas in Puget Sound*. Olympia, WA: Puget Sound/Georgia Basin International Task Force from the Puget Sound Water Quality Action Team.

Paddack, M. J., and J. A. Estes. 2000. Kelp forest fish populations in marine reserves and adjacent exploited areas of central California. *Ecological Applications* 10:855–870.

Palsson, W. A., and R. E. Pacunski. 1995. The response of rocky reef fishes to harvest refugia in Puget Sound. In *Puget Sound Research '95 Proceedings*, Vol. 1, 224–234. Olympia, WA: Puget Sound Water Quality Authority.

Parrish, R. H., J. Seger, and M. Yoklavich. 2001. *Marine Reserves to Supplement Management of West Coast Groundfish Resources—Phase 1: A Technical Analysis*. Portland, OR: Pacific Fishery Management Council.

Pinnegar, J. K., N. V. C. Polunin, P. Francour, F. Badalamenti, R. Chemello, M.-L. Harmelin-Vivien, B. Hereu, M. Milazzo, M. Zabala, G. D'Anna, and C. Pipitone. 2000. Trophic cascades in benthic marine ecosystems: Lessons for fisheries and protected-area management. *Environmental Conservation* 27:179–200.

Polacheck, T. 1990. Year around closed areas as a management tool. *Natural Resource Modeling* 4:327–353.

Prince, J. D., T. L. Sellers, W. B. Ford, and S. R. Talbot. 1987. Experimental evidence for limited dispersal of haliotid larvae (genus *Haliotis;* Mollusca: Gastropoda). *Journal of Experimental Marine Biology and Ecology* 106:243–263.

Rowley, R. J. 1994. Marine reserves in fisheries management. *Aquatic Conservation: Marine and Freshwater Ecosystems* 4:233–254.

Schroeter, S. C., D. C. Reed, D. J. Kushner, J. A. Estes, and D. S. Ono. 2001. The use of marine reserves in evaluating the dive fishery for the warty sea cucumber (*Parastichopus parvimensis*) in California, USA. *Canadian Journal of Fisheries and Aquatic Sciences* 58:1773–1781.

Shanks, A. L., B. A. Grantham, and M. H. Carr. 2003. Propagule dispersal distance and the size and spacing of marine reserves. *Ecological Applications* 13(1, suppl):S159–S169.

Sladek Nowlis, J., and C. M. Roberts. 1999. Fisheries benefits and optimal design of marine reserves. *Fishery Bulletin* 97:604–616.

Swearer, S. E., J. E. Caselle, D. W. Lea, and R. R. Warner. 1999. Larval retention and recruitment in an island population of a coral-reef fish. *Nature* 402:799–802.

Tegner, M. J. 1992. Brood-stock transplants as an approach to abalone stock enhancement. In Shepherd, S. A., M. J. Tegner, and S. A. Guzman del Proo, eds. *Abalone of the World: Biology, Fisheries, and Culture,* 461–473. Oxford: Blackwell Scientific.

Tegner, M. J., and P. K. Dayton. 2000. Ecosystem effects of fishing in kelp forest communities. *ICES Journal of Marine Science* 57:579–589.

Tegner, M. J., and L. A. Levin. 1983. Spiny lobsters and sea urchins: Analysis of a predator–prey interaction. *Journal of Experimental Marine Biology and Ecology* 73:125–150.

Tegner, M. J., P. K. Dayton, P. B. Edwards, and K. L. Riser. 1995. Sea urchin cavitation of giant kelp (*Macrocystis pyrifera* C. Agardh) holdfasts and its effects on kelp mortality across a large California forest. *Journal of Experimental Marine Biology and Ecology* 191:83–99.

Wallace, S. S. 1999. Evaluating the effects of three forms of marine reserve on northern abalone populations in British Columbia, Canada. *Conservation Biology* 13:882–887.

NINE

# Bahamian Marine Reserves—Past Experience and Future Plans

CRAIG P. DAHLGREN

As an archipelago composed of over 700 islands and cays, the Commonwealth of the Bahamas (Fig. 9.1) is reliant upon its marine resources in a variety of ways. The white sand beaches, crystal clear waters, coral reefs, and abundant marine life attract millions of tourists to the Bahamas each year, making marine-based tourism one of the largest sectors of the Bahamian economy.

Marine fisheries in the Bahamas provide the country with one of its chief exports, the Caribbean spiny lobster (*Panulirus argus*), with exports valued at over $60 million in four of the past five years. Similarly, fisheries for queen conch (*Strombus gigas*) and Nassau grouper (*Epinephelus striatus*), which have suffered economic collapse elsewhere in the Caribbean, remain an important part of the Bahamian economy (Fig. 9.2). Fishing provides full employment to 6.8 percent of the Bahamian workforce and partial employment to an additional 8.8 percent (MacAllister Elliot and Partners 1998).

More than just a source of economic gain, the marine environment of the Bahamas is an integral part of Bahamian culture. Many Bahamian legends, traditions, and symbols are closely connected with the sea. This is evident in the Bahamian coat of arms, postage stamps, and money, all of which are adorned with fish and other sea creatures. Based on the close bond between Bahamian economics, history, culture, and the sea, it is only fitting that one of the oldest marine protected areas in the world is located in the Bahamas, the Exuma Cays Land and Sea Park.

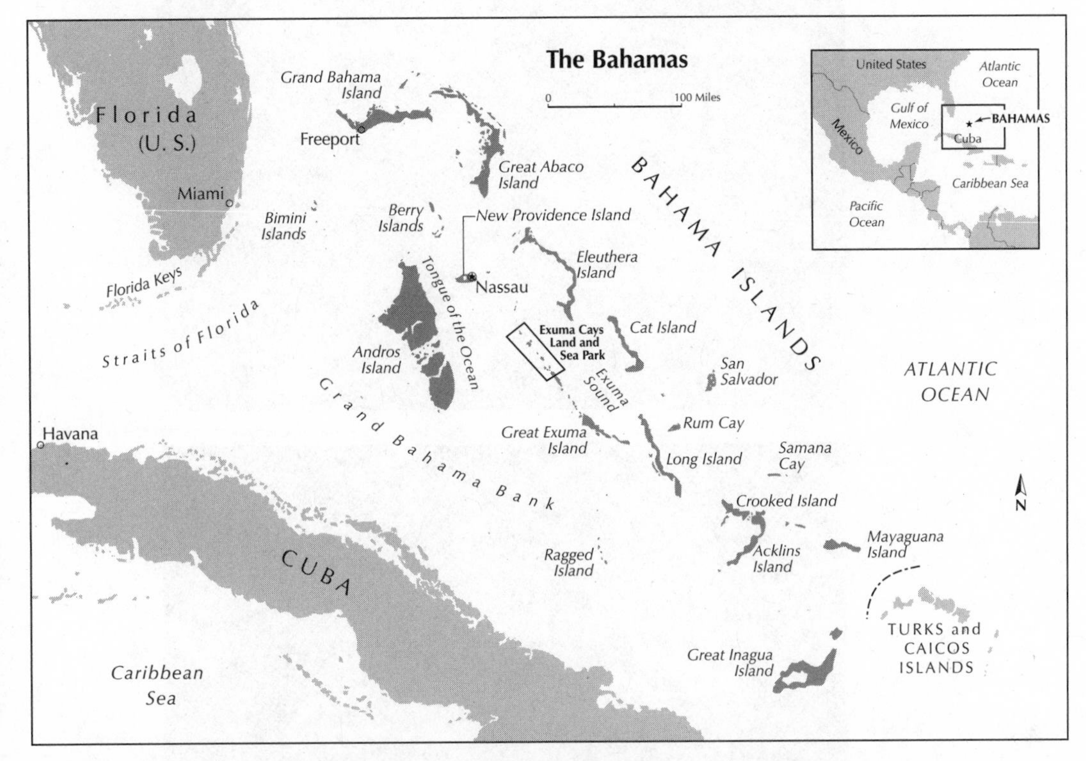


**FIG. 9.1 Map of Bahamian Archipelago, Showing its Neighbors and Geographic Context.** The Bahamas consists of over 700 islands on large carbonate platforms separated by deep basins such as Exuma Sound and the Tongue of the Ocean. This geological heterogeneity creates a diversity of marine habitats within the archipelago.

## THE EXUMA CAYS LAND AND SEA PARK

Although the Bahamas has an abundance of marine life, relatively healthy coral reefs, and high water quality when compared to many other parts of the Caribbean, these resources face threats from a wide range of human impacts (e.g., Dahlgren 1999; Ray 1999; Sullivan-Sealey 2000). As long ago as the 1890s there was concern in the Bahamas over the destruction of coral reefs, as evident in the Sea Gardens Protection Act of 1892, which prohibited dredging or the removal of coral, sea fans, or other organism from the seabed (Mascia 2000). Concern over decreases in the abundance of marine life, particularly species targeted in fisheries, were increasing by the 1940s (Ray 1998). This concern mobilized efforts in both the conservation and the scientific communities to increase protection of Bahamian marine environments. As a result of these efforts, the Exuma Cays Land and Sea Park (Fig. 9.3) was officially created by the Bahamas National Trust Act in 1959.

**FIG. 9.2 The Three Most Important Fishery Species of the Bahamas.** (A) Caribbean spiny lobster or "crawfish," *Panulirus argus*; (B) the queen conch, *Strombus gigas*; and (C) the Nassau grouper, *Epinephelus striatus*. These three species account for approximately 90 percent of the value of Bahamian fisheries. Local populations of all three have benefited from creation of the Exuma Cays Land and Sea Park.

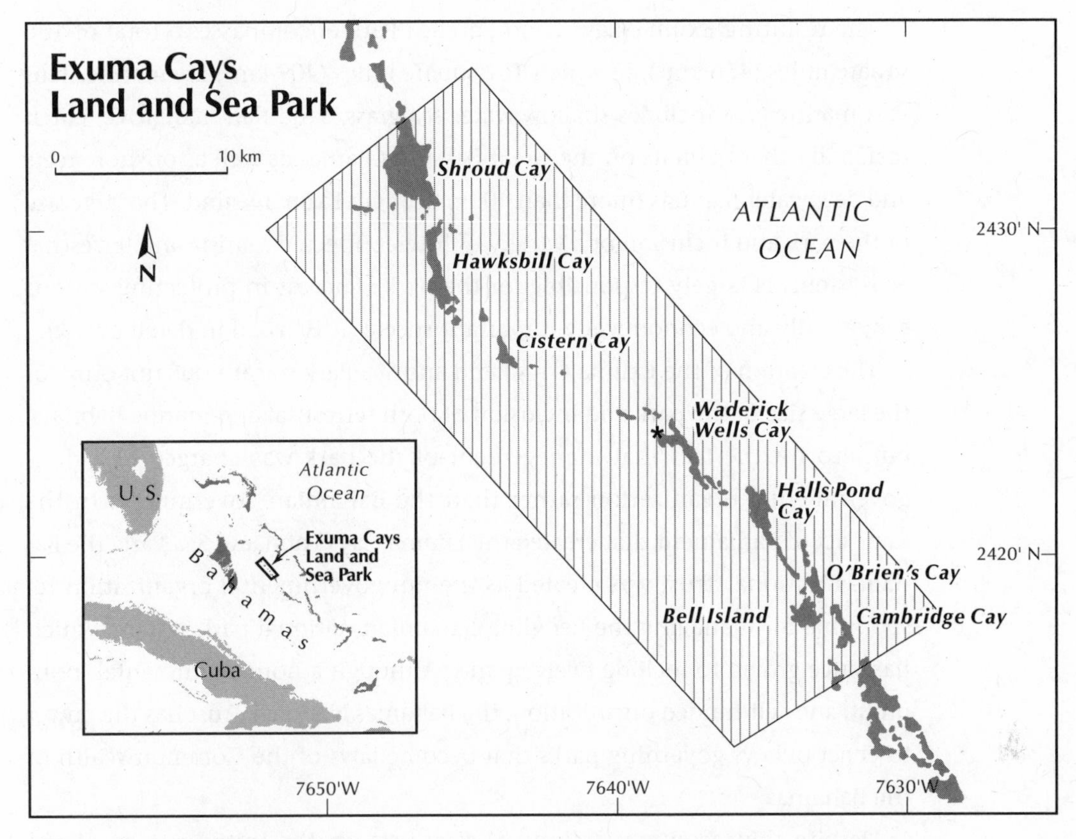


**FIG. 9.3 The Exuma Cays Land and Sea Park (map).** This park is one of the largest (456 km$^2$) and oldest (circa 1958) marine protected areas in the wider Caribbean region. Since 1986, it has been managed as a no-take marine reserve. The park includes terrestrial protection to many of the islands within it and encompasses extensive marine areas, including sand and seagrass areas along the Bahama Banks to the west of the islands; mangroves, creeks, sand flats, hardbottom and coral reef systems around the islands; and deepwater environments of Exuma Sound to the east of the islands. *Park headquarters located on Waderick Wells Cay.

Although the creation of the Exuma Cays Land and Sea Park may have been the result of opportunism and several serendipitous events (see Ray 1998 for a history of the Exuma Cays Land and Sea Park), the creation of the park was characterized by foresight into the scientific basis for ecosystem conservation and marine protected area design. In a report assessing the suitability of the area for protection, scientists identified the importance of including interrelated marine and terrestrial ecosystems within a single unit for conservation, emphasized the importance of preserving ecological processes, stressed the role that protected areas can play in conservation education, and even suggested the importance of setting aside reefs "free of fishing pressure" for study in their "primordial state" (Ray 1998).

The resulting Exuma Cays Land and Sea Park encompasses a total of 186 square miles (456 $km^2$), of which 167 square miles (409 $km^2$) are marine. This vast marine area includes shallow water sea grass, sand flat, mangrove, patch reef, and other habitats on the Great Bahamas Banks, as well as offshore reefs and deepwater habitats (more than 400m deep) in Exuma Sound. The large size of the park and inclusion of a variety of interconnected marine and terrestrial ecosystems is largely responsible for the park's success in protecting various ecologically and economically important species (discussed in detail below).

The creation of the Exuma Cays Land and Sea Park was unique not only for the large size of the park and inclusion of both terrestrial and marine habitats, but also for the fact that management of the park was charged to a nongovernmental organization rather than the Bahamian government. In the same act of Parliament that created the Exuma Cays Land and Sea Park, the Bahamas National Trust was created as the nongovernmental organization responsible for managing the fledgling Bahamian national park system, which has since grown to include twelve parks. Although a nongovernmental, nonprofit, and self-funded organization, the Bahamas National Trust has the power to enact bylaws governing parks that become laws of the Commonwealth of the Bahamas.

Despite the recommendations of scientists on the initial survey of the Exuma Cays Land and Sea Park, the original bylaws of the park allowed some fishing (Mascia 2000). However, an increase in fishing pressure in the 1980s forced the Bahamas National Trust to change the park's bylaws, so that all fishing and other extractive uses were prohibited from the entire park area in 1986 (Sluka et al. 1996). This made the Exuma Cays Land and Sea Park one of the first no-take marine reserves in the tropical western Atlantic, and one of the biggest no-take marine reserves in the world (at the time of writing, it is still one of the largest no-take marine reserves in the world).

Its large size, no-take status, and inclusion of a wide variety of marine and terrestrial habitats make the Exuma Cays Land and Sea Park an excellent example of effective marine reserve design, yet it is not without its problems. Perhaps the greatest of these problems is ensuring adequate compliance with park restrictions on fishing and other activities. Despite its large size, the park has only a single warden charged with enforcing park regulations. The relatively central location of the park headquarters on Waderick Wells Cay allows for maximum efficiency for a single individual to patrol the park, but the large size of the park limits the ability of one warden to effectively patrol the entire area, particularly near the northern and southern boundaries where fast "dayboat" fishing vessels can move into and outside of the park undetected (see Fig. 9.3). Thus poaching became a significant problem in the 1990s. To effectively enforce park regulations over such a large area, the park warden often relies on visitors to the park reporting any poaching that they observe. Furthermore, the recent addition of armed Royal Bahamian Defense Force (RBDF) personnel to assist the park warden has greatly improved compliance with park regulations. For example, the RBDF boarded and impounded a 72-foot fishing vessel that was found fishing in the park during the summer of 2000.

Despite the increased capacity for enforcement, poaching remains a challenge to management of the Exuma Cays Land and Sea Park, particularly by local dayboats at times when there is a reduced enforcement mechanism in the park (e.g., when the park's primary patrol boat is not functioning). Although the Exuma Cays Land and Sea Park apparently attracts poachers from hundreds of miles away, poaching by local residents and tourists presents a greater problem. One reason why poaching by local residents may occur within the park is that there was little consultation with inhabitants of the small settlements that surround the park when it was created, and again when it was made a no-take reserve. Because there is a growing disparity between the abundance of fishery resources in the park and decreasing resources of surrounding areas (see Fig. 9.4), there is an economic incentive for dayboat fishermen to poach within the park. Furthermore, local resource users were not involved in decision making with respect to the designation of the park, or making the park no-take. Thus many local fishermen do not agree with park management. In addition to poachers from nearby settlements, yachtsmen that frequently cruise the Exuma Cays in the winter and spring also fish within the park. This occurred particularly during the park's early stages as a no-take marine reserve. Effective outreach and education efforts, however, have turned this user group into strong supporters of the park, who often serve as volunteers in various capacities there.

In addition to consumptive threats like poaching, the Exuma Cays Land and Sea Park also faces problems resulting from land use within the park. Although protecting the small cays that made up the terrestrial component of the park has always been a goal of the Bahamas National Trust, several of the privately owned islands escaped protection during the initial creation of the park. Most of these islands have remained largely undeveloped and uninhabited, but a few have been sold and built upon by their new owners in recent years. Much of the development within the park has progressed without following any acceptable environmental protection guidelines. For example, extensive networks of unpaved roads were cleared throughout one small island without any thought to erosion control. In addition to any adverse impact that terrestrial development has to the marine environment of the park, it has also contributed to resentment from much of the local Bahamian community who feel that the foreigners who own islands are getting preferential treatment when it comes to resource use within the park. At the time of writing, however, the Bahamian government and the Bahamas National Trust are addressing several issues related to terrestrial development within the park and its impact on the marine environment.

## MARINE RESERVE RESEARCH IN THE BAHAMAS

Although science played an important role in its designation, scientific studies of marine populations, communities, and ecosystems within the Exuma Cays Land and Sea Park since its creation remain limited, and our knowledge is far from complete. This may be due to the limited resources and capacity of the Bahamian government and Bahamas National Trust to support research, as well as the park's relatively remote location. The nearby Caribbean Marine Research Center provides a remote field station in the vicinity and contributes significant supplementary capacity for research within the park. Nonetheless, results from several studies conducted within the park demonstrate dramatic effects of reserve protection for some of the most important fishery species in the Bahamas: Nassau grouper, Caribbean spiny lobster, and queen conch.

Comparisons of Nassau grouper populations within the Exuma Cays Land and Sea Park to those up to 30 km north and south of the park indicate that grouper size, density, and biomass (see also chapter 4, Fig. 4.6) differed between the park and unprotected areas outside it (Sluka et al. 1996, 1997). Within the park, mean density of Nassau grouper was 75 to 100 percent greater and fish were over 20 percent larger on average than outside the park (Sluka et al. 1996). These differences resulted in Nassau grouper biomass being three times greater within than outside the park. Observed differences in density, size, and biomass are believed to result from the protected status of the park rather than other factors (e.g., habitat) that may differ between the park and surrounding areas because there were no significant differences in Nassau grouper abundance among different habitat types (Sluka et al. 1996). Nassau grouper biomass within the park also compares very favorably with more heavily fished areas elsewhere in the Bahamas with up to an order of magnitude difference (See Fig. 9.4)

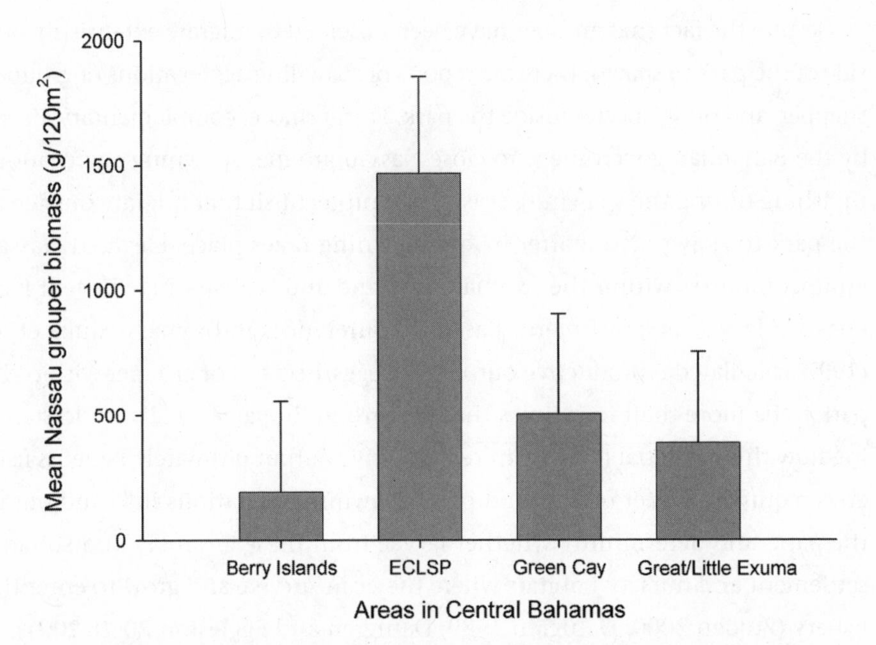


**FIG. 9.4 Comparison of Nassau Grouper Biomass, Observed in Transect Surveys on Coral Reefs within the Exuma Cays Land and Sea Park (ECLSP), to Similar Reefs in Other Locations throughout the Central Bahamas.** These other locations are all open to fishing and include areas near (Great Exuma, Berry Islands) and far (Green Cay) from human population centers. Several of these other sites include some areas that are proposed as marine fishery reserves and national parks.

There is also indirect evidence that the elevated Nassau grouper biomass within the Exuma Cays Land and Sea Park supports fisheries outside the park. A closer examination of the spatial distribution of Nassau grouper as a function of distance from the center of the park indicates that Nassau grouper biomass immediately outside the park boundaries is more similar to biomass within the park than farther away from the park (see also Fig. 4.6 and Sluka et al. 1997). This provides indirect evidence of spillover of biomass from the park to fished areas (an alternative hypothesis is that poaching within the park may be causing these patterns; however, this explanation does not adequately explain why abundance immediately outside the park is higher than farther outside). Further evidence that Nassau grouper periodically leave the park and enter the fishery comes from studies of movement by tagged fish. Although most tagged grouper exhibited movement over relatively small areas, at least two grouper tagged within the Exuma Cays Land and Sea Park were caught by a fisherman at spawning aggregations hundreds of kilometers away (Bolden 2000; Dahlgren, unpublished data).

Despite the fact that grouper have been observed to migrate extensively outside of the park to spawn, there are reports of spawning aggregations of grouper, snapper, and other species inside the park. Furthermore, complementary efforts by the Bahamian government to close Nassau grouper spawning aggregations to fishing during the spawning season can protect fish that migrate outside of the park to spawn. No matter where spawning takes place, elevated Nassau grouper biomass within the Exuma Cays Land and Sea Park may benefit fisheries via larval replenishment. Based on differences in biomass, Sluka et al. (1997) calculated reproductive output (no. eggs/hectare) of grouper within the park to be more than four times that of surrounding areas. Fully understanding how this potential increase in reproductive output ultimately benefits fisheries requires a better understanding of spawning migrations into and out of the park, and determining whether larvae from these groupers find suitable settlement and nursery habitats where they can survive and grow to enter the fishery (Bolden 2000; Dahlgren 1999; Dahlgren and Eggleston 2000, 2001).

Studies within and outside the Exuma Cays Land and Sea Park have also found the park to be effective at protecting spiny lobster stocks. Lipcius et al. (1997) compared adult spiny lobster abundance within the park to other areas throughout Exuma Sound. Despite the fact that the Exuma Cays Land and Sea Park did not have particularly high rates of postlarval supply and did not contain the most suitable habitat, densities within the park were significantly higher than the other sites examined around Exuma Sound, a difference that appears to be increasing as lobster stocks outside the park are subject to increasing fishing effort. Furthermore, appropriate juvenile and adult lobster habitat covers less area within the Exuma Cays Land and Sea Park than other locations within Exuma Sound, suggesting it may not be ideally located to protect lobsters. Lobster abundance throughout Exuma Sound also decreased on an annual basis from the closed season to the open fishing season and the decrease was similar for all sites, including the Exuma Cays Land and Sea Park. Given these factors, the high abundance of adult lobsters within the park compared to outside is especially noteworthy. Models based on observed lobster abundance and hydrodynamics of Exuma Sound predict that although some reproductive output from the park may be carried to inappropriate settlement habitats, potentially limiting the park's ability to support regional fisheries, the location of the park may result in significant local retention of larvae and support local fisheries (Lipcius et al. 2001). Empirical investigations into the effects of the park on local and regional fisheries, and tests of model assumptions will continue to improve our understanding of how the park affects lobster fisheries in surrounding areas. Studies should also address the indirect effects of increased lobster abundance on ecosystems within the park.

As with Nassau grouper and spiny lobster, queen conch populations also benefit from protection within the Exuma Cays Land and Sea Park. Unlike lobster, however, there is greater evidence that the park benefits conch fisheries throughout the central Bahamas. Within the park, conch densities were significantly higher than in most fished areas throughout the Exuma Sound region (Stoner and Ray 1996; Stoner et al. 1998). Conch densities in only one fished area, the unprotected site around the Schooner Cays off Cape Eleuthera had densities similar to that of the park. High adult conch densities in the Schooner Cays area, however, are likely to be the result of extremely productive juvenile conch populations there, as suggested by the presence of large juvenile conch aggregations in the area and the fact that most adults appeared to be young (Stoner et al. 1998). Recent anecdotal reports from the Schooner Cays area, however, suggest that this is no longer the case and conch stocks there have suffered a dramatic reduction due to high levels of fishing. Although the Exuma Cays Land and Sea Park contains some juvenile conch aggregations, the density and area of these are not much higher than other parts of Exuma Sound, so high adult densities are likely to result from reserve protection. Within the Exuma Cays Land and Sea Park, mean density of adult conch exceeded 49 conch per hectare at all depths from 5 to 30 m, and had a maximum mean density of 270 conch per hectare, the highest observed density anywhere in the region. With the exception of the Schooner Cays, adult conch densities rarely exceeded forty-nine per hectare outside of the Exuma Cays Land and Sea Park (Stoner et al. 1998). Because conch reproduction is density dependent, with spawning not observed at densities lower than forty-eight conch per hectare in Exuma Sound (Stoner and Ray-Culp 2000), high densities maintained within the park may be essential for the maintenance of the population throughout the region.

Further evidence for the ability of the Exuma Cays Land and Sea Park to support conch populations and possibly conch fisheries throughout the region comes from sampling larval conch (veliger) distribution within and outside of the park. Plankton tows throughout Exuma Sound indicate that larval production (density of early-stage larvae) is significantly greater within the Exuma Cays Land and Sea Park than in other parts of Exuma Sound. Early-stage larval densities within the park ranged from 159 to 929 veligers per 100 $m^3$, but did not exceed 100 veligers per 100 $m^3$ anywhere else in Exuma Sound (Stoner et al. 1998). Later stage larval conch, however, were not concentrated around the park, but were dispersed throughout Exuma Sound by regional current patterns. Thus, the high densities of conch within the Exuma Cays Land and Sea Park are likely to serve as a source of conch recruits to the fishery throughout the Exuma Sound region.

Based on these studies of the effect of the Exuma Cays Land and Sea Park on important fishery species, we can draw several conclusions. It is clear that the park provides protection against exploitation for conch, lobster, and grouper. This is particularly important for queen conch and Nassau grouper, which have been fished to such an extent that they are threatened throughout most of their range. There is also evidence that the park provides some fisheries benefits on local (e.g., less than 30 km from the park) and regional (e.g., throughout Exuma Sound) scales. These benefits are derived despite the lack of quantitative studies of marine resources of the Exuma Cays prior to the creation of the park, and despite the fact that the park may not protect the best habitat in the region for important fishery species. Thus, despite any ecological and socioeconomic shortcomings, the Exuma Cays Land and Sea Park is an effective tool for marine conservation.

## THE FUTURE OF MARINE RESERVES IN THE BAHAMAS

At present, the Bahamas has few marine protected areas and virtually no effective no-take marine reserves aside from the Exuma Cays Land and Sea Park. Other marine protected areas that exist are essentially "paper parks" due to a lack of enforcement, limited protective measures, and/or small size (i.e., Pelican Cays Land and Sea Park near Abaco Island, Union Creek Reserve, and Black Sound Cay National Reserve off Great Inagua Island; Fig. 9.5). However, the Bahamian government is actively building on the Bahamian experience with no-take marine reserves. In November 1999, the Bahamian government announced an ambitious plan to create a network of no-take marine reserves that is intended to encompass at least 20 percent of the Bahamian marine environment. Although achieving this eventual goal may take years, in January 2000, the government identified sites for the first five new marine reserves in this network and began the designation process. The first of these new fishery reserves are expected to be designated in late 2004, and are expected to include areas near Bimini and Southern Berry Islands and Great Exuma Island (see Fig. 9.5). In addition to these new reserves that will fall under the management of the Department of Fisheries, the Bahamas National Trust is creating a number of new parks that will include marine areas zoned for various uses, including several no-take zones. The first of these new National Parks were announced

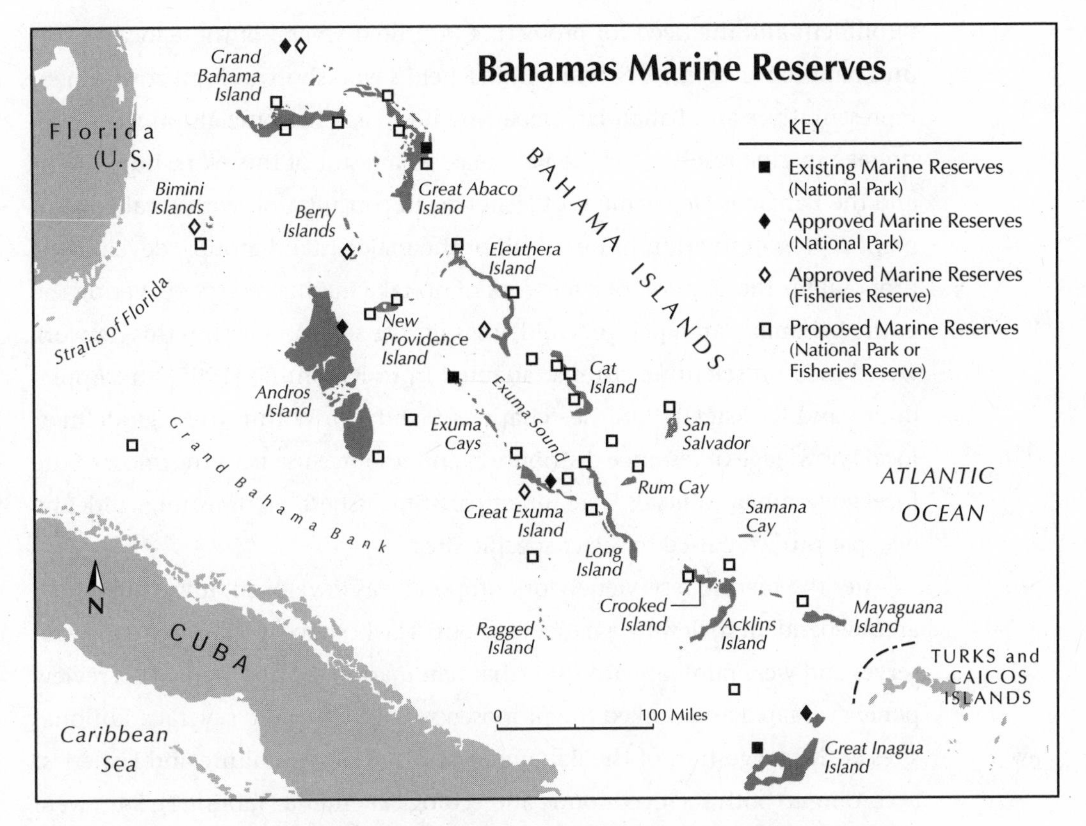


**FIG. 9.5 Existing and Proposed Marine Reserves of the Bahamas.** Sites shown on this map have either been formally selected or proposed as marine fishery reserves or national parks. Sites that have been selected for protection as fishery reserves are currently subject to public hearings and discussions prior to their boundaries and management plans being finalized.

in April 2002 and include marine areas off Walker's Cay; Central Andros; Little Inagua; and Moriah Harbour Cay, Exuma (see Fig. 9.5).

Initiatives by the Bahamian government and Bahamas National Trust are, in part, a response to vocal support for Bahamian marine reserves from a variety of sources. For a number of years, dive operations and marine research laboratories within the Bahamas have established what amount to "voluntary" no-take areas to reduce user conflicts and have requested official government designation for these areas. Local communities have also formed committees such as the Andros Conservancy and Trust (ANCAT) on the Island of Andros, and the Exuma Tourism and Environment Advisory Committee (TEAC) in the Exuma Cays, to address environmental concerns and petition the government for the establishment of marine protected areas.

The Bahamas Reef Environment Educational Foundation (BREEF), a nonprofit organization dedicated to educating Bahamians about their marine environment and the need for protecting it, united several of these local efforts on a national level in 1998, when BREEF held a workshop for local government representatives and Bahamian Department of Fisheries officials on the use of no-take marine reserves for the Bahamas. As a result of this workshop, BREEF and the Bahamas Department of Fisheries, in consultation with local government representatives from nearly all of the major island groups, developed a proposal for the creation of a network of no-take marine reserves covering the entire Bahamian archipelago. Guidelines for site selection within this network were based on scientific criteria (adapted from Ballantine [1995] and Appledoorn and Recksiek [1998]; see chapters 5 and 11). Within these guidelines, local knowledge of resource distribution and resource use patterns (most of the local government officials were full or part-time fishermen or fishing guides by occupation) were used to select specific sites.

After the marine reserve network proposal was developed, it was subject to an independent review by a panel of scientists who have worked on marine reserves and were familiar with the Bahamian marine environment. This review panel evaluated and ranked the proposed reserve sites (and several additional sites at the suggestion of the Bahamian Minister of Agriculture and Fisheries) according to both socioeconomic and ecological criteria (Table 9.1). Sites were ranked on a scale of 1 to 3 for various criteria, and individual scores were averaged for all socioeconomic and all ecological criteria independently. Overall site ranks were then determined by adding socioeconomic and ecological scores together (Stoner et al. 1999). Similar site selection criteria are also used by the Bahamas National Trust to select National Park sites (Table 9.2).

Based on the marine reserve network proposal and its subsequent review by scientists, the Bahamian government announced that it would be creating a marine reserve network with the ultimate goal of protecting 20 percent of Bahamian marine habitats, and that the first five reserves in this new network were created in 2000. The five sites chosen for protection included the area of Walker's Cay, Bimini, the Berry Islands, Cape Eleuthera, and Lee Stocking Island/Great Exuma (see Fig. 9.5). Although few of these sites were among the highest ranked sites, all were considered to be acceptable marine reserve locations by the scientific review panel (Stoner et al. 1999).

Since the first five new sites were selected for the marine reserve network, the Bahamian Department of Fisheries has held several public meetings on a national level and in communities near the new reserve sites. The purpose of these meetings is to explain the need for marine reserves to the public and get public input into drawing reserve boundaries and developing reserve manage-

Table 9.1 Site Evaluation Criteria Used by the Scientific Review Panel to Rank Sites Proposed as Marine Fishery Reserves

Socioeconomic Criteria

A. Fishing Impact—the degree to which fishing will be displaced due to the creation of the MPA
 1 point = Major displacement of fishing activity
 2 points = Minor displacement of fishing activity
 3 points = Negligible displacement of fishing activity
B. Community Management—the ability of local communities or existing organization to participate in management of the MPA
 1 point = No community nearby and no existing park (or other management organization)
 2 points = Community nearby but support uncertain
 3 points = Supportive community nearby or existing park
C. Community Benefits—likelihood that marine reserve would provide benefits to local communities
 1 point = Both nonconsumptive benefits and spillover effect negligible
 2 points = Minor nonconsumptive benefits and/or spillover effect likely
 3 points = Major nonconsumptive benefits and/or spillover effect likely

Ecological Criteria

A. Habitat Diversity—the diversity of marine habitats important for supporting Bahamian fisheries and marine biodiversity. These habitats included sea grass, mangroves, and coral reefs.
 1 point = habitat sparse or degraded by human activities
 2 points = Healthy reef or seagrass/mangrove (not both) habitats present
 3 points = Healthy reefs, seagrass, and mangrove habitats present
B. Regional Importance—the potential importance of the area for supporting fisheries throughout the Bahamas. Because the prevailing currents in the Bahamas run from SE to NW, larval retention within the Bahamas is likely to be greatest for sites in the SE half of the country.
 1 point = Negligible potential source of larvae for the Bahamas.
 2 points = Minor potential source of larvae for the Bahamas
 3 points = Major potential source of larvae for the Bahamas

Stoner et al. 1999.

For each proposed site, scores for socioeconomic criteria and ecological criteria were averaged separately then added together to give an overall score for each site. The resulting priority scores ranged from 2 to 6 and were used to rank sites.

ment plans. Thus local resource users are having a large amount of input into the planning process in terms of deciding the exact location and boundaries of individual reserves, as well as the development of management plans for the reserves. Because the resources for enforcement of the new reserves are limited, community participation in reserve designation and management is essential for the success of these marine reserves. At the time of writing, local response to marine reserves in the five areas has been mixed. In Bimini, for example, the proposed marine reserve was relatively well received. This may be due to the fact that many residents of Bimini feel that their natural resources are under

Table 9.2 Parks and Protected Area Site Selection Criteria Used by Bahamas National Trust.

| Criteria | Subcriteria |
|---|---|
| Biogeographic importance | Presence of rare biogeographic qualities or representativeness of a biogeographic type |
| | Unique or unusual geological features |
| | Characteristic of the biogeographic province or region |
| Ecological importance | Essential part of ecological process or life support systems |
| | Area's integrity encompasses a complete ecosystem |
| | Variety of ecosystem |
| | Habitat for rare or endangered species |
| | Nursery or juvenile area |
| | Feeding, courtship, breeding, rest, or migration areas |
| | Rare or unique habitat for species |
| | Genetic diversity |
| | High level of primary and/or secondary production and attendant higher trophic level communities |
| Biodiversity importance | Variety and number of life forms and communities that occur within the specified habitat type or within the biogeographic province or region |
| | Representative variety of species or an important sample of the diversity of ecosystems, communities, species, populations, and gene pools found within the region or habitat |
| Naturalness/ habitat structure | Extent to which area has been protected from human-induced change |
| | Unique rare or unusual chemical, physical, geological, and/or oceanographic features, structures, or conditions |
| Economic importance | |
| Social importance | |
| Scientific importance | |
| International/national importance | |
| Practicality/feasibility | Degree of insulation |
| | Social or political acceptability |
| | Accessibility for education, tourism, recreation |
| | Compatibility with existing uses |
| | Ease of management |

For each criteria and subcriteria, sites are scored on their value using a scale of low, medium, and high.

increasing threat from habitat loss and exploitation, and local stakeholders have united to protect their resources from these outside threats. Acceptance of a reserve in this area may also be related to the fact that the greatest resource use in the proposed area is catch and release bonefishing, which (after consultation with local stakeholders) will be allowed within the reserve, so the establishing the reserve is likely to have little adverse impact on the local economy. Bonefish guides may even be involved in community-based management and enforcement of the reserve.

In other proposed reserve areas, there has been some opposition to the proposed reserve area, but a general acceptance of the need for marine reserves. In south Eleuthera, where poor economic conditions leave local residents few options other than fishing in nearby areas to support their families, there is concern that creating a reserve there would create undue economic hardship. As in the case of the Florida Keys and other places, conflict among user groups has slowed the marine reserve designation process in some parts of the Bahamas. In Great Exuma, for example, local fishermen objected to the creation of a large marine reserve around Lee Stocking Island because they believed that it would only benefit a marine research lab in the area and actually hurt local fishing since several important fishing areas would be closed. These fishermen, however, recognized the importance of marine reserves for conserving marine resources and suggested an alternative site near Great Exuma that was included in the Department of Fisheries marine reserve network proposal and was ranked high by the scientific review panel. Local support for this reserve largely stems from the fact that commercial fishermen from other parts of the Bahamas use this area and are thought to practice destructive fishing with high rates of bycatch that result in reduced landings for local fishers. The Bahamas Department of Fisheries is currently considering both areas for protection, with the area around Lee Stocking Island likely to be reduced to areas immediately around research sites, but a much larger reserve is being considered for the other proposed Great Exuma reserve site.

To help direct future marine reserve designations, there is increasing interest in forming a national marine reserve advisory panel composed of representatives from the scientific community, marine education and conservation nongovernmental organizations, the government of the Bahamas, and the Bahamas National Trust. This panel will provide further advice to the Bahamian government on the selection of sites for marine reserves, the design of those reserves, and the marine reserve designation process, as well as the development and implementation of marine reserve management plans. The continued interest of this advisory panel will greatly facilitate the creation of an effective network of marine reserves in the Bahamas. On a local level, the Department of Fisheries has encouraged the creation of local advisory panels comprised of various stakeholders, including fishing guides, commercial fishers, recreational and subsistence fishers, as well as conservationists. These local panels are used to make recommendations about the designation and management of individual reserves.

## LESSONS LEARNED FROM THE BAHAMIAN EXPERIENCE WITH NO-TAKE MARINE RESERVES

The Bahamas has a long history with the use of marine protected areas and no-take marine reserves, but the Bahamian experience with no-take marine reserves is really just beginning. Based on the example provided by the Bahamas, we can draw several conclusions about marine reserves. First and foremost, the Bahamas provides an excellent example of a large marine reserve that effectively protects some of the most important and most vulnerable fisheries species in the wider Caribbean region and the coral reef ecosystems of which they are an integral component. The Exuma Cays Land and Sea Park effectively protects Nassau grouper, Caribbean spiny lobster, and queen conch populations, despite the fact that it may not contain the greatest abundance of high quality habitat for any of these species. The effectiveness of the park is likely to be the result of its large size and inclusion of a variety of habitats that these species utilize during various life stages. Conservation of these species results in fisheries benefits on local and regional scales for some species, but may not for others with different life-history characteristics.

Based on the Bahamian experience, we can also draw several conclusions from a socioeconomic perspective. The mixed reaction to marine reserves in the Bahamas suggests that acceptance of marine reserves requires consultation with local communities during planning stage of the designation process and ensuring that access remains to at least some traditional fishing grounds. Similarly, public support also depends on ensuring that certain user groups (e.g., foreign scientists and island owners) are not perceived as receiving preferential treatment. Thus all resource users must be treated openly and fairly in the designation process.

Finally, the new reserve network being created in the Bahamas provides an example of how a combination of bottom-up and top-down approaches can be used to create marine reserves. The development of the no-take marine reserve network proposal also involved a balance of socioeconomic and ecological considerations by combining local knowledge of resource use and other socioeconomic factors with scientific criteria. Similarly, the review and ranking process evaluated sites on equally weighted socioeconomic and ecological criteria. Although primarily driven by a top-down approach, involving consultation with both scientists and local representatives, there is much bottom-up participation in the marine reserve planning process by local resource users. Consequently, several new reserve sites being considered by both the Department of Fisheries and the Bahamas National Trust have evolved from local proposals.

Only time will tell how effective these efforts will be, and whether or not a functioning marine reserve network that encompasses more than 20 percent of the Bahamian marine environment will be protected. Fortunately, Bahamian fisheries and marine ecosystems remain relatively healthy, but threats are increasing. Unlike many Caribbean nations and tropical island nations around the world, the Bahamas still has an opportunity to create marine reserves in order to prevent declines in fisheries species and ecosystem health, rather than using reserves to rebuild collapsed stocks and restore severely degraded ecosystems. Although the degree of urgency may not be present, the relative health of Bahamian marine resources may help with the creation of marine reserves. To use an analogy, it is much easier to set aside a piece of pie for the future when everyone can still eat their fill, than when people are fighting to pick up every last crumb. By creating an effective marine reserve network before marine resources are depleted and marine ecosystems degraded, the Bahamas may have sustainable fisheries and healthy marine ecosystems for the economic and cultural benefit of present and future generations of Bahamians.

## REFERENCES


Appledoorn, R. S., and C. W. Recksiek. 1998. Guidelines for the design of marine fisheries reserves, with emphasis on Caribbean reef systems. Presented at the 2nd William R. and Lenore Mote Symposium on Essential Fish Habitat and Marine Reserves.

Ballantine, W. J. 1995. Networks of "no-take" marine reserves are practical and necessary. In Shackell, N. L., and J. H. M. Willison, eds. *Marine Protected Areas and Sustainable Fisheries,* 13–20. Wolfville, Nova Scotia: Science and Management of Protected Areas Association.

Bolden, S. K. 2000. Long-distance movement of a Nassau grouper (*Epinephelus striatus*) to a spawning aggregation in the central Bahamas. *Fisheries Bulletin* 98:642–645.

Dahlgren, C. P. 1999. The biology, ecology, and conservation of the Nassau grouper, *Epinephelus striatus,* in the Bahamas. *Bahamas Journal of Science* 7(1):6–12.

Dahlgren, C. P., and D. B. Eggleston. 2000. Ecological processes underlying ontogenetic habitat shifts in a coral reef fish. *Ecology* 81:2227–2240.

———. 2001. Spatio-temporal variability in abundance, size and microhabitat associations of early juvenile Nassau grouper, *Epinephelus striatus* in an off-reef nursery system. *Marine Ecology Progress Series* 217:145–156.

Lipcius, R. N., W. T. Stockhausen, D. B. Eggleston, L. S. Marshall, Jr., and B. Hickey. 1997. Hydrodynamic decoupling of recruitment, habitat quality and adult abundance in the Caribbean spiny lobster: source-sink dynamics? *Marine and Freshwater Research* 48:807–815.

Lipcius, R. N., W. T. Stockhausen, and D. B. Eggleston. 2001. Marine reserves for

Caribbean spiny lobster: Empirical evaluation and theoretical metapopulation recruitment dynamics. *Marine and Freshwater Research* 52:1589–98.

MacAllister Elliott and Partners. 1998. *Fisheries Management Action Plan for the Bahamas.* A report to the Bahamas Department of Fisheries.

Mascia, M. B. 2000. *Institutional Emergence, Evolution, and Performance in Complex Common Pool Resource Systems: Marine Protected Areas in the Wider Caribbean.* Ph.D. diss., Duke University, Durham, North Carolina.

Ray, G. C., 1998. Bahamian protected areas part I: How it all began. *Bahamas Journal of Science* 6(1):2–10.

———. 1999. Bahamian protected areas part II: Developing a marine protected area system. *Bahamas Journal of Science* 6(2):2–9.

Sluka, R., M. Chiappone, K. M. Sullivan, and R. Wright. 1996. *Habitat and Life in the Exuma Cays, the Bahamas: The Status of Groupers and Coral Reefs in the Northern Cays.* Arlington, VA: The Nature Conservancy.

———. 1997. The benefits of a marine fishery reserve for Nassau grouper *Epinephelus striatus* in the Central Bahamas. *Proceedings of the 8th International Coral Reef Symposium* 2:1961–1964.

Stoner, A. W., and M. Ray. 1996. Queen conch, *Strombus gigas,* in fished and unfished locations of the Bahamas: Effects of a marine fishery reserve on adults, juveniles, and larval production. *Fisheries Bulletin* 94:551–565.

Stoner, A. W., and M. Ray-Culp. 2000. Evidence for Allee effects in an over-harvested marine gastropod: Density-dependent mating and egg production. *Marine Ecology Progress Series* 202:297–302.

Stoner, A. W., N. Mehta, and M. Ray-Culp. 1998. Mesoscale distribution patterns of queen conch (*Strombus gigas* Linne) in Exuma Sound, Bahamas: Links in recruitment from larvae to fishery yields. *Journal of Shellfish Research* 17:955–969.

Stoner, A. W., M. H. Hixon, and C. P. Dahlgren. 1999. *Scientific Review of the Marine Reserve Network Proposed for the Commonwealth of the Bahamas by the Bahamas Department of Fisheries.* Report to the Bahamas Ministry of Agriculture and Fisheries.

Sullivan-Sealey, K. 2000. Tourism and water quality in Elizabeth Harbour. *Bahamas Journal of Science* 5:2–19.

TEN

# Belize's Evolving System of Marine Reserves

JANET GIBSON, MELANIE MCFIELD, WILL HEYMAN, SUSAN WELLS, JACQUE CARTER, AND GEORGE SEDBERRY

The nation of Belize stretches along the Caribbean coast of Central America between Mexico and Guatemala and is home to the Belize (Meso-American) Barrier Reef system, second in length only to Australia's Great Barrier Reef. This national landmark, along with other natural and cultural treasures, has transformed the former British colony of British Honduras into a mecca for tourism, especially ecotourism, since it declared its independence in 1981. Despite a land area of just less than 23,000 km$^2$ and a population of about a quarter of a million people, Belize possesses a spectacular coastline nearly 400 km long and is home to a trove of coastal and offshore resources. Three largely undeveloped and awe-inspiring "atolls" (Turneffe, Lighthouse, and Glover's Reefs) lie offshore of the internationally acclaimed barrier reef. Belize continues to develop one of the world's most advanced and visionary systems of marine protected areas (MPAs).

MPAs, including marine no-take reserves, are an essential component of Belize's Coastal Zone Management Strategy, consistent with its overarching goal of "improved management of coastal resources to ensure economic growth is balanced with sound environmental management" (CZMAI 2001a). As in many countries, MPAs are being established in Belize with a variety of objectives, including tourism management, biodiversity protection, and fisheries management. Management of tourism is particularly important given that tourism, much of it marine oriented, is one of the most important sources of foreign exchange in Belize and that the diversity and comparatively healthy condition of reefs and other marine ecosystems are of recognized international importance. MPAs are therefore being established with zoning schemes designed to benefit various stakeholders.

Most of the scientific work conducted within MPAs in Belize has focused on measuring the effectiveness of the no-take zones for enhancing commercial fish stocks within MPAs as a means of fisheries management. Less work has been focused on measuring the effectiveness of MPAs to maintain ecosystem function. Stakeholders and the general public readily acknowledge the tourism and educational values of the MPAs. However, some commercial fishing interests have recently raised concerns about and opposition to MPAs, illustrating that their value as a fisheries management tool is not universally recognized within this sector in Belize.

An existing network of thirteen MPAs, designed to encompass the range of marine habitat diversity in Belize and conserve overall ecosystem functions, was already developed and approved by 2002. This MPA network includes a number of no-take areas or marine reserves (as defined in this book). In 2002, additional legislation was enacted strengthening this MPA framework by creating eleven new no-take marine reserves to protect known spawning aggregation sites for Nassau grouper and other reef fish and providing additional nationwide protection for Nassau grouper during its spawning season. This case study emphasizes four MPAs that have had active management for several years, but also discusses other sites and recent developments. Belize's system of MPAs is an important component of the wider integrated coastal zone management program that is guiding marine conservation efforts in Belize.

## THE BELIZEAN FISHING INDUSTRY

The fishing industry in Belize is still characterized by small-scale commercial interests involving about 800 boats, some of them sailing boats, and employing between 2000 and 3000 fishermen. Most fishers are members of the five active fishing cooperatives, which purchase, process, and export fishery products. The fishery is essentially "open access," although there appear to be some traditional areas and grounds, the use of which warrants further study. Fisheries production is the country's third largest foreign exchange earner (Iioka 2001), with about 80 percent being exported, of which most goes to the United States. Exports of lobster, conch, finfish, wild-caught marine shrimp and shark were worth over $12 million in 2000; farmed shrimp has expanded so rapidly that it now surpasses the capture fisheries and generated $23 million in 2000.

Table 10.1 Exports (lbs. x 1000) of Marine Products

| Commodity | Average 1995–1998 | 1999 | 2000 |
|---|---|---|---|
| Whole fish | 240.7 | 60.9 | 73.9 |
| Fish fillet | 3.2 | 0.0 | 0.0 |
| Shark fins | 3.5 | 4.8 | 12.7 |
| Lobster tails | 488.5 | 566.0 | 646.1 |
| Lobster meat | 88.4 | 89.8 | 26.7 |
| Whole lobster | 61.3 | 0.0 | 0.0 |
| Aquarium fish | 100.3 | 49.9 | 6.1 |
| Pink sea shrimp | 49.8 | 33.7 | 33.9 |
| Conch | 421.3 | 392.9 | 517.4 |
| Crab | 8.3 | 11.8 | 4.0 |
| SUBTOTAL | 1465.3 | 1209.8 | 1320.8 |
| White farmed shrimp | 1786.6 | 4658.0 | 4945.6 |
| TOTAL | 3251.9 | 5867.8 | 6266.4 |

The full extent of the direct and indirect contribution of fishing to the economy is not known because fish sold in local markets are not recorded, and a growing volume of sea food products are sold directly to restaurants and hotels or illegally across international borders. Subsistence catch and consumption are also significant and similarly not recorded. Sport and recreational fishing are becoming increasingly important activities that target a range of species, including permit, tarpon, bonefish, snook, marlin, and tuna.

The principal fisheries are for conch and lobster, which constituted 90 percent of all fisheries exports (excluding farmed shrimp) in 2000 (Table 10.1; CZMAI 2000, 2001b). Lobster is the most lucrative and the 2000 exports, totaling 672,800 pounds were worth over $9 million. Conch exports have in general declined since the 1970s but in 2000 they reached a high of over 0.5 million pounds. Exports of finfish and marine shrimp have declined since 1989. Of the finfish that are exploited, grouper (Serranidae) and snapper (Lutjanidae) are of greatest commercial value and are fished mainly at the traditional spawning banks. Two of the four licensed aquarium fish collectors were active in 2000 with 6000 aquarium fish and invertebrates exported with a total value of over $46,000.

The status of commercial fish stocks in Belize is not known with any great certainty because catch and effort data have not been systematically collected and analyzed. Although Koslow et al. (1994) suggested that some elements of Belize's finfish fishery are underutilized, there is general consensus within the country that overfishing is occurring, at least for certain species (Carter et al.

1994b; McField and Wells 1996), particularly aggregating species like the Nassau grouper (Heymen 2001; Paz and Grimshaw 2001; Sala et al. 2001).

The pattern of peaks and troughs in lobster production over recent years suggests that this fishery is being exploited at, or just above, its maximum sustainable yield. The conch fishery has declined dramatically from the 1970s, when Belize was one of the largest producers in the world, exporting over 1.2 million lbs. in 1972. This fishery is presumed to be overexploited. The grouper spawning banks are heavily overfished, as elsewhere in the Caribbean, and at some sites such as Rocky Point it appears that spawning migrations have ceased (Carter et al. 1994b). Recent conservation efforts have focused attention on these spawning banks, several of which are already within protected areas. The Gladden Spit Marine Reserve was created primarily to conserve its unparalleled fish spawning and whale shark aggregations. The recently enacted legislation provides additional protection for the remaining viable spawning banks.

The fisheries sector is currently managed largely by traditional methods. All vessels and fishermen must be licensed, but at present the fee is so low ($12.50 per annum) that it plays an insignificant role in regulation of fishing effort. There are a number of gear restrictions, including the prohibition of scuba gear, poisons, and explosives for any form of fishing. The conch and lobster fisheries are regulated through minimum size limits, closed seasons, and, in the case of lobsters, protection of soft-shelled and berried individuals. There is a closed season for shrimp and efforts are under way to restrict the size of the fleet. Guidelines have been prepared for the aquarium fish industry, and the Fisheries Department has an in-house policy by which it regulates this. Professional sport fishing guides voluntarily adhere to a catch and release policy for tarpon, permit, bonefish, and snook, but this does not extend to billfishing tournaments. Bonefish is protected as a sport fish species by a ban on its sale and purchase. A similar law is being drafted for tarpon, permit, and snook.

The Conservation Compliance Unit of the Fisheries Department enforces fishery regulations. When first established, with funding from the United States Agency for International Development (USAID), was well equipped and provided an enforcement capability considerably greater than that found in most other Caribbean countries. Although government resources have proved insufficient to maintain this original level, the Unit remains active, conducting patrols while staying within its severely limited budget. Nevertheless, there is illegal fishing of undersized conch and lobster, and fishing also occurs in the closed seasons; illegal foreign fishermen are a major problem particularly in the south of the country.

## MARINE PROTECTED AREAS (MPAS) IN BELIZE

In Belize, MPAs can be designated under two pieces of legislation, the Fisheries Act and the National Parks Systems Act. Use of the term *marine reserve* in Belize is not consistent with the definition used elsewhere in this book and is not synonymous with *no-take*. Belizean marine reserves are designated under the Fisheries Act, zoned for multiple use, and administered by the Fisheries Department, sometimes in partnership with nongovernmental organizations (NGOs) via formal comanagement agreements. For clarity, throughout this chapter we will try to avoid confusion on this by adding *no-take* where appropriate, by capitalizing *Marine Reserve* when used in the Belize sense, and by limiting such use to proper names or references to actual named areas. There are currently eight such designated Marine Reserves: Hol Chan, Glover's Reef, Bacalar Chico, Caye Caulker, South Water Caye, Sapodilla Cayes, Port Honduras and Gladden Spit.

National parks, wildlife sanctuaries, natural monuments, and nature reserves are strictly no-extraction and are administered by the Forest Department or designated comanaging organizations. There are two natural monuments, Half Moon Caye and Blue Hole; two national parks, Laughing Bird Caye and Bacalar Chico; and two wildlife sanctuaries, the Corozal Bay (Manatee) Sanctuary and the Swallow Caye (Manatee) Sanctuary. In some cases, adjacent marine and terrestrial protected areas were designated and are managed jointly, as with the Bacalar Chico National Park and Marine Reserve, and the Caye Caulker Forest Reserve and Marine Reserve. The Caye Caulker Forest Reserve was declared under the Forests Act. Table 10.2 provides details of the thirteen MPAs (CZMAI 2001); the locations of both marine and coastal protected areas are shown in Figure 10.1.

In December 1996, seven of the MPAs were declared the Belize Barrier Reef Reserve System World Heritage Site, under the United Nations Educational, Scientific, and Cultural Organization's (UNESCO's) World Heritage Convention. These include Bacalar Chico Marine Reserve and National Park, Blue Hole Natural Monument, Half Moon Caye Natural Monument, Glover's Reef Marine Reserve, South Water Caye Marine Reserve, Laughing Bird Caye National Park, and Sapodilla Cayes Marine Reserve.

The Hol Chan Marine Reserve was especially successful in terms of visibly increasing fish populations and attracting overseas visitors, leading to widespread enthusiasm for establishing MPAs in communities with tourism potential. Given the small size of the country and the limited management resources

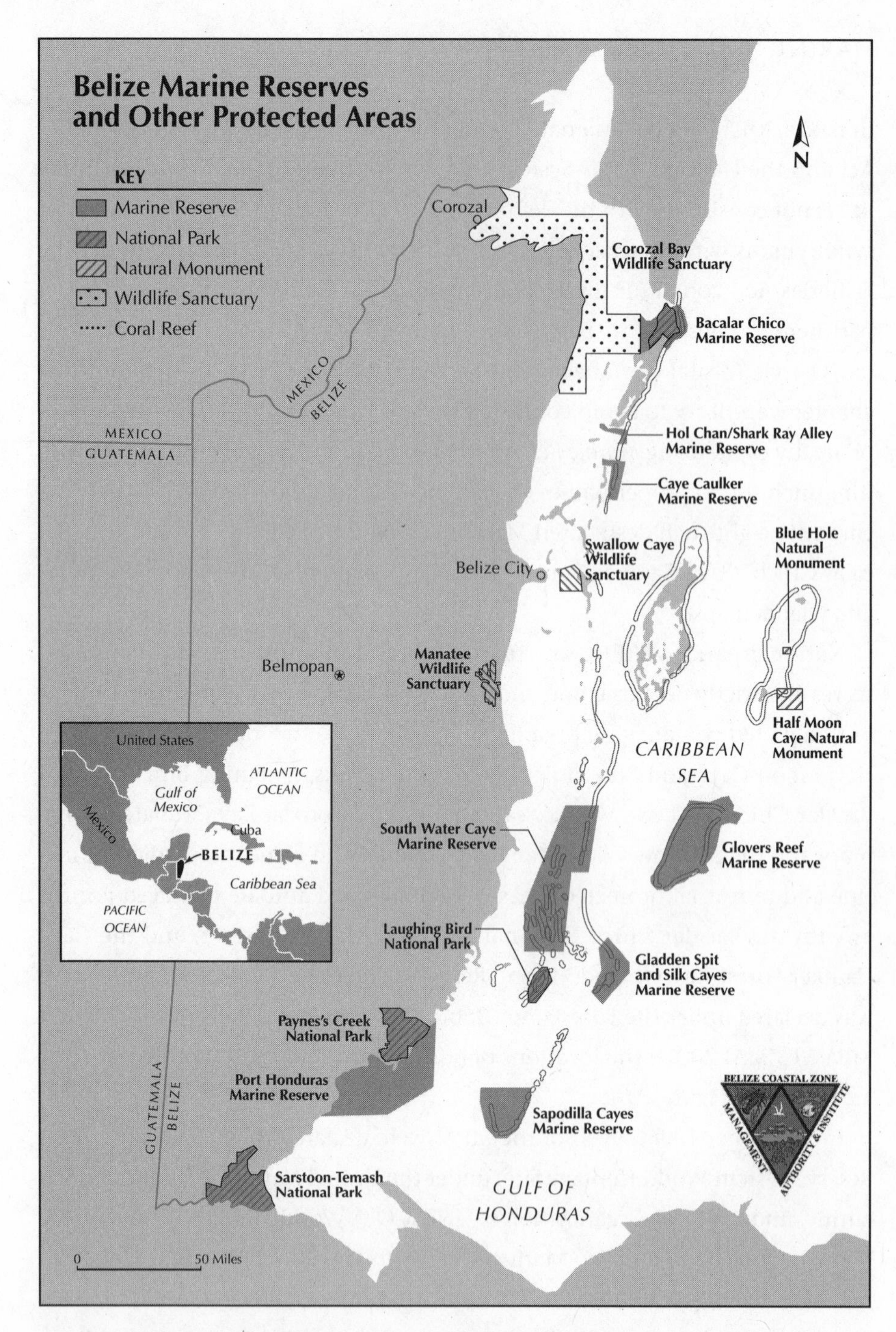


**FIG. 10.1 Belize Possesses a Network of 13 Marine Protected Areas (Not Including 11 Spawning Aggregation Sites Declared in 2002).** These include several marine reserves (In Belize, marine reserves are not all no-take), which have been established under the Fisheries Act and are administered by the Fisheries Department, and also wildlife sanctuaries, national parks, and natural monuments, which were declared under the National Parks System Act and are the responsibility of the Forest Department. Seven of the marine protected areas comprise the Belize Barrier Reef Reserve System World Heritage Site. Source: Belize Coastal Zone Management Authority & Institute.

available, Belize has a remarkably ambitious MPA program that has resulted in a network of protected sites along the length of the Barrier Reef, on two of the three atolls, and along the mainland coast. To date the existing MPA network system is believed to include virtually all the main and unique habitats, as recommended in the National Protected Areas Systems Plan (Programme for Belize 1995), with the exception of Turneffe Atoll.

Although most of the MPAs have a management presence, with basic infrastructure and equipment in place, management plans need to be completed or updated, and the MPA staff need training in all aspects of MPA management because many of them are newly appointed and inexperienced. A recent evaluation of management effectiveness of four MPAs (Hol Chan, Glover's, Bacalar Chico, and Half Moon Caye) found that overall management effectiveness was "moderately satisfactory," with common problems regarding administration and some management programs, like environmental education and research (McField 2000).

Funding support for the MPA system is presently being provided by donor agencies such as the Global Environment Facility/United Nations Development Programme Coastal Zone Management Project, the European Union, the GEF/World Bank Mesoamerica Barrier Reef System Project, The Nature Conservancy, World Wildlife Fund, and the Summit Foundation. However, a long-term financial strategy for the MPAs needs to be developed and implemented as soon as possible.

## DESCRIPTION OF SITES AND STUDIES IN BELIZE MPAS

### Hol Chan Marine Reserve

Hol Chan Marine Reserve, lying on the northern part of the Barrier Reef off the southern tip of Ambergris Caye, was established in 1987 with the support of the Wildlife Conservation Society and World Wildlife Fund U.S. It was established at the request of the local community in San Pedro to reduce conflict between fishermen and dive operators, and covers a transect across the barrier reef from the mangrove cayes at the southern end of Ambergris Caye to the outer reef. Before protection, it was one of the most productive and heavily fished cuts on the Barrier Reef (Carter et al. 1994a). Covering just 1545 ha (3816 acres), this protected area has experienced the longest period of full management and successfully enforces its regulations by having a warden on duty throughout each day. The reserve's overall management effectiveness was rated as satisfactory (76 percent) with the primary problems being administration

Table 10.2 List of Marine Protected Areas

| Name of MPA | Total Area (ha) | Marine Area (ha) | Marine No-Take Area (ha) | Year Established | Responsible Agency | Comanagement Organization |
|---|---|---|---|---|---|---|
| Bacalar Chico Marine Reserve and National Park* | 11,487 | 6303 | 1699 | 1995 | Fisheries Dept. and Forest Dept. | — |
| Blue Hole Natural Monument* | 414 | 414 | 414 | 1996 | Forest Dept. | Belize Audubon Society |
| Caye Caulker Marine and Forest Reserve | 3951 | 1545 | (zoning under preparation) | 1998 | Fisheries Dept. and Forest Dept. | Forest and Marine Reserve Association Association of Caye Caulker (FAMRACC) |
| Corozal Bay Wildlife Sanctuary | 73,050 | 72,350 | — | 1998 | Forest Dept. | — |
| Gladden Spit Marine Reserve | 10,523 | 10,463 | 153 | 2000 | Fisheries Dept. | Friends of Nature |
| Glover's Reef Marine Reserve* | 32,876 | 32,834 | 7226 | 1993 | Fisheries Dept. | — |
| Half Moon Caye Natural Monument* | 3954 | 3921 | 3921 | 1982 | Forest Dept. | Belize Audubon Society |
| Hol Chan Marine Reserve | 1545 | 1452 | 273 | 1987 | Fisheries Dept. | — |
| Laughing Bird Caye National Park* | 4095 | 4077 | 4077 | 1991 | Forest Dept. | Friends of Nature |
| Port Honduras Marine Reserve | 40,457 | 39,748 | 1277 | 2000 | Fisheries Dept. | Toledo Institute for Development and Environment (TIDE) |
| Sapodilla Cayes Marine Reserve* | 15,619 | 15,591 | (zoning under preparation) | 1996 | Fisheries Dept. | Toledo Association for Sustainable Tourism and Empowerment (TASTE) |
| South Water Caye Marine Reserve* | 47,703 | 46,833 | (zoning under preparation) | 1996 | Fisheries Dept. | — |
| Swallow Caye Wildlife Sanctuary | 3630 ha | 3230 ha | — | 2002 | Forest Dept. | Friends of Swallow Caye |

*Denotes the marine protected areas that make up the World Heritage Site.

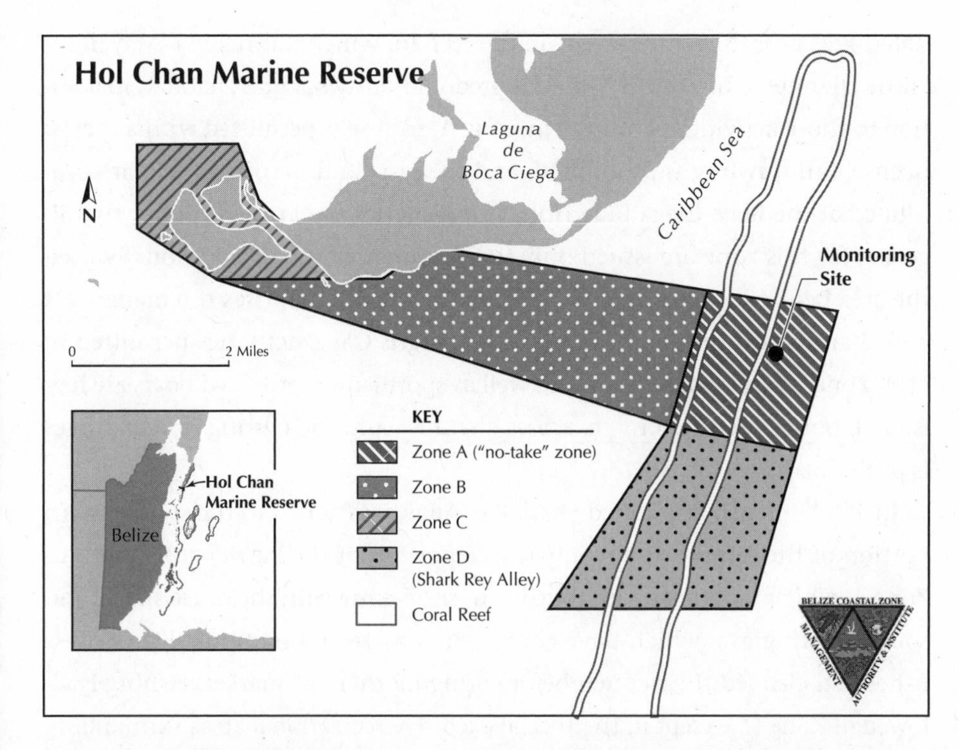


**FIG. 10.2 The Marine Reserve Has Been Zoned to Allow for Multiple-Use.** Zone A, the coral reef zone, is the fully protected area where no extraction is permitted. Traditional fishing is allowed in Zones B, C, and D, but there are some restrictions on gear and other special regulations. Zone D, added in 1999, provides for the protection for the Shark Ray Alley area and the management of visitors to this attraction. Source: Belize Coastal Zone Management Authority & Institute.

and threats (McField 2000). Threats include natural disasters, pollution, development, dredging, illegal fishing, and so forth. It is run by a "semiautonomous" staff within the Fisheries Department, which is financed primarily by a trust fund established through visitor fees. Despite its small size, Hol Chan has attracted much international attention as a "successful" MPA. Its principal appeal lies in the large fish (including groupers and snappers) that can be seen there, even by snorkelers. Hol Chan receives nearly 40,000 visitors a year, each paying an entrance fee of $2.50 and a separate fee for zone D of $3.50. which sustains basic operating costs. However, there are concerns about its carrying capacity, particularly given the concentration of visitors in the cut area.

The Marine Reserve originally included three zones, which reflect the main habitat divisions as well as different uses (Fig. 10.2). Zone A is almost 300 ha (741 acres) in area and comprises the segment of barrier reef that lies within the reserve. It is a no-take zone in which nonextractive recreational activities, such as snorkeling and diving, are permitted, and boats may moor at designated sites near the natural "cut" in the reef, for which it is named (Mayan for "little channel"). In Zone B, which is predominantly sea grass habitat, in addition to those activities permitted in Zone A, fishing is permitted with a special license, but trawling is prohibited, and spearing and netting of fish are prohibited at the Boca Ciega Blue Hole. It is Fisheries Department policy that licenses for this zone are issued only to fishermen who had traditionally used the area before it was protected. In Zone C, which comprises the mangroves and channels off the southern tip of Ambergris Caye, activities permitted in both Zones A and B are allowed, as well as sportfishing provided boats are registered, but the setting of nets across channels and the cutting of mangroves is prohibited.

In 1999 Zone D, also called Shark Ray Alley, was added to the southeastern portion of the reserve. It encompasses a portion of the barrier reef, south of Zone A, which is popular with tourists for snorkeling with the nurse sharks and southern stingrays, which have congregated there at a traditional site where fishermen cleaned their conch before bringing them to market. Fishing is allowed in Zone D, except in the special exclusive recreational areas surrounding the Shark Ray Alley and Amigos del Mar Dive Wreck sites. Scuba diving and feeding of the fish by tourists are not permitted in the Shark Ray Alley recreational area, although the snorkeling guides are allowed to feed the fish in this zone.

In 1988 and 1989, one to two years after the reserve had been established, Sedberry et al. (1992) compared Hol Chan cut with the Tres Cocos cut, a heavily fished cut about 5 miles to the north, as well as two other unprotected sites, Mexico Rocks and Rocky Point. Hol Chan had significantly more fish per observation than Tres Cocos; species diversity was also higher but not significantly so. The snappers, *Lutjanus griseus, L. mahogani,* and *Ocyurus chrysurus* were among the most abundant species at Hol Chan but were rare at Tres Cocos, as were groupers. The results were complex, but total abundance was higher at the protected site than at the unprotected sites. This was predominantly due to Nassau and black groupers (*Mycteroperca bonaci*), although graysby (*Cephalopholis cruentatus*) and snappers were also found to be more abundant in the protected area than in unprotected areas. Conversely, Acanthuridae, Scaridae, and small coney (*Cephalopholis fulvus*) were more abundant in unprotected areas, this abundance of herbivorous species suggesting that a prey-release effect had occurred.

A second study was carried out in 1991–1992 at Hol Chan, four years after protection was initiated (Polunin and Roberts 1993; Roberts and Polunin 1993). Sites at the Hol Chan cut, at two different depths, were compared with similar sites at three fished but similar cuts to the north (Mata Cut, San Pedro Cut, and Basil Jones Cut). Significant differences were found in fish abundance and diversity between habitat types and cuts at different depths for some but not all species. In Hol Chan, 45 percent of the target species showed abundance, size, or biomass increases. Several commercial species, such as school master snappers (*Lutjanus apodus*) had a higher frequency of very large individuals within the protected area. At the family level, differences were much clearer, with predatory families (groupers, snappers, and grunts) having greater biomass in Hol Chan.

Data were also collected by the reserve staff on lobster (*Panulirus argus*) and conch (*Strombus gigas*) populations about four to five years after protection (Hol Chan 1991a, 1991b, 1992a, 1992b). Data for lobsters, collected in June 1992 near the end of the lobster closed season, using timed-swimming counts of individuals per hour on the back reef, showed a significantly larger population in Hol Chan (30 lobsters/hr) than in similar habitats at Mexico Rocks (1.6 lobsters/hr) and Robles Point (2.4 lobsters/hr), which are unprotected areas to the north. Data for conch, collected in July–September 1991, gave similar results: averaging across all habitats, there were counts of 33 conch/$km^2$ in Hol Chan and 4 conch/$km^2$ at Mexico Rocks. The size distribution of conch also differed between the two sites, with mature conch (with flared, thick lips) accounting for 42 percent of the conch population on the reef flat and reef crest in Hol Chan (where this habitat lies in Zone A and is totally protected) and only 14 percent in the same habitat at Mexico Rocks. At both sites, the lagoonal and sea grass bed populations were dominated by juvenile conch, with only 10 percent mature conch at Hol Chan and 2 percent mature conch at Mexico Rocks.

The Hol Chan studies lack strong baseline data from before the reserve was established, which would further strengthen their results, although Sedberry and Carter (1993) studied the fish population of the mangrove habitat in Zone C before it was protected. Nonetheless, local people claim that fish populations in the reserve have greatly increased since protection (Carter et al. 1994a), and the scientists who have worked in the area have made qualitative observations that confirm this.

In 1998, Hol Chan was included as one of the protected area sites in a regional study to investigate the causes of the widespread increase in macroalgal cover on reefs (Williams and Polunin 2001). Findings of the study showed that there is a higher biomass of herbivorous fish inside MPAs like Hol Chan, and there is also a negative correlation between the macroalgal cover and the biomass of herbivorous fish in shallow reefs. However, at middepth reefs, which often have low cover of corals even inside protected areas, herbivorous fish populations seem unable to exclude macroalgal cover, which may have been previously controlled via sea urchin (*Diadema antillarum*) grazing, prior to its Caribbeanwide die-off (see earlier discussion).

### Half Moon Caye Natural Monument

The first marine habitat to be included in a protected area in Belize was at Half Moon Caye, on Lighthouse Reef atoll, which was designated as a natural monument in 1982 (a portion of the caye itself having been protected since 1928 due to its booby colony). Its total area is 3954 ha (9766 acres), of which 3921 ha (9685 acres) is marine habitat in which fishing is prohibited. Presently, there are no provisions for zoning of the marine habitat. The Belize Audubon Society (BAS) manages Half Moon Caye on behalf of the Forest Department. In the past, limited funding restricted active enforcement by BAS. However, regular visiting by dive boats and other tourists has tended to discourage fishing activities within the protected area. Volunteers from BAS and other organizations have carried out management activities such as beach cleanups, bird counts, marine resource monitoring, and construction of visitor facilities. Since 1996, however, with funding support from GEF/UNDP and the Summit Foundation, active management has been under way. Overall management effectiveness has been evaluated as satisfactory (77 percent), with the main problems being legal issues and threats such as hurricanes, coral bleaching, water pollution, visitor impact, and illegal fishing (McField 2000).

Between 1988 and 1990 Sedberry et al. (1992) compared sites in Half Moon Caye (six to eight years after legislative protection) with similar sites at Glover's Atoll, which at that time had not been designated a marine reserve. Forereef sites at Half Moon Caye, legally protected for six to eight years, were found to have significantly more fish per observation than similar unprotected sites at Glover's Reef. However, unlike Hol Chan, fish populations at Half Moon Caye were not strikingly abundant to the casual observer, and individuals were not unusually large. Less effective enforcement and lack of compliance at the time are among possible explanations for the failure to see more striking differences. Other sites on Lighthouse Reef atoll that are not protected, such as the Aquarium dive site, have a greater reputation for fish size and abundance, probably due to fish feeding by dive guides.

On-site management by BAS began in 1997, with increased enforcement in 1999. A survey by park staff in November–December 1999 found approximately 250 percent more lobster inside (five/hr searching) versus outside (two/hr searching) the park boundaries, with all large (greater than 30 cm) lobsters found inside the park (Belize Audubon 2001). Likewise, a July 1999 survey found approximately 650 percent more conch inside (30.3 conch/150 $m^2$ belt transect ) versus outside (4.6 conch/150 $m^2$ transect) the park boundaries (Belize Audubon 2001).

A study conducted in 2001 comparing herbivorous fish populations (Acanthurids and Scarids) at shallow reefs inside and outside Half Moon Caye Natural Monument found that the biomass of Acanthurids and total herbivorous fish surveyed was significantly higher inside the MPA (Sandman 2001). Inside the MPA there was also a higher proportion of large fish (Sandman 2001). A negative correlation between percent algal cover and fish biomass was also demonstrated (Sandman 2001). Although this study was rather limited in scope, it represents additional evidence supporting the theoretical ecological benefits (reduced algal cover) resulting from no-take zones and increased fish populations.

### Glover's Reef Marine Reserve

Glover's Reef Marine Reserve, only part of which is no-take, was established in 1993 with the assistance of Wildlife Conservation Society and the UNDP/GEF Coastal Zone Management Project. It has an area of 32,876 ha (81,204 acres), covering the entire atoll, and is zoned for multiple uses. There are six privately owned cayes on the atoll, on which are situated three resorts, a marine laboratory and the reserve headquarters, and a few small private houses. These are excluded from the reserve statutes, which do not apply to private land. The reserve is managed by the Fisheries Department and has "moderately satisfactory management" (68 percent), with the main problems related to its administration and management programs (McField 2000). The atoll suffered serious overfishing in the 1970s, by both Belizean and foreign fishermen, largely from Honduras. With increased human presence on the atoll and public awareness of the Marine Reserve, illegal fishing by aliens has declined, although poaching remains a problem. Several dive boats and commercial and recreational fishermen also visit the atoll.

The regulations for the reserve were gazetted in 1996 and then revised in 2001 in response to new data (Acosta 1998). Changes to the boundaries of the zones were made that reflect more accurately the location of the Seasonal Closure Area to encompass the grouper spawning bank, standardize the shape of

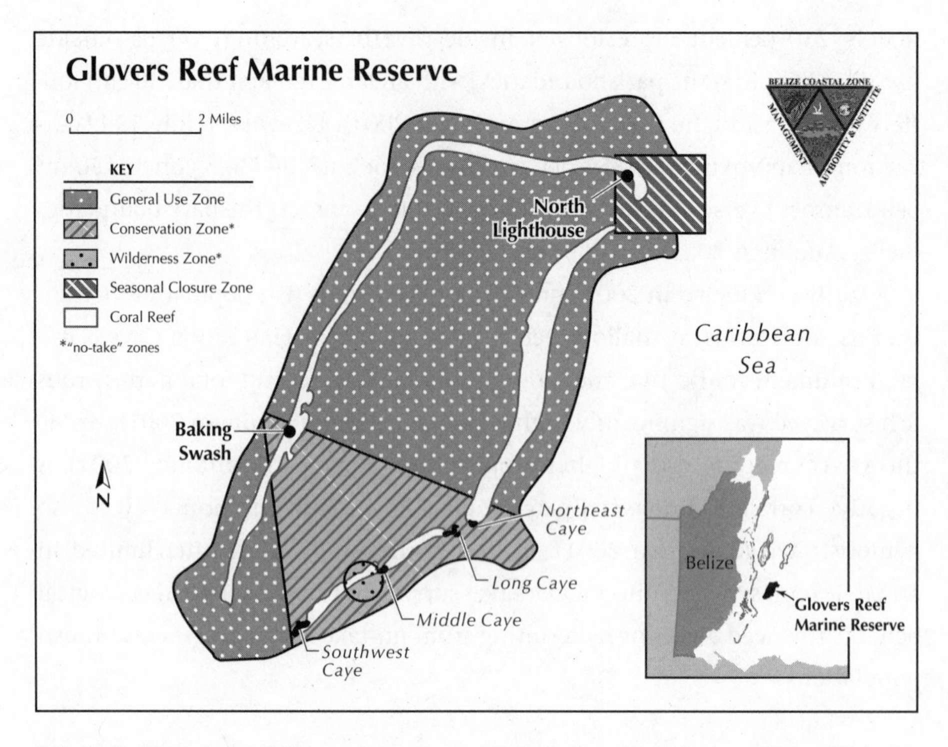


**FIG. 10.3 Glover's Reef.** The southernmost of Belize's three atolls is zoned for multiple use and possesses the largest fully protected area in Belize's marine protected area system. This no-take area includes the wilderness and conservation zones. Source: Belize Coastal Zone Management Authority & Institute.

the Conservation Zone for ease of demarcation and enforcement and to more effectively protect its populations of lobster and conch and place the Wilderness Zone in a more effective area, away from its previous location in a highly used operational zone centered around the north end of Middle Caye (Fig. 10.3). Fishing of any kind as well as diving and other water activities are prohibited in the wilderness zone or no-take area. In the Conservation Zone, which covers much of the southern third of the atoll, nonextractive recreational activities such as snorkeling and diving are permitted, and boats are required to use designated moorings and to register with the reserve. Subsistence fishing by island residents (these include staff of the two resorts, and marine reserve and research station personnel) and sportfishing (catch-and-release only) are also permitted under special license. In the General Use Zone, fishing, tourism, and other activities are permitted, but commercial fishing is restricted to those fishermen who currently use the area, a special license is required, and fish traps and nets are prohibited. The Seasonal Closure Area, which covers the grouper spawning bank, has been replaced with a no-take area, closed to all fishing.

A study designed to test the effectiveness of the no-take area of the reserve was initiated in 1994 (Carter 1994). Researchers and park staff used visual census methods to assess abundance and diversity of nineteen diurnally active fish families, including economically valuable species (e.g., snappers, groupers, and porgies), site-pecific species (e.g., damselfishes), transient species (e.g., jacks, barracudas), and different trophic levels (e.g., filefishes, grunts, surgeonfishes, parrotfishes, barracudas). The censuses, which also included conch and lobster, were carried out in three habitats (patch, backreef, and forereef) at complementary sites in the Wilderness Zone (no-take) and outside the boundary of the marine reserve. Local fishers assisted in the collection of commercial fishing data. With the reserve staff and the researchers, they helped to devise a data collection kit (consisting of waterproof data collection sheets with fish names in English and Spanish, a handheld spring balance, and a measuring tape) for each boat that uses the atoll. The legislating of the reserve regulations was critical to the progress of the study because it was based on a comparison of populations within the fished and nonfished areas. However, because this action was delayed by a couple of years, the investigation ultimately served as baseline data for the premanagement condition of fishery stocks.

In 1996, Acosta (1998) initiated a study on the movements of lobster and conch, and the tidal transport of recruits within the reserve. This was carried out to evaluate how the original boundaries of the no-take areas affected the function of the reserve. He concluded that the configuration of the Conservation Zone was only effective in protecting the more sedentary species, such as conch. However, the more mobile adult lobsters were capable of dispersing into the General Use Zone in a relatively short time. To make the protection more effective, he recommended some changes to the boundaries. The southwestern portion of the atoll was recommended for inclusion in the Conservation Zone to protect the highly productive conch grounds and an important area for larval recruits due to the southwesterly current regime. In addition the inclusion of the "cuts" (tidal or current passes) in the fringing reef and a reduction in the perimeter:area ratio of the Conservation Zone were recommended. The data collected during this study were used to develop a predictive framework to evaluate reserve effectiveness in protecting target species.

Acosta and Robertson (2001) followed up this initial study with further investigations in Glover's Reef and South Water Caye Marine Reserves and the Half Moon Caye and Blue Hole Natural Monuments. During his initial study, enforcement in the no-take areas was minimal at Glover's Reef, and the density of lobster and conch in the protected areas were not different from the fished areas. However, with increased enforcement starting in late 1998, he noted an increase in the density of conch and lobster in the protected area. Density of spiny lobsters increased by 300 percent, and biomass by over 700 percent, due mainly to growth of large adults in the area. The density of conch increased by over 200 percent. Key fish species were also monitored, with black grouper, mutton snapper, and hogfish twice as abundant in the protected area than in the fished area.

In preliminary comparisons between the MPAs, Acosta and Robertson (2001) noted that Blue Hole and Half Moon Caye had substantially higher densities as well as larger average sizes of lobsters and conch than in the adjacent fished habitats. In the case of South Water Caye, however, which has no enforcement yet of no-take areas, the populations of conch and lobsters consisted of small juveniles only, although their densities were high. This is indicative of severe recruitment overfishing. These comparative studies are providing valuable information for the development of a protocol of rapid ecological assessments of marine reserves, which will assist in economic cost–benefit analyses to evaluate no-take areas.

A year long study of the interactive role of MPAs and direct macroalgal reduction was conducted in 1997–1998 on the lagoonal patch reefs (McClanahan et al. 2001). This study found that the no-take zone had higher densities for thirteen of thirty fish species, and higher rates of fish herbivory than the general use zone. However, the two management zones did not differ in the effects of the experimental algal reduction. It was hypothesized that the higher fish populations in the no-take zone would prevent the return of high algal cover to these patch reefs. Between 1970 and 1996 these patch reefs experienced a community shift, with the coral:algal ratio changing from 4:1 to 0.25:1, with continued coral declines in recent years (McClanahan and Muthiga 1998; McClanahan et al. 2001). Thus this study indicated that the benefits of MPAs to the benthic reef community may require a longer timeframe to take effect or may be influenced by other uncontrolled confounding factors, such as the coral bleaching event and catastrophic hurricane that intervened during the course of this study. Results of more recent experiments indicate that high nutrient levels negatively influence brown frondose algal cover. They also suggest that coral mortality, due to disease and bleaching, and low herbivory due primarily to the *Diadema* die-off and over fishing of grazers, are the most likely factors responsible for the high levels of brown fleshy algae on the patch reefs of Glover's Reef (McClanahan et al. 2003).

## Bacalar Chico

The Bacalar Chico National Park and Marine Reserve lies 10 to 15 miles north of San Pedro, Belize's most popular tourist destination, and has long been used by commercial fishermen (from both San Pedro and Xcalak, a small Mexican town 5 miles to the north) and for sport and recreational fishing. Rocky Point was once a major spawning bank for Nassau and black grouper and is now a popular spot for sport fishermen. The reserve was established in 1996, with support of the International Tropical Conservation Foundation/European Union. The fisheries department manages the reserve, along with a local advisory committee that has remained active over the past seven years. The reserve has "moderately satisfactory management" (61 percent), with the main problems being administration, threats, and illegal uses (McField 2000).

The marine reserve's zoning plan was legally implemented in 2001 and includes a Preservation Zone lying immediately south of the Mexican border, which is a no-take area and in which recreational activities are also prohibited; a Conservation Zone 1, also a no-take area but where recreational activities are permitted; a Conservation Zone 2, which includes the Rocky Point area, and in which only catch-and-release sportfishing and other recreational activities are allowed; and General Use zones throughout the remainder of the reserve where fishing is permitted, but the use of gill nets, spearguns, and traps (other than beach traps established before or during 1996) is banned (Fig. 10.4).

Studies of fish, conch, and lobster populations are ongoing, using methods similar to those used in the other reserves. Although the no-take areas have just recently been legally established, reserve staff members have been informally enforcing the zone regulations for several years. Results of lobster surveys conducted at the opening of the lobster season have consistently shown that densities are higher in the protected areas compared to the fished areas (Alegria 2000; Gomez 1998). Catch per unit effort (CPUE) is being measured for beach traps. In 1995, over 16,000 pounds of fish were caught by nine beach traps (Gibson et al. 1996). In 1997, over 22,000 pounds were caught by eight traps, with the catch consisting of mainly yellow fin mojarra (*Gerres cinereus*), grey snapper (*Lutjanus griseus*), great barracuda (*Sphyraena barracuda*), and the bluestriped grunt (*Haemulon sciurus*; Gomez 1998).

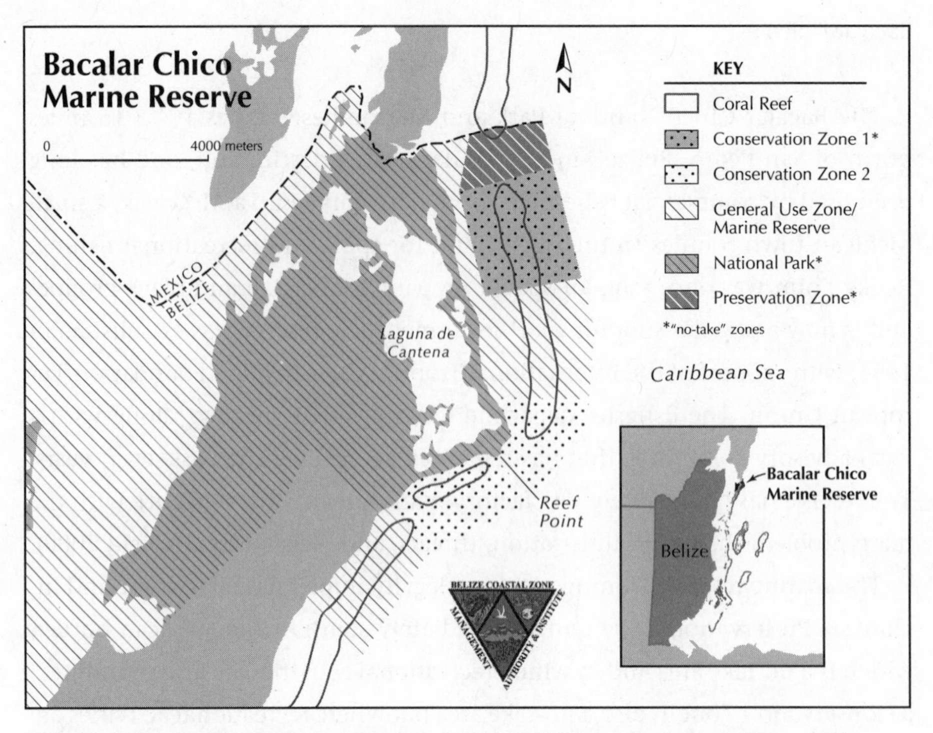


**FIG. 10.4 Bacalar Chico Marine Reserve, Located Offshore Ambergris Caye on the Border with Mexico, Is Zoned for Conservation, Recreational Activities, and Traditional Fishing.** The northern end of the caye comprises the Bacalar Chico National Park. Conservation Zone 1 and the Preservation Zone are both no-take. Source: Belize Coastal Zone Management Authority & Institute.

## Gladden Spit and Silk Cayes Marine Reserve

Fishermen of southern waters have fished at "the Elbow" or Gladden Spit since the 1920s, capitalizing on the healthy spawning aggregations of mutton snapper (*Lutjanus analis*) in April and May each year. These fishermen have also recognized that whale sharks (*Rhincodon typus*) frequent the area during spawning time. The area attracts the most predictable and dense aggregation of whale sharks in the world; whale sharks are attracted to the site to feed on the freshly released spawn from cubera (*Lutjanus cyanopterus*) and dog snapper (*Lutjanus jocu*; Heyman et al. 2001). It was further recognized that over twenty-five species aggregate at the site for spawning at various times of year (Heyman and Requena 2002). One mutton snapper tagged at Gladden traveled over 200 miles in less than three weeks, and another returned to the site two years after tagging, indicating long migration distance and site fidelity for spawning (Heyman, unpublished data).

A local community-based organization, the Friends of Nature (FoN), held a

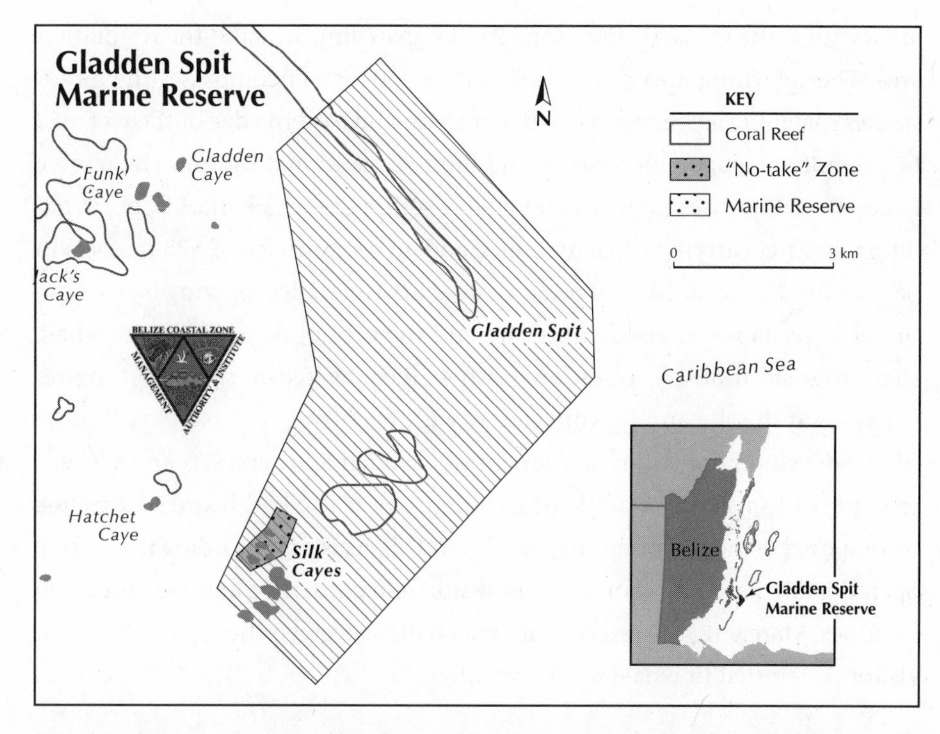


**FIG. 10.5 Gladden Spit and Silk Cayes Reserve and Spawning Aggregation Site Closures.** This marine reserve, spanning a large opening in the central portion of the barrier reef, Gladden Entrance, encompasses an important spawning aggregation site for mutton snapper and many other species. Only the marked area near the Silk Cayes is no-take. Source: Belize Coastal Zone Management Authority & Institute.

series of community consultations over a two-year period that culminated in the designation of the reserve on May 18, 2000 (Government of Belize SI [Statutory Instrument] #68 of 2000). FoN subsequently signed a Memorandum of Understanding with the Fisheries Department for the comanagement of the reserve, and the organization assumed management of Gladden Spit in January 2002. FoN now operates two vessels and has a biologist and four rangers on staff for the marine reserve. A number of local and international organizations, such as the Coastal Zone Management Authority and Institute (CZMAI), The Nature Conservancy (TNC), World Wildlife Fund (WWF), the Oak Foundation, and others, have assisted FoN with institutional strengthening, community consultations, planning, and reserve management.

The total reserve area is 10,523 ha (25,992 acres) of which 153 ha (378 acres), Conservation Zone 1 surrounding the Middle Silk Caye was designated as no-take, while the remainder is designated as a General Use Zone (Fig. 10.5). FoN has developed a draft management plan for the reserve and is sharing the plan with communities via public consultations. The plan includes two additional areas within the Conservation Zone 2—the spawning area and the restoration area. The spawning area of whale shark zone, which encompasses the multi-species spawning aggregation on the northern and eastern edge of the reef, will be carefully managed for tourism and limited traditional fishing. The reserve issues special licenses for boats and guides within the whale shark zone, in addition to strict carrying capacity limits, and an entrance fee of $25 per person per day for diving during the peak whale shark times. Recognizing its international importance as a spawning site and the economic value of the whale shark tourism industry, traditional fishers have agreed to severe fishing restrictions at the site and to work closely with scientists.

The development and management of Gladden Spit serves as an excellent example of comanagement by local communities. The FoN board of directors is composed of local community leaders, and all decisions regarding the reserve operation are based on thorough discussion and consensus amongst the communities. Managing the reserve will be a challenge, given the rapid increase in visitors interested in whale shark watching.

## PROTECTION OF SPAWNING AGGREGATION SITES

Most larger reef fish species aggregate to spawn at specific times and locations. This has long been recognized by fishers from Belize who have traditionally fished Nassau grouper (*Epinephelus striatus*) during the December and January full moon period at a variety of traditional spawning sites throughout Belize (Carter et al. 1994b). Glover's Reef Marine Reserve included seasonal protection for one such site, and Bacalar Chico had the area around its spawning bank at Rocky Point zoned to allow for only sport fishing and recreational activities (see earlier discussion). Investigations at Glover's Reef demonstrate that fishing during the spawning season by even a limited number of traditional hand-line fishermen can easily remove more than 10 percent of the spawning population; tag returns also show that spear fishing over the balance of the year takes a further 14 percent of the population (Sala 2002). Fisheries models indicate that this level of fishing is unsustainable; if continued, the population will disappear by 2013, with commercial extinction even sooner (Sala et al. 2001).

In 1998, investigations were initiated by TNC on the spawning aggregations of snapper at Gladden Spit, which led to year-round assessments at that site for the period between March 1998 and May 2002 (Heyman 2001). Recognizing the need for protection of the Nassau grouper, Green Reef Environmental Institute developed and implemented the first national synoptic assessment of spawning aggregations during the 2000–01 Nassau grouper season (Paz and Grimshaw 2001). This project also contained a major education and advocacy program, culminating in a national workshop, the creation of a Spawning Aggregation Working Group, and ultimately policy recommendations for national legislation. A similar national study was sponsored in January 2002, along with spot checks at various reef promontory spawning sites at varying times of year, with data from these studies compiled into a status report on multispecies spawning aggregations in Belize (Heyman and Requena 2002). It appears from the report that many reef promontory sites serve as multispecies spawning aggregation sites through much of the year.

Based on the need to protect Nassau groupers, a coalition of seven national and international NGOs ( BAS, FoN, Green Reef Environmental Institute, TNC, Toledo Institute for Development and Environment [TIDE], Wildlife Conservation Society, and WWF) jointly advocated for new legislation to protect Nassau grouper spawning aggregation sites, and in November 2002, two new laws were enacted. The first protects the Nassau grouper from December through March each year, the time when this species is known to breed; the second created eleven new no-take marine reserves at Nassau grouper spawning aggregation sites, which are also known to host multiple species of spawning aggregations throughout the year. Taken together, these two pieces of legislation provide unprecedented protection for aggregating finfish and serve as a model for other countries. Similarly, the development of the legislation represents unprecedented collaboration between conservation groups, the commercial fishing industry, and the government of Belize. To evaluate the effectiveness of the closures and the possible migration to and from these sites, the coalition of NGOs and the participating MPAs will continue to monitor the spawning aggregation sites in January and April each year.

## ECONOMIC ALTERNATIVES

Perhaps Belize's most effective tool for the development of marine reserves has been the use of economic alternatives for fishers. Marine reserves, by definition, restrict fishing activity. Although it is understood that the reserves will contribute to the long-term management of the fishery, fishermen are often more supportive of the reserves when offered alternative livelihoods in the short term. TIDE and FoN have been particularly cognizant of this fact and have included extensive economic alternatives training in their conservation programs. The most effective of these have been fly-fishing guide training and scuba diving guide training, which have led to the inclusion of many ex-fishermen into the guide business.

## LESSONS LEARNED

Conservation of ecosystem functions and biodiversity, and management of tourism and recreational activities, are the major roles of MPAs in Belize, and to achieve these objectives strict no-take zones may not be essential. However, the potential importance of marine refugia to fisheries management is recognized in Belize's fisheries legislation, with its provisions for the designation of marine reserves. Designation of MPAs often comes under the general legislation for establishment of protected areas. In the past, marine reserves were viewed as a response to tourism, and fisheries management policy focused primarily on enforcement of traditional management practices and diversification to different species and to aquaculture to reduce pressure on stocks that are being overexploited. However, recently there has been greater recognition of the role of marine reserves in fisheries management (Carter and Sedberry 1997), although some commercial fishing interests are not in full agreement with the concept. In the recently revised Fisheries Act (2003), the primary purposes for declaring marine reserves are to maintain commercial fishing, encourage research, and protect fish habitat (section 47 (2)).

Carter and Sedberry (1997) recommend that 30 percent of the coastal zone of Belize should be closed to fishing, the remainder to be managed by traditional methods. This recommendation is based on the work of the Plan Development Team (1990), which concluded that if 20 percent of the southern U.S. coastal waters were to be closed to fishing, a sustainable fishery in the remaining 80 percent of the waters would be achievable. The total marine area with statutory protection in Belize is about 238,761 ha (587,610 acres), but only about 19,040 ha (47,028 acres), or 8 percent, of these protected areas are strictly no-take. Thus less than 4 percent of the estimated 500,000 ha (1,235,000 acres) of reef and sea grass habitat, or 1 percent of the country's territorial waters (1,881,400 ha [19,467,058 acres]), are in no-take zones.[1] Closure to fishing of 30 percent of just the reef and sea grass habitats (as opposed to the entire territorial waters) would involve the creation of no-take zones over a total area of some 150,000 ha (370,500 acres).

[1] These figures do not include the recently protected spawning aggregation sites.

Traditional fishery management methods will continue to play an important role. It will be important to have a range of management methods, and certain aspects of the fishery will still need the species approach to management. MPAs, including no-take marine reserves, should be viewed not as a replacement for such methods and approaches, but rather as complementary. The recent revision of the legislation previously discussed recognizes this. The act requires the preparation of management plans for prime commercial species, and newly passed legislation creates a seasonal closure to protect Nassau grouper and creates a network of small no-take areas to protect key reef fish spawning sites. Since the snapper and grouper (reef fish) fishery is seasonal and site specific, these new seasonal and area closures should be highly effective, particularly where such areas can be incorporated within an MPA and thus have regular surveillance. This was investigated and some boundary adjustments were made to ensure inclusion of entire spawning sites. These closures will be most effective if implemented in conjunction with other traditional fishery management approaches. The shrimp fishery and sportfishing industries may also benefit from specific regulations such as those on gears and closed seasons.

The data available at present are insufficient to conclusively demonstrate that the existing marine reserves contribute significantly to the sustainable management of Belizean fisheries. However, they indicate several potential benefits. The research at Hol Chan, Glover's Reef, and Half Moon Caye illustrate that MPAs protected from fishing enhance fish stocks, and significant differences can appear within two to four years after closure of the no-take zones. In these protected sites, abundance and size of certain species of fish, conch, and lobster were higher than in other sites (Acosta 1998; Carter and Sedberry 1997; Hol Chan Marine Reserve 1991a, 1991b, 1992a, 1992b; McClanahan et al. 2001). Such populations, particularly given the larger sizes and resulting greater reproductive output, could act as breeding stock and result in improved recruitment outside the no-take zones. It can also be presumed that total protection in MPAs of habitats important to certain life stages of commercial species will also benefit the fishery. Once Belize's network of MPAs is completely established and enforced, a significant area of mangroves, sea grass beds, and other habitats important for juvenile commercial species will be protected and could be expected to increase populations. However, studies confirming this or showing higher yields in adjacent areas have not been collected yet, and more research is required on the extent to which populations, particularly of relatively sedentary species such as conch and lobster, migrate out of protected areas. The role of no-take zones in the management of the shrimp and aquarium fisheries and in sport and recreational fishing, other than the general direct benefits described following here, will need further study.

It is also expected that MPAs may lead to a reduction in fishing effort by providing alternative sources of income. This is already being seen in San Pedro, where tourism has increased dramatically since the establishment of Hol Chan Marine Reserve, which now receives almost 40,000 visitors a year. The large fish in the reserve are undoubtedly one of the main attractions and can be equated with elephants in Africa because visitors have an almost guaranteed certainty of seeing large fish at a location such as Hol Chan.

The MPAs in Belize can greatly facilitate Conservation Compliance Unit (CCU) operations and the enforcement of fisheries regulations, as has already been demonstrated with Hol Chan, Glover's Reef, and Bacalar Chico. The resources of CCU are insufficient to police the entire Belize exclusive economic zone. MPA rangers have certain enforcement powers, and fisheries officers may only need to be called out in special circumstances. Furthermore, because of the focus on one area, reserve staff can develop closer relationships with the local fishermen who use the area, helping to educate them in the rationale for the MPA and additional fisheries regulations and thus ultimately reducing the need for surveillance and enforcement. MPAs adjacent to international borders can also play an important role in reducing illegal harvesting by foreign fishermen. This has already been noticed at the Bacalar Chico Marine Reserve, where the presence of reserve staff has reduced incursions by Mexican fishermen. Similarly, the Glover's Reef and Port Honduras Marine Reserves near the southern border have helped reduce illegal fishing by Honduran and Guatemalan fishermen. MPAs must be effectively managed and enforced to fully attain this potential benefit, and some parks have not yet attained satisfactory management. However, the enforcement or protection programs generally received higher scores, although there was wide variability between MPAs (McField 2000).

MPAs may also play an important educational role in the context of fisheries management. The establishment of MPAs in Belize involves holding public forums, which are invariably attended by fishermen whose livelihoods are potentially affected by the reserve. Representatives of the fishing community also sit on the advisory or management committee, which is generally established to assist with the development and management of the MPA. These processes are designed to ensure that fishermen are generally well informed about the purpose of the reserve and that they develop an understanding of the rationale behind various fishery management strategies. They are also encouraged to participate in the preparation of the management plan for the area, a process that generally ensures much greater cooperation and willingness to obey regulations once these are in force. However, there may not be adequate flow of information from the representatives on these committees to the active fishermen in the reserves. Recently there have been several instances of groups of fishermen calling for the dereservation of MPAs or reductions of their no-take zones.

Education of other marine users, such as tourists and residents, may also have a long-term indirect benefit, enhancing the conservation ethic, and potentially reducing demand for out-of-season or undersized fisheries products. Over the past couple of years there has been a growing move toward comanagement of the MPAs by NGOs and community-based organizations. Presently, comanagement agreements are in place for seven MPAs (see Table 10.2). This type of comanagement arrangement provides a mechanism for community involvement not only in the planning for MPAs but also in their day-to-day management. Such comanagement arrangements were also found to be more effective than direct governmental management of the reserves, although this was based on only four MPAs (McField 2000).

The hypothesized benefits of MPAs, particularly the no-take zones on benthic reef community structure, have not been adequately addressed by research. McField (2001) analyzed the influence of environmental and management-linked variables (including protected areas status) on reef community structure for twelve moderate-depth forereef sites, including seven sites within MPAs under various levels of active management. This work applied a multivariate analysis technique of Clarke and Ainsworth (1993) that links the matrices of biotic community data and the environmental and/or management linked variables to determine the selection of variables that best fit or explain the community patterns. The results indicate that the MPA status was among the most influential factors shaping reef community structure prior to the catastrophic disturbances of 1998, after which the best fit was attained with only the fluvial variable, indicating that run-off associated with Hurricane Mitch had a greater role in shaping the postdisturbance communities. Although more testing of this method is recommended, this approach and others are needed to holistically examine the ecological effects of the MPAs system. The few other site-specific studies of benthic MPA effects that have been conducted have produced somewhat mixed results. Sandman (2001) and Williams and Polunin (2001) reinforced the hypothesized benefit of reduced macroalgal abundance correlated with increased herbivore biomass at shallow reef sites in Half Moon and Hol Chan, respectively. However, Williams and Polunin (2001) found that the results did not hold for intermediate-depth forereef sites. Furthermore, McClanahan et al. (2001) did not find any "beneficial" effect of the no-take zone in shallow patch reefs in the Glover's Reef Marine Reserve, although there were confounding disturbances. Clearly, this is an important area of research that requires further attention.

Research in Belize has reinforced the priority areas for further research that have also been identified in the literature (Roberts, 1995; Roberts and Polunin 1991; Wells 1995), including the following:

- The movements of fish and invertebrates in and out of reserves and the potential of reserves to export fish to adjacent fished areas, enhancing yields or catches.
- The level of fish catch per unit effort inside MPAs and in adjacent areas over time following enforcement of regulations.
- Larval dispersal in relation to prevailing currents (to identify the sources of recruits to reserves and to fishing stocks) and the fate of progeny spawned in reserves; this may influence the choice of location of no-take zones and provide some guidance as to how effective the proposed network of MPAs will be in Belize and in the wider Mesoamerican reef system; this may require a search for genetic markers and work on both local and large-scale oceanographic processes.
- Optimum sizes for no-take zones; in the absence of relevant information, recommendations have been made that, where a protected area is large enough, two or more no-extraction zones should be established as replenishment areas, and that these should be as large as possible.
- Impact of fish feeding and diver tourism within the MPAs, particularly at locations such as Hol Chan where both are important issues.
- Impact of changing the trophic structure of fish communities; for example, if carnivorous species increase in MPAs at the expense of herbivores, including sea urchins, will this result in an increase in algal cover and deteriorating corals?
- Comparison of genetic diversity in fished and unfished areas.
- Relationship of no-take zones and increased fish abundances on benthic community structure, particularly on the hypothesized reduction in macroalgal cover.

Equally important is the establishment of long-term monitoring and data-gathering programs, including the compilation of locational catch per unit effort data on important fisheries. Despite several efforts, such a fisheries data management program has not become fully successful. The recent opposition to the enforcement of additional no-take zones within established MPAs attests to the importance of working closely with commercial fishermen to adequately demonstrate the effectiveness of these zones at enhancing fisheries yields outside the MPAs. The importance of collecting accurate catch and effort data cannot be overestimated, and methods that are simple to use, easily replicated, and involve local fishermen and park staff should be used. Belize's network of MPAs provides a unique opportunity to carry out a long-term, large-scale study on the role of marine refugia in fisheries management, using a variety of methods. Finally, given the numerous threats to coral reefs emanating from local, regional, and global anthropogenic sources, including from global climate change, the potential ecological benefits of MPAs are paramount and should be more adequately addressed through scientific investigations.

## ACKNOWLEDGMENTS

We are extremely grateful to Dylan Gomez, Gilbert Richards, Julie Robinson, Sergio Hoare, and Ian Gillett for their assistance in providing data for this chapter.

## REFERENCES

Acosta, C. A. 1998. *Population Dynamics of Exploited Species and "Reserve Effects" in the Glover's Reef Marine Reserve, Belize.* Annual Progress Report to the Wildlife Conservation Society Project, Belize.

Acosta, C. A., and Robertson, D. N. 2001. *Population Assessments of Exploited Species at Glover's Reef Marine Reserve, Belize, with Preliminary Comparisons to the South Water Cay Marine Reserve and Lighthouse Reef Natural Monuments.* Annual Progress Report to the Wildlife Conservation Society, Belize.

Alegria V. 2000. *Bacalar Chico National Park/Marine Reserve Annual Report 2000.* Belize: Fisheries Department.

Belize Audubon Society. 2001. Presentation of research findings to the Lighthouse Reef advisory committee. Belize City, Belize.

Carter, J. 1994. A test of the effectiveness of designated fisheries reserves as tools to enhance and/or sustain yields of tropical marine resources in Belize. Research Project Document to GEF/UNDP Coastal Zone Management Project, Belize.

Carter, J., and G. R. Sedberry. 1997. The design and use of marine fishery reserves as tools for the management and conservation of the Belize Barrier Reef. *Proceedings of the 8th International Coral Reef Symposium* 2:1911–1916.

Carter, J., J. Gibson, A. Carr, and J. Azueta. 1994a. Creation of the Hol Chan Marine Reserve in Belize: A grass-roots approach to barrier reef conservation. *Environmental Professional* 16:220–231.

Carter, J., G. Marrow, and V. Pryor. 1994b. Aspects of the ecology and reproduction of Nassau Grouper, *Epinephalus striatus,* off the coast of Belize, Central America. *Proceedings of the Gulf and Caribbean Fisheries Institue* 43:65–11.

Clarke, K. R., and M. Ainsworth. 1993. A method of linking multivariate community structure to environmental variables. *Marine Ecological Progress Series* 92:205–219.

Coastal Zone Management Authority and Institute (CZMAI). 2000. *State of the Coast Report 1999.* Belize: CZMAI.

———. 2001a. Draft national integrated coastal zone management strategy for Belize. Belize: CZMAI.

———. 2001b. *State of the Coast 2000 Belize.* Belize: CZMAI.

Gibson, J., J. Azueta, D. Gomez, and M. Somerville. 1996. *Bacalar Chico Final Report to International Tropical Conservation Foundation.*

Gomez, D. 1998. *Bacalar Chico 1997 Report to the International Tropical Conservation Foundation.*

Heyman, W. D. 2001. *Spawning Aggregations in Belize.* Belize: The Nature Conservancy.

Heyman, W., and N. Requena. 2002. *Status of Multispecies Spawning Aggregations in Belize.* Technical Report. Punta Gorda, Belize: The Nature Conservancy.

Heyman, W. D., R. T. Graham, B. Kjerfve, and B. Johannes. 2001. Whale sharks, *Rhincodon typus,* aggregate to feed on fish spawn in Belize. *Marine Ecological Progress Series* 215:275–282.

Hol Chan Marine Reserve. 1991a, 1991b, 1992a, 1992b. Semi-annual staff reports to World Wildlife Fund, Washington, D.C., and the Belize Fisheries Department, Belize City.

Iioka, C. 2001. *Aquaculture Development in Belize.* Japan International Cooperation Agency. Belize City.

Koslow, J. A., K. Aiken, S. Auil, and A. Clementson. 1994. Catch and effort analysis of the reef fisheries of Jamaica and Belize. *Fisheries Bulletin* 92:737–747.

McClanahan, T. R., and N. A. Muthiga. 1998. An ecological shift in a remote coral atoll of Belize over 25 years. *Environmental Conservation* 25:122–130.

McClanahan, T. R., M. McField, M. Huitric, K. Bergman, E. Sala, M. Nyström, I. Nordemer, T. Elfwing, and N. A. Muthiga. 2001. Responses of algae, corals and fish to the reduction of macro algae in fished and unfished patch reefs of Glover's Reef Atoll, Belize. *Coral Reefs* 19:367–379.

McClanahan, T. R., E. Sala, P. A. Stickels, B. A. Cokos, A. C. Baker, C. J. Starger, and S. H. Jones IV. 2003. Interaction between nutrients and herbivory in controlling algal communities and coral condition on Glover's Reef, Belize. *Marine Ecology Progress Series* 261:135–147.

McField, M. 2000. *Evaluation of Management Effectiveness Belize Marine Protected Area System.* Prepared for Coastal Zone Management Authority and Institute, Belize City. August 2000.

———. 2001. A multivariate approach to determining key environmental and management-linked influences on coral reef community structure in Belize. In *The Influence of Disturbance and Management on Coral Reef Community Structure in Belize*, chap. 6. Ph.D. diss., College of Marine Science, University of South Florida.

McField, M., and S. M. Wells. 1996. *The State of the Coastal Zone Report, Belize 1995*. Coastal Zone Management Programme, Government of Belize, Belize City.

Paz, G., and T. Grimshaw. 2001. *Status Report on Nassau Groupers for Belize, Central America*. Scientific report of the Green Reef Environmental Institute. San Pedro, Ambergris Caye, Belize.

Plan Development Team (PDT). 1990. The potential of marine fishery reserves for reef fish management in the U.S. southern Atlantic. NOAA Technical Memorandum NMFS-SEFC-261.

Polunin, N. V. C., and C. M. Roberts. 1993. Greater biomass and value of target coral-reef fishes in two small Caribbean marine reserves. *Marine Ecolological Progress Series* 100:167–176.

Programme for Belize. 1995. *National Protected Areas System Plan*. Report to Government of Belize and Natural Resources Management Program.

Roberts, C. M. 1995. Effects of fishing on the ecosystem structure of coral reefs. *Conservation Biology* 9(5): 988–993.

Roberts, C. M., and N. V. C. Polunin. 1991. Are marine reserves effective in management of reef fisheries? *Rev. Fish. Biol. Fish.* 1:65–91.

———. 1993. Marine reserves: Simple solutions to managingcomplex fisheries? *Ambio* 22(6):363–368.

Sala, E. 2002. *Conservation of Grouper Spawning Aggregations in Belize*. Scientific report to the Belize Fisheries Department. July 2002.

Sala, E., E. Ballesteros, and R. M. Starr. 2001. Rapid decline of Nassau grouper spawning aggregations in Belize: Fishery management and conservation needs. *Fisheries* 26:23–30.

Sandman, A. 2001. *The Value of Grazers for Fisheries and Tourism at the Lighthouse Reef Atoll, Belize*. Degree project thesis, vol. 13, Department of Systems Ecology, Stockholm University.

Sedberry, G. R., and Carter, J. 1993. The fish community of a shallow tropical lagoon in Belize, Central America. *Estuaries* 16(2):198–215.

Sedberry, G., J. Carter, and P. Barrick. 1992. A comparison of fish communities between protected and unprotected areas of the Belize Reef ecosystem: Implications for conservation and management. *Proceedings of the Gulf and Caribbean Fisheries Institute* 45: 95–127.

Wells, S. M. 1995. Science and management of coral reefs: Problems and prospects. *Coral Reefs* 14:177–181.

Williams, I. D., and N. V. C. Polunin. 2001. Large-scale associations between macroalgal cover and grazer biomass on mid-depth reefs in the Caribbean. *Coral Reefs* 19:358–366.

ELEVEN

# Global Review: Lessons from around the World

> There is an emerging consensus among fisheries scientists and managers throughout the world that marine [no-take] reserves, if well-placed and of the appropriate size can achieve many of the goals that fishery management has failed to achieve using conventional methods. Particularly, there is overwhelming evidence from both temperate and tropical areas that exploited populations in protected areas will recover following the cessation of fishing and that spawning biomass will be rebuilt. Also, there is widespread recognition throughout the world that loss of biodiversity is largely driven by ecosystem modifications and the habitat loss that ensues. Hence, preserving biodiversity implies the maintenance or reestablishment of the natural ecosystems as in marine reserves in which no extractive anthropogenic effects are allowed or are minimized.
>
> —Roberts et al. 1995

Since the preceding conclusions were reached by a distinguished group of international scientists several years ago, consensus continues to emerge. Though the verdict is not yet unanimous, an increasing number of scientists, managers, stakeholders, plain folks, conservationists, and others are looking at the same basic facts, reviewing the global experience, applying common sense, and, inevitably, reaching the same conclusions. The more information people have, the stronger this consensus will become. Marine reserves have now been implemented around the globe, in many countries, at numerous locations. Most are small, many extremely so, and the percentage of the oceans they cover remains a tiny fraction of a percent. Nonetheless, documentation of their many and diverse benefits continues to proliferate.

In this chapter, we provide an overview of the global experience with lessons from sites scattered throughout the world. The examples discussed are by no means intended to be all-inclusive or comprehensive, but rather to highlight key points, trends, and issues. Many of these sites, examples, and points have been briefly discussed earlier in the book, but are discussed more fully here and with a greater place-based focus. They complement, amplify, and reinforce lessons from the preceding, more comprehensive and in-depth case studies.

## NEW ZEALAND

The first clear proposal to develop marine reserves in New Zealand, dating back to 1965, centered initially on the waters surrounding the University of Auckland's Leigh Marine Laboratory (Ballantine 1991). In response, the New Zealand Parliament enacted the Marine Reserves Act of 1971 a few years later providing a vehicle to create such reserves to protect areas in a natural state for scientific studies (NZDOC 2002). In the three decades since, New Zealand created the Leigh Marine Reserve, established fifteen additional marine reserves, and emerged as a global leader with respect to marine reserve thinking, development, creation, and assessment. Much can be learned from New Zealand's thirty-year experience with reserves that can be adapted and applied elsewhere.

According to the noted anthropologist, Margaret Mead, "Never believe that a few caring people can't change the world. For, indeed, that's all who ever have." In New Zealand, a small core group of committed people, focused initially on the Leigh reserve and later on marine reserves more generally, shaped the future of marine reserves there and influenced their development around the world. Among the most influential of this group was Bill Ballantine, a new recruit to and professor at Leigh when the reserve was initially proposed. Ballantine (1991) argued that New Zealand should lead with respect to marine reserves, given its position at the center of the "water [Southern] Hemisphere," long coastline (15,000 km [9300 mi]), rich marine traditions, and other relevant factors. Consistent with this, and despite the Leigh reserve's small size and slow development, the principles, lessons, and thinking that emerged from its creation greatly impacted the development of reserves not only throughout New Zealand but also around the globe.

A dozen years passed from the initial proposal to create a marine reserve near Leigh until the reserve, formally known as the Cape Rodney to Okakari Point Marine Reserve, was gazetted in 1977. Located near the Hauraki Gulf, 100 km north of Auckland by road, the reserve covers just 5.5 $km^2$ of rocky reef, kelp

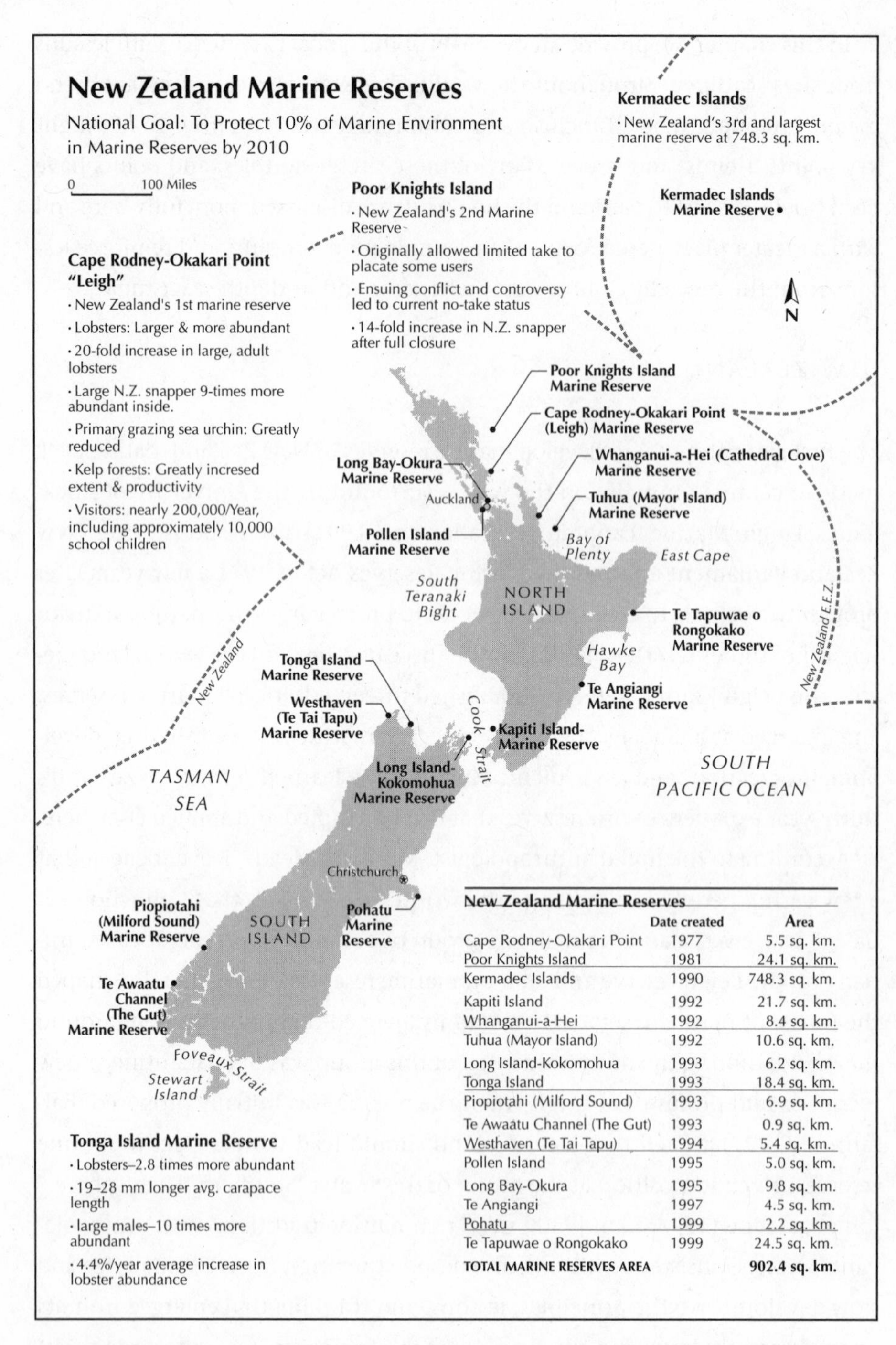


**New Zealand Marine Reserves**

| | Date created | Area |
|---|---|---|
| Cape Rodney-Okakari Point | 1977 | 5.5 sq. km. |
| Poor Knights Island | 1981 | 24.1 sq. km. |
| Kermadec Islands | 1990 | 748.3 sq. km. |
| Kapiti Island | 1992 | 21.7 sq. km. |
| Whanganui-a-Hei | 1992 | 8.4 sq. km. |
| Tuhua (Mayor Island) | 1992 | 10.6 sq. km. |
| Long Island-Kokomohua | 1993 | 6.2 sq. km. |
| Tonga Island | 1993 | 18.4 sq. km. |
| Piopiotahi (Milford Sound) | 1993 | 6.9 sq. km. |
| Te Awaatu Channel (The Gut) | 1993 | 0.9 sq. km. |
| Westhaven (Te Tai Tapu) | 1994 | 5.4 sq. km. |
| Pollen Island | 1995 | 5.0 sq. km. |
| Long Bay-Okura | 1995 | 9.8 sq. km. |
| Te Angiangi | 1997 | 4.5 sq. km. |
| Pohatu | 1999 | 2.2 sq. km. |
| Te Tapuwae o Rongokako | 1999 | 24.5 sq. km. |
| **TOTAL MARINE RESERVES AREA** | | **902.4 sq. km.** |

**FIG. 11.1 New Zealand Marine Reserves Map, Table, and Key Points.** This map and table show New Zealand's growing marine reserves network as of 2003. Two additional sites were recently added bringing the total number of reserves to eighteen. Key points highlight noteworthy aspects of specific reserves. Sources: NZDOC (2003), NZMSS (2002), Davidson et al. (2002), Kelly et al. (2000), Babcock et al. (1999), Ballantine (1999, 1991).

forest, and other habitats; includes several kilometers of coastline; and extends about 0.5 km offshore (Fig. 11.1). It is now home to increasingly large, abundant, and diverse marine life and expanding kelp forests that are extremely popular with a broad range of human visitors. It is also a hotspot for scientific research, monitoring, and education. Furthermore, it is strongly supported by the local community and many local fishers who believe it benefits adjacent fisheries.

The Leigh reserve's long gestation period, with many labor pains, and its need for sustained commitment along the way, are attributes shared by many successful marine reserves, especially the first ones in an area. The return on a successful delivery is shared by most. Ballantine (1991) details the road traveled and events leading to the establishment of this reserve well. Among the most critical to success were the building of strong, broad, and essential public support; developing appropriate legislation and the political will to enact it; and the decision by proponents not to compromise their principles by allowing some fishing within the reserve (see also following text), even though this required additional public education and process and delayed the designation.

The story of New Zealand's second marine reserve around the Poor Knights Islands (see Fig. 11.1), designated in 1981, takes a different route and provides a different lesson, but eventually reaches a similar conclusion. The Poor Knights Islands Marine Reserve now fully protects a remote 24.1 $km^2$ scenically spectacular and biologically unique area surrounding these islands from all extractive activities, including fishing, but this was not always the case. The initial approach taken in establishing this marine reserve differed markedly from that taken to protect the relatively ordinary stretch of coast within the Leigh reserve. This approach, similar to that used by a majority of marine protected areas (MPAs) worldwide, focused on avoiding controversy, limiting public debate, and regulating only activities determined by consensus to be demonstrably destructive. Since many believed that some types of fishing were not causing harm, complex regulations allowed some fishing within the reserve when it was created. This seemed to work at first, but serious conflicts emerged over time between visitors who expected a protected reserve and others who wanted to catch more and bigger fish. The initial regulations became difficult to explain, justify, and enforce. In 1997, sixteen years after its creation, Poor Knights became a no-take marine reserve, following multiple public processes, a court injunction, and heated political controversy (Ballantine 1999, 1991). Within three years of implementation, adult snapper abundance increased over 14-fold, average size also increased, and other changes in fish community composition were observed (Denny et al. 2003).

Although it took some time and hard knocks, New Zealand seems to have gleaned the following from the experience with its first two marine reserves and those created subsequently: (1) marine reserves are good public policy; (2) their primary goal should be to protect biodiversity and restore and maintain a natural marine environment; (3) consequently, marine reserves should be fully no-take and no-fishing; (4) such marine reserves provide numerous public benefits in areas of conservation, scientific research, recreation, and education in addition to benefiting fisheries; (5) New Zealand needs more and larger marine reserves; (6) a national system or network of reserves is desirable and should include sites representative of both typical and rare or unique ecosystems and communities; (7) a minimum of 10 percent of New Zealand's marine environment should be protected by 2010; (8) people should be free to enter and enjoy marine reserves subject to ensuring their natural values are not harmed; (9) the process for establishing reserves should be improved, streamlined, open, and transparent; and (10) the public should be involved in marine reserve creation and management (Ballantine 1999; NZDOC 2004).

New Zealand currently has eighteen marine reserves ranging in size from several less than a $km^2$ to the Kermadec Islands Marine Reserve covering nearly 7500 $km^2$ (see Fig. 11.1). Roughly twenty additional sites are under various levels of development or consideration. The New Zealand Department of Conservation (NZDOC), created in 1987, now has primary management responsibility for marine reserves. Protecting 10 percent of New Zealand's marine environment by 2010 is listed as a priority action in New Zealand's National Biodiversity Strategy (NZBS). New Zealand's Parliament is in the process of reviewing and reauthorizing the Marine Reserves Act of 1971. The government's Marine Reserve Act of 2002 largely reflected the lessons learned (as already listed here), several of which are explicitly stated in the Act. Furthermore, the government has stated its intent to protect 10 percent of New Zealand's marine environment in marine reserves by 2010 (Ballantine 1999). Details on the current status of the marine reserve legislation and marine reserves in New Zealand are available from NZDOC (2004).

The Marine Reserves Act of 1971 focused on protecting areas in their natural state for scientific study, although existing marine reserves in New Zealand are widely acknowledged to provide a much broader range of benefits than were recognized in the Marine Reserves Act of 2002. New Zealand's existing marine reserves, most notably the initial Leigh reserve, have been remarkably successful with respect to the original statute's intent and much more. Creese and Jeffs (1993) reviewed nearly 150 papers based on research conducted to that point in time within the Leigh reserve and concluded that most of these were wholly dependent on or benefited from its protected status. Such research would have been more difficult or impossible without Leigh and other New Zealand reserves, has expanded greatly since then, and has produced major findings with respect to marine reserve effects, fishing impacts, spiny lobster, marine fish, animal behavior, kelp forest, and other ecosystem dynamics (NZMSS 2002).

Studies of the spiny lobster (*Jasus edwardsii*) initially focused on the Leigh reserve (MacDiarmid 1989; MacDiarmid and Breen 1992) and later involving other New Zealand reserves (Babcock et al. 1999; Kelly et al. 2000), documented the recovery of lobster populations within these reserves compared to similar areas outside and provided other valuable information about the biology of this species (see Fig. 11.1). Density and size of lobsters increased quickly and substantially within reserves and continued to steadily increase for female lobsters but were more variable over time for male lobsters, probably due to migrations outside reserves. Larger lobsters, more legal-size lobsters, greater size distribution, and higher densities of lobsters were found within reserves versus similar areas outside. Lobsters within two reserves were 1.6 to 3.7 times more abundant than outside, and mean length of lobsters was nearly 20 percent greater within reserves than outside. An analysis of lobsters from within four marine reserves and similar sites outside concluded that mean length within reserves increased by more than 1 mm/year of protection and that density, biomass, and egg production each increased by between 3.9 and 5.4 percent per year of protection for shallow water sites and between 9.1 and 10.9 percent per year for deep water sites. Perhaps most impressive is the twenty-fold increase in large lobster reported within the Leigh reserve (Ballantine 1991, 1999). Research has also revealed some idiosyncrasies of lobster reproductive behavior and raised additional concerns about fishing impacts on overall reproductive output and success due to such impacts. Much of this research was conducted in northeastern New Zealand reserves and would have been impossible without them (NZMSS 2002).

Likewise, studies of marine fish within and adjacent to New Zealand's marine reserves, especially the Leigh reserve, documented the recovery of marine fish species within these reserves compared to similar areas outside and provided other valuable information about their biology. Adults of the highly prized and most common demersal predatory fish in northeastern New Zealand, the New Zealand snapper, *Pagrus auratus,* discussed previously in this book, were found to be 5.75 to 8.7 times more abundant within reserves than outside and also to be much larger on average within reserves, 316 mm (12 in.), compared to

186 mm (7 in.) in adjacent fished areas (see Fig. 11.1; Babcock et al. 1999). Another recent study on the blue cod (*Parapercis colias,*) the most common edible reef fish in the Marlborough Sounds, within and adjacent to the Leigh reserve, showed that it too was considerably larger and somewhat more abundant within the reserve, provided direct evidence of limited dispersal to adjacent fished areas, and concluded that this species will grow larger in reserves, and via spillover become available to fishers in adjacent areas (Cole et al. 2000). Research at the Leigh reserve also provided insights into less prized, less commercially valuable species. Spotties (*Pseudolabrus celidotus*) are a nearshore species of limited interest to most commercial and recreational fishers, but are commonly caught by novice shore bound anglers, especially children. Research possible only within the reserve enabled discoveries about this species, its biology, and especially its territoriality, aggressiveness, courtship, and sex change behaviors, that have proven fascinating to young and old alike (Ballantine 1991).

Research results related to overall changes in community structure, habitat, and ecosystem dynamics within New Zealand marine reserves versus adjacent areas are likely of even greater value and importance than the single species findings already discussed, though they are inextricably linked. Babcock et al. (1999) utilized the Leigh and Tawharanui marine reserves in northeastern New Zealand to assess cascading effects due to fishing impacts outside the reserves and the reciprocal effects of recovering predator populations within the reserves on grazer and algal communities. They related the changes in lobster (*Jasus edwardsii*) and predatory fish (*Pagrus auratus*) already discussed to resulting changes in the primary grazing sea urchin (*Evechinus chloroticus*) and dominant kelp (*Ecklonia radiata*). In the Leigh reserve, predator recovery reduced abundance of this urchin in barren rocky reef areas formally dominated by it from 4.9 to 1.4 per $m^2$ between 1978 and 1998. Consequently, kelp forests were far more extensive within the reserve in 1998 than when it was created and urchin-dominated barrens made up only 14 percent of available reef substrate inside the reserves compared to 40 percent of similar substrate outside (see Fig. 11.1). The observed changes to reserve community structure demonstrate higher trophic complexity than expected for this region's ecosystems, and increased primary and secondary production within reserves as a result of protection. These changes also indicate large-scale reduction of benthic primary production as an indirect result of fishing in unprotected areas outside reserves.

Far beyond the preceding summary of marine reserve research results, New Zealand's thirty-five plus years of experience with marine reserves has altered the way it thinks about, looks at, and uses its oceans. More than 10,000 school children and 200,000 total visitors come to the small Leigh reserve to view and enjoy the reserve, but remove virtually nothing from it. Such heavy visitation can have costs, as demonstrated by Brown and Taylor (1999) and may require limitation or areas where access is further restricted, but the real lesson learned from this is that such reserves meet real and legitimate public needs and more should be created to both fill these needs and avoid damage to existing sites from overuse.

New Zealand's relatively long experience with reserves has also contributed greatly to the continuing development of the reserve concept and principles related to it. Ballantine (1999) does an excellent job of reviewing this experience and attempts to distill key principles for successful reserve establishment. Ballantine (1997) also provides a good New Zealand perspective on marine reserve system or network design principles. This remote island nation and its mostly small marine reserves have sent key and clear messages heard around the globe. The recently created no-take marine reserve networks off California's Channel Islands (discussed earlier) and Victoria, Australia (discussed later) drew, for example, on the lessons learned and principles derived from New Zealand's experience.

## THE PHILIPPINES

Another Pacific island nation and two of its small reserves have sent similarly loud and important messages across the world's oceans. The Philippine Archipelago includes more than 7000 islands surrounded by extensive coral reef, sea grass, and mangrove ecosystems, and stretches more than 1000 km. It is home to tremendous marine and terrestrial biodiversity and is among the hottest of the hot spots. Over 2000 species of marine fishes can be found in the Philippines, among the highest diversity of reef fish anywhere (Sale 1980). In addition, Philippine waters contain over 500 species of corals, 1000 species of algae, 13 species of sea grass, and over 1700 species of mollusks (Rubec 1988). This island nation is also home to two of the best-studied and most renowned marine reserves in the world, those adjacent to Apo and Sumilon Islands (Fig 11.2).

Coral reef ecosystems of the Philippines possess tremendous social and economic value. The fishery sector of the economy directly employs over 1 million people, including over 600,000 small-scale fishers (Barber and Pratt 1997). Reef fish are the most important source of animal protein in the Filipino diet (ADB 1993). Another key aspect of Filipino fisheries is the export of fish for food and the aquarium trade. Roughly 75 to 80 percent of tropical marine

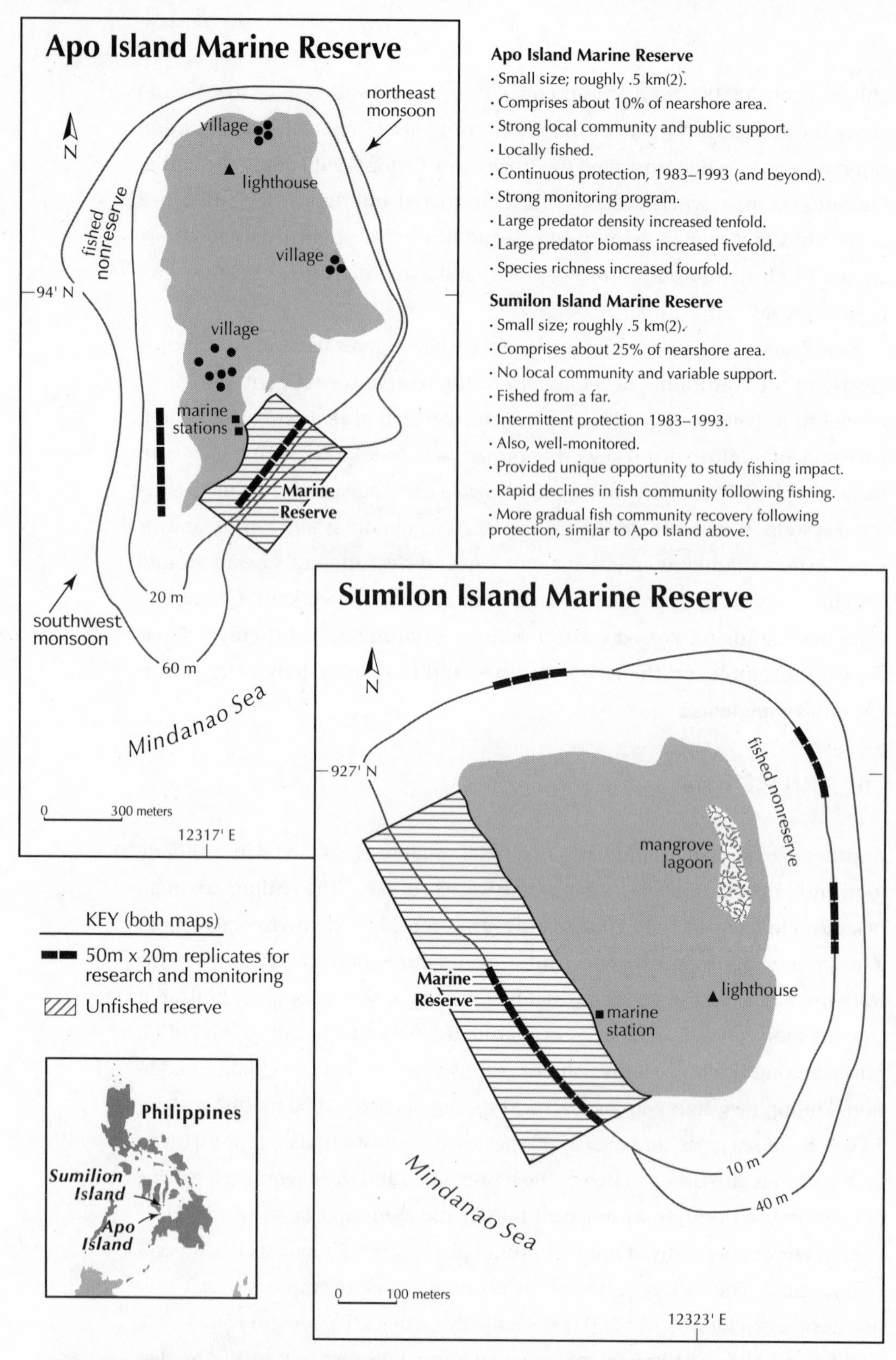


**FIG. 11.2 Philippines Apo and Sumilon Marine Reserve Maps and Key Points.** Small map shows location of each island within the Phillipine Archipelago. Individual reserve maps provide island geography, depth contour, reserve, nonreserve, and research and monitoring site locations. Key points highlight noteworthy aspects of each reserve. Source: Garry Russ and Angel Alcala; Material adapted with their permission.

aquarium fish sold worldwide are from the Philippines (Barber and Pratt 1998). The export of live food fish to the fish markets of Hong Kong is also a major component of fisheries in the Philippines. Coral reefs also provide Filipinos with a source of income from tourism and the ornamental and curio trades.

Despite their biological and socioeconomic significance, the coral reefs of the Philippines continue to be destroyed by multiple anthropogenic stresses. Gomez et al. (1981) estimated that over 70 percent of the reefs in the Philippine Archipelago were already in only poor to fair condition. A wide variety of human impacts have contributed to coral reef degradation (reviewed by Rubec 1988). Some of the greatest threats to Philippine reefs are fisheries related. Intense fishing pressure has driven stocks of many fish to extremely low levels. Overfishing not only impacts reef fish themselves but also has secondary or cascading effects on entire reef ecosystems. Extensive overfishing and especially destructive fishing techniques, including illegal coral harvesting, blast or dynamite fishing, muro-ami, and cyanide fishing, are commonplace in the Philippines. These techniques indiscriminately kill and waste target and nontarget species alike while destroying the reef framework itself. Because habitat complexity is often important in determining the abundance and composition of reef fish assemblages, areas of reduced complexity following blast fishing are not likely to recover quickly (Jennings and Lock 1996; Rubec 1988; Russ 1991). The array and scale of human impact in the Philippines continue to escalate, and the coral reefs and their associated fish and invertebrate communities continue to decline (Wilkinson 1998, 2000, 2002).

In response to overfishing, destructive fishing, and overall reef degradation various MPAs have been established throughout the Philippines to preserve and restore relatively healthy reef ecosystems. The United Nations Environmental Program/International Union for the Conservation of Nature (UNEP/IUCN; 1988) listed about thirty MPAs of various size, importance, and level of protection in the Philippines. By 1995, the number of nationally and locally designated MPAs had grown to 260. These MPAs have been created by a variety of means at the national level and by local municipalities, with the support of universities, nongovernmental organizations, and stakeholder groups. They also serve a variety of purposes from protecting endangered species (e.g., dugong and marine turtles in the Calauit reserve) to promoting tourism (e.g., the EI Nido resort) (Alcala 1988). The few Philippine MPAs that are fully no-take provide excellent examples of their ability to protect and restore reef fish and fisheries, and demonstrate how social factors can influence reserve efficacy.

The Philippine's Sumilon and Apo marine reserves' inordinate importance and impact relative to their size derives largely from (1) their unique, complex, and interesting management histories; (2) the strong research, monitoring, and documentation of these histories and their effects; and (3) the value, status, and concerns regarding Philippine coral reefs and the level of fishing occurring on them. Russ and Alcala (1996; 1999) and White (1989) provide detailed descriptions of the environment and management histories for both reserves. A condensed version drawn from these sources follows.

Sumilon and Apo Islands are two of many small islands (less than 1 $km^2$) located close to one another in the central Visayas region of the Philippines and surrounded by fringing coral reef systems (see Fig 11.2). Both are within two hours by boat from Dumaguete, home to Silliman University. The university's marine lab and associated researchers initiated community-based marine conservation, research, and education programs focused on Sumilon (1973) and Apo (1976), based on concerns that observed degradation of their reefs required proactive efforts to protect them. Although reef systems around both islands were experiencing heavy fishing pressure and being severely impacted by it, a key difference was that Sumilon was uninhabited and fished by off-island communities, whereas Apo had about 600 residents that were totally and directly dependent on its surrounding waters for their livelihood. The university's efforts eventually led to the creation of marine reserves off each island, but with very different results.

In 1974, the town of Oslob, Cebu, in cooperation with the university created the Sumilon marine reserve (see Fig 11.2), providing no-take protection for 25 percent of the marine area (0.5 $km^2$) adjacent to a 750 meter section of shoreline on the western side of the island, as part of a larger MPA surrounding the entire island via local government action. The remainder of the MPA along the rest of the shoreline remained open to traditional fishing methods only (predominantly trap, net, and hook-and-line fishing). Initially things went well and Sumilon Island became the first well-managed marine park in the Philippines. Outreach, education, research, and monitoring programs were developed and implemented, fishermen generally cooperated with the restrictions, and the reserve was protected from fishing. In 1980, the election of new local leaders opposed to the reserve initiated friction and began to erode reserve protection. An appeal to the remote national government resulted in some additional federal protection, but reserve protection collapsed completely in 1984 (the reserve caretaker left the island for his own safety) with widescale poaching, including the use of destructive fishing methods, such as muro-ami techniques and explosives. Effective reserve protection was reinstated in 1987 with support from local leaders and lasted from 1988 to 1992, but broke down again in 1992 and remained ineffective at least through 1998.

By contrast, the evolution of the Apo marine reserve took longer, but benefited from the existence of a true local community on the island with a direct stake in protecting its marine environment and the slower, but ultimately longer lasting, development of that community's support for the reserve. The idea for the reserve evolved as a result of the university's long-term, community-based reef protection efforts initiated in 1976 but was not endorsed by the local community and government until 1982 and did not receive formal legislative protection until 1986. Nonetheless, the local endorsement, local government action, and carefully built community support effectively created the 0.45 $km^2$ Apo marine reserve along a 500-meter stretch of shoreline in 1982, and has consistently and effectively provided reserve protection for 10 percent of the island's marine area since (see Fig. 11.2). The Apo marine reserve remains one of three community-managed marine reserves in the southern Philippines totally maintained by the people who live on the island since the mid-1980s.

The creation of the Apo and Sumilon marine reserves and their divergent management histories, especially the opening and closing of the Sumilon reserve versus the continuous protection afforded at the Apo reserve, provided an interesting opportunity for examining the effects of reserve protection on fish populations and communities (Russ and Alcala 1998). Long-term research and monitoring programs developed in association with reserve development enabled analysis of these effects. A series of papers by Russ and Alcala (1989, 1990, 1994, 1996, 1998) document how fish abundance, biomass, and community composition changed within and outside of the reserves as a function of the status of the reserves (open or closed to fishing). These results are summarized following here and in Figures 4.3 and 11.2.

The most remarkable and important findings with respect to fishing impacts and marine reserve effects related to Sumilon and Apo Islands and their respective reserves concern observed changes to their "large predator" fish communities, which underwent striking changes attributable to both fishing impacts and, conversely, reserve protection. Gradual, but significant, increases in large predator density were documented a total of three times on the two islands following cessation of fishing activities for periods of five or more years. Density and biomass of large predators increased steadily while they were protected from fishing at both Sumilon and Apo and were still increasing after nine and eighteen years of protection, respectively (Russ and Alcala 1996; 2003).

At Sumilon, the density and biomass of large predators were highest in the reserve in 1983 after nine years of continuous protection. Predator density increased fivefold and biomass eightfold during this period. But when protection broke down, predator density decreased rapidly and significantly, wiping out the gains from reserve protection in less than two years. A similar pattern was also observed in the nonreserve area, when it too was closed to fishing in 1988 (Russ and Alcala 1996). In addition, the opening of the reserve area to destructive fishing resulted in changes in the composition of the reef fish community (Russ and Alcala 1989, 1998), with large predators (characterized by low growth, natural mortality, and recruitment) and species subject to intense fishing pressure (e.g., Caesionidae) being most affected (Russ and Alcala 1998).

Apo's large predator community responded similarly to continuous reserve protection, but some additional changes were documented. Large predator density and species richness within the reserve increased significantly and directly from shortly after the reserve was created in 1983 until at least 2000. Predator density and species richness within the reserve increased roughly tenfold and fourfold, respectively, during the period from 1983–1995. Predator biomass inside the reserve also increased significantly, but with a more curvilinear response of about fivefold. Outside of the Apo Island reserve, the density and species richness of large predatory fish were much lower than inside the reserve, but both density and species richness also increased consistently following reserve creation, showing a strong positive correlation with the number of years of reserve protection. The ratio of large predators inside the reserve to fished nonreserve areas outside of it peaked several years after its creation, with density inside the reserve more than ten times greater than outside. The linear increase in predator density and curvilinear response of predator biomass continued through at least 2000. By then predator density had increased more than 15-fold and biomass more than 20-fold (Russ and Alcala, 2003a).

Research results clearly show that the Apo reserve is protecting and rebuilding its own fish community and may also be responsible for increasing large predatory fish density and diversity in adjacent fished areas. If the reserve is positively impacting adjacent fished population due to juvenile or adult spillover from the reserve, the enhancement effect should be greatest close to the reserve. This was not obvious during the first eight years following reserve protection, but a strong density gradient for large predators decreasing with distance from the reserve developed subsequently (Russ and Alcala 1996a). These results are consistent with the prediction that, as biomass and abundances increased within the reserve over time, spillover to adjacent areas enhanced the population outside of the reserve. The time lag also suggests that some fishery benefits may take a relatively long time to be experienced locally. By 2000, additional evidence of spillover was accumulating for specific large predators (Russ and Alcala 2003b).

Opening the Sumilon reserve to destructive fishing in 1984 also resulted in a decline in species richness and density of smaller coralivores and benthic carnivores, such as butterfly fish (Cheatodontidae; Russ and Alcala 1989). Reserve protection prevented both direct and indirect negative impacts of fishing on several species of reef fish and allowed more natural reef fish populations and communities to develop. Other fish species such as caesionids, showed periodic increases in abundance over time, but their populations were more affected by natural fluctuations, such as recruitment variability, than the length of time protected (Russ and Alcala 1998).

The disruption of Sumilon reserve protection also appears to have adversely impacted adjacent fisheries. Alcala and Russ (1990) compared fisheries yields in terms of catch per unit effort (CPUE), from the area around Sumilon Island ten years after the reserve was established, to yields eighteen months after the breakdown of reserve protection. CPUE in the fishery was significantly higher during the period when the reserve was in place, despite 25 percent of the reef area being closed to fishing (Alcala and Russ 1990). Total catches dropped 54 percent following the breakdown of reserve protection. Spillover of juveniles or adults resulting from increased survivorship within the reserve and subsequent movement into the fishery is among the potential mechanisms by which reserve protection could be enhancing local fisheries (Alcala and Russ 1990; Russ et al. 1992). Thus the historical record for both Apo and Sumilon Island suggests the buildup of populations within their reserve areas may contribute to those outside of reserve borders, and provides evidence that the reserve can contribute to higher fishery yields. White et al. (2000) provides a case study of the Apo Island reserve that combines estimates of such improved yields with other reserve generated values, such as increased local diving tourism, and concludes that the initial $75,000 investment in the reserve now provides an annual return between $31,900 and $113,000.

Despite their small size (less than 1 $km^2$), the Apo and Sumilon Island reserves still generate significant ecological and economic benefits and insights into their efficacy. They provide strong evidence that (1) marine reserves are highly effective at rebuilding density, biomass, and even species richness of fish, especially those heavily targeted and highly vulnerable including large predators that may be easily endangered and difficult to protect using traditional means; (2) the effects of reserve protection will be variable among species and dependent on both life history and level of exploitation; (3) reserve benefits, particularly those related to fisheries, may accrue more slowly than the rapid declines in reef fish populations and communities due to fishing and may take several decades to reach their full potential; (4) marine reserves may improve fishery yields outside of their borders via juvenile and adult spillover and larval dispersal, but proving this can be extremely difficult; and equally significant, (5) marine reserve success is largely dependent on public support and respect for reserve regulations, and (6) community-based programs can be essential to obtaining these, especially in developing and dispersed island archipelagos like the Philippines.

The local community continuously supported the Apo Island reserve, and benefits continued to build over time. In contrast, local politics in communities near Sumilon Island resulted in breakdown of reserve protection and subsequent loss of reserve benefits. Lessons learned from Sumilon and Apo are being broadly applied with respect to other marine conservation efforts throughout the Philipines, especially the value of community-based approaches.

## LESSONS FROM THE MEDITERRANEAN RESERVES (FRANCE, SPAIN, AND ITALY)

Despite Europe's long maritime history, the extensive development and historical importance of its fisheries, and the continuing importance of its fishing industries, or possibly because of these factors; Europe's development of marine reserves has been relatively slow and modest. The best-known and most visited European marine reserves dot the Mediterranean coastline spanning the coasts of France, Spain, Monaco, and Italy. Currently, they include at least a dozen sites within a small but growing set of MPAs and terrestrial/MPAs that are either entirely no-take or have no-take areas zoned within them. These no-take zones or marine reserves are generally small, mostly less than 1 km$^2$. This loosely knit group resulted from a mélange of national authorities and legislation applied largely at a local level. Not all of these have dedicated staff, even though such staffing has been identified as the single factor most critical to their success (Francour et al. 2001; Ramos-Espla and McNeil 1994). Although currently limited in size and scope, a brief review of these reserves is provided here for geographic representation, to demonstrate or reinforce specific lessons, and in consideration of their potential to develop as a more integrated and extensive network.

The stated purposes or reasons for the creation of existing individual Mediterranean reserves generally mirror the full range of potential benefits provided earlier and include: (1) protection or conservation of general marine biodiversity; diverse or representative marine communities, habitats, and species; and specific habitats and species of special conservation or commercial value; (2) development, enhancement, or improvement of tourism or ecotourism, especially diving and fishing; (3) scientific research; (4) general public and marine-oriented education; and (5) rebuilding, restocking, and replenishing overexploited species, even including aquaculture and artificial reef development. However, they also reflect specifics related to the northwestern Mediterranean and local environments, and consequently include: (1) a heavy emphasis on rocky shore, rocky reef, and sea grass (*Posodonia oceanica*) habitats; (2) considerable focus on larger, high-profile, heavily targeted, local fish species native to these habitats, such as the large groupers (*Epinephelus marginata* and *E. costae*), dentex (*Dentex dentex*), amberjack (*Seriola dumerili*), and brown meagre (*Sciaena umbra*), as well as the more abundant sea breams (*Diplodus* spp.), large wrasses (*Labrus* spp.); and (3) the Islas Chafarinas site near Morocco designed to protect the last colony of the highly endangered Mediterranean monk seal (*Monachus monachus*) in Spain, but only extending to mean low water. Also, depending on local circumstances, zoning regulations in and adjacent to these reserves may alternatively favor diving or commercial or recreational fishing interests (Francour et al. 2001; Ramos-Espla and McNeil 1994).

Given their relatively small size and limited research efforts, a surprising number of marine reserve benefits and effects have been documented for these northwestern Mediterranean reserves. Common to sites around the world, several studies here confirm greater density, size, and diversity of targeted and vulnerable species (see preceding list) and also overall fish numbers within versus outside several of these reserves. In some cases, spectacular overall increases of fishes, especially targeted ones, have been observed and reported in the shallow waters of these reserves. A so-called buffer effect, in which seasonal variation of fish communities within reserves was damped resulting in a more stable year round fish community, was first reported from France's Scandola Marine Reserve. Researchers documented and reported cascading effects in Spain's Medes Marine Reserve similar to those discussed earlier in this chapter for New Zealand reserves. In Medes, an increase in large Sparid fishes within the reserve led to increased predation on sea urchins, reduced grazing, and ultimately the return of large, erect algae. Another interesting effect is the documented recovery of large adult populations of the dusky grouper (*E. marginata*) in several

Mediterranean reserves following their closures and the movement of fish into them from nursery areas to the south. Finally, the existence of these reserves in combination with adjacent MPAs and management zones has helped to tease apart the impacts of commercial fishing, recreational angling, and spearfishing and demonstrates that both angling and spearfishing impacts can equal or exceed commercial impacts in some settings (Francour et al. 2001; Ramos-Espla and McNeil 1994).

The successful experience to date with the small but growing number of marine reserves along the northwest Mediterranean coast suggests the potential to develop a more extensive and integrated network and that such development is warranted. Significant reserve benefits accrue as a result of the existing reserves, despite their extremely limited spatial extent. These have clearly included conservation, education, scientific, and nonconsumptive benefits. Some direct external fisheries benefits have also been suggested or reported, but these are less clear. At a minimum, fisheries may benefit from improved knowledge resulting from the reserves. Broader conservation and more significant fisheries benefits will require more and larger reserves. Even the maintenance and sustainability of the existing reserves will require a more extensive network.

The small size of existing no-take areas in this region reflects the strong influence of extractive user groups and decisions to compromise conservation needs and accommodate their views to ensure acceptance. This approach has achieved some success, but overcoming this precedent to develop a more sustainable network will now require engaging a broader and more supportive cross section of the public and working with, exchanging information, and changing the views of such interests. Ongoing professional networking among scientists and managers among reserves in all three countries raises the prospects for such development. A number of proposals could expand the existing system and further a future network, including sites in each country and a joint French–Spanish effort. Several European Union (EU) initiatives could help foster this, but such efforts will require support at a more local level. Similarly, EU members recently voted to create networks of marine reserves for the Northeast Atlantic and Baltic Sea, but actual protection will require development of stronger public and stakeholder support and input.

## KENYA AND ADJACENT TANZANIA

Compared to the status of marine no-take reserves in Europe already described, East Africa possesses a remarkably well-developed system of such reserves that is among the best studied in the world. Kenya's coastline extends about 450 km along the Indian Ocean just south of the equator between Somalia and Tanzania. The continental shelf is narrow here and supports fringing reefs, patch reefs, mangrove forests, sea grass meadows, and other nearshore tropical marine communities (UNEP/IUCN 1988). These nearshore communities cover an area of about 800 km$^2$, most of which is intensively fished by roughly 8000 artisanal (small-scale) fishers who land more than 80 percent of Kenya's coastal catch, or between 8000 and 16000 tons per year in 1990. Several marine national parks, fitting the no-take marine reserve terminology used in this book (Kenyan "marine reserves" allow fishing), constitute nearly 5 percent the nearshore area off Kenya and adjacent Tanzania and have effectively prohibited fishing for varying lengths of time (Fig. 11.3). Three of these, Malindi, Watamu, and Kisite have been protected for more than twenty-five years, while Mombassa and Chumbe have each been protected for about a decade. Kenyan nearshore fished reefs and their associated fish communities are highly degraded relative to their protected counterparts (McClanahan 1995; McClanahan and Arthur 2001).

Earlier in this book, we mentioned that researcher Tim McClanahan, who authored "Are Conservationists Fish Bigots?" (1990), relocated from Florida to Kenya to study coral reefs fully protected from extractive activities. Over the past two decades, extensive reserve-oriented research by him and others has revealed an incredible story of ecosystem linkages, fishing impacts, and the efficacy of marine reserves that would likely have remained hidden were it not for the reserves set up more than a quarter century ago. Among the lessons learned from this research to date are: (1) fishing impacts appear to be the primary driver behind both major declines in East African nearshore fish communities and serious degradation of coral reef ecosystems; (2) certain large fish, including keystone predators and herbivores, are especially vulnerable to fishing pressure, and their removal can profoundly alter urchin populations and degrade reefs; (3) Kenya's no-take marine national parks have been remarkably successful in protecting, restoring, and reversing declines with respect to both fish and coral reefs; (4) no-take marine reserves are likely the best and may be the only way to protect Kenya's rare and vulnerable fish species and precious coral reef systems; (5) protection from fishing impacts also provides resistance and resilience to other stresses; and (6) marine reserves can provide real fisheries, tourism, and overall economic benefits, but ensuring or optimizing these may be complicated and design-dependent (McClanahan and Arthur 2001; McClanahan and Obura 1995).

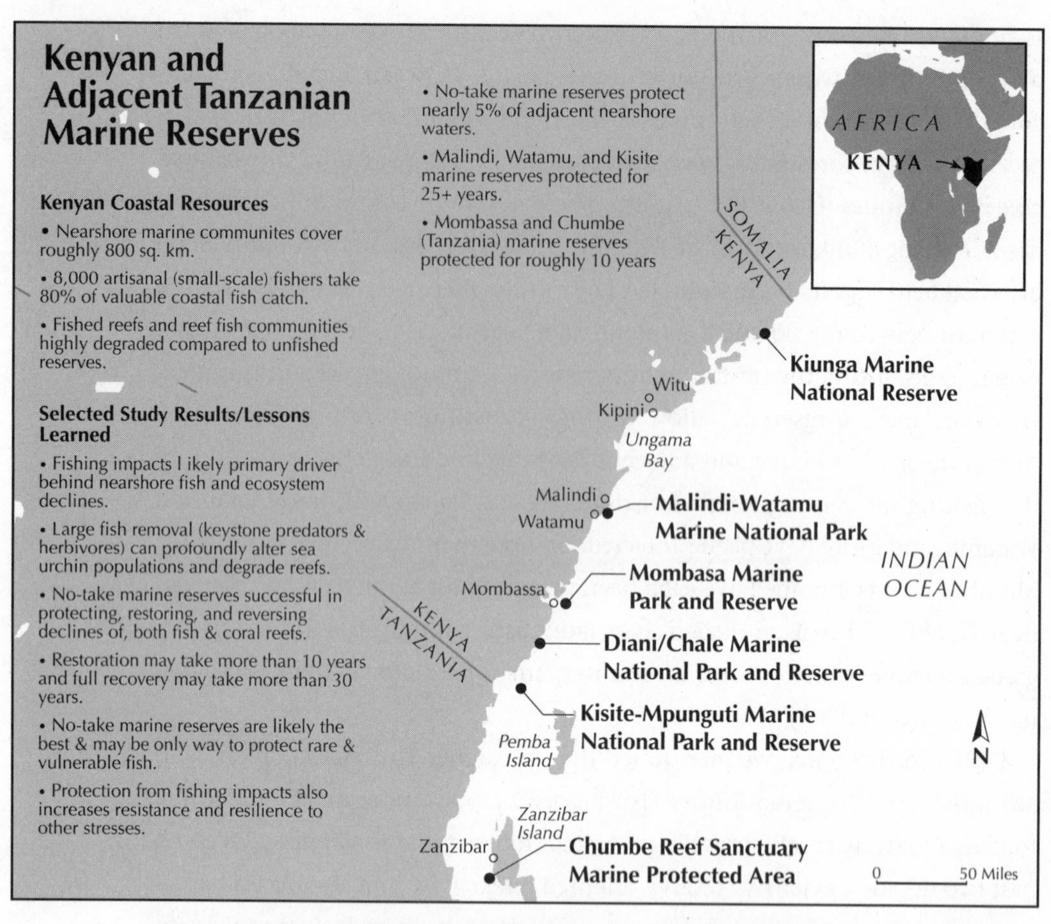


**FIG. 11.3 Kenyan and Adjacent Tanzanian Marine Reserves Map and Key Points.** Map shows section of East African coast and locations of five marine reserves in Kenya and the adjacent Chumbe Reef site in neighboring Tanzania. Key points highlight information about these reserves, local coastal resources, and what's been learned about both. Sources: McClanahan and McClanahan et al. (see references); UNEP/IUCN (1988).

Multiple stresses and cumulative impacts are almost certainly responsible for recent declines in global coral reefs, and their relative contribution probably varies by location. More than two decades of research on Kenyan and other East African coral reefs, enabled in part by Kenya's no-take marine national parks, provides strong evidence that fishing is probably the primary driver of declines in Kenyan reef fish and coral reefs. McClanahan and Obura (1995) reported on the status of Kenyan coral reefs and research results related to it. They summarized earlier work, provided additional evidence, and concluded that (1) removal of fish was the primary impact on unprotected Kenyan coral reefs and had cascading impacts on other faunal groups and ecological processes; (2) removal of sea urchin predators on unprotected reefs was causing high densities of sea urchins (see Fig. 11.3); (3) high urchin density in turn led to reduced coral cover, simpler reef structure, and slower reef deposition; (4) creation of the Mombasa Marine National Park resulted in the rapid recovery of fish abundance, diversity, and coral cover; (5) some gastropods are directly affected by shell-collecting, but even more are affected by removal of their finfish predators; and (6) river sedimentation and discharges were of secondary importance. They suggested that protection from fishing might provide resistance and resilience to these secondary impacts.

More recently, McClanahan and Arthur (2001) extended and expanded prior work, examining the effects of fishing, length of protection, and reef characteristics to determine their ability to predict reef fish communities. They estimated population density, species richness, and rarity for over one hundred fish species on more than twenty reefs along 400 km of East African coastline, including sites inside and outside of the five marine national parks previously mentioned. They found that fish diversity was positively correlated with coral and coralline algae cover and negatively correlated with sea urchin and algal turf abundance; protection from fishing was the strongest single factor affecting fish abundance and diversity; the habitat correlations likely resulted from fishing impacts to reef ecology where heavy fishing increased sea urchin and algal turf abundances and decreased coral and coralline algae cover; and protected areas had much higher densities and species richness for commercially important fish species (see Fig. 11.3). They also found that older reserves contained more and rarer species than young reserves or fished reefs, suggesting that maintaining marine reserves older than ten years may be necessary to protect the full diversity of fishes. Several targeted fish species were an order of magnitude more abundant in areas protected from fishing than those heavily fished, but not all fish were more common in the reserves. Small fish such as damselfish and small wrasses were the most abundant species overall and were more common on recently protected or moderately fished reefs than on protected ones, but least common in heavily fished areas. When these species were excluded from the analysis, overall fish density was an order of magnitude higher in the protected reserve areas than in heavily fished ones. Fish density, species richness, and rarity all declined along a gradient from unfished (protected for more than twenty-five years) to newly protected, moderately fished, and heavily fished reefs (see Fig. 11.3).

The story of Kenya's coral reef systems, the effects of fishing on them, and the role of its marine reserves in protecting and restoring them took many years to piece together, remains incomplete, and is certainly far from over. The full story is as complex and diverse as reef systems themselves and space precludes telling it in detail here. However, we highlight a few aspects following here to illustrate some key points.

Research results, published in a series of scientific papers by McClanahan and others since the late 1980s, detail the critical importance of fishing impacts, especially on a single keystone predatory fish; the resulting release of a key destructive grazing urchin; the cascading effects that reverberated across Kenya's reefs; and the prevention and reversal of these changes within its marine reserves (see Fig. 11.3). McClanahan (2000) also details the recovery of the keystone predator, the red-lined triggerfish (*B. undulates*)and resulting control of the grazing urchin (*E. mathaei*) in Kenya's five marine reserves based on protection of up to thirty years. His results suggest that *B. undulatus* dominance and *E. mathaei* control return in five to ten years following protection, but that full recovery of *B. undulates,* reduction of *E. mathaei,* and reef restoration may take more than thirty years (see Fig. 11.3).

Several published papers also analyze or discuss the potential for direct fisheries, tourism, and economic benefits from Kenya's marine reserves (McClanahan 1995; 2000; McClanahan and Kaunda-Arara 1996; McClanahan and Mangi 2000, 2001). The results of these studies generally recognize potential for and importance of such benefits from existing or proposed reserves, but remain incomplete and are inconclusive about the degree to which they have been achieved to date and how to optimize them in the future. With respect to fisheries, they have been limited, focusing largely on direct short-term spillover effects and confounded by other fisheries and demographic changes. Some spillover has been suggested, but concern has also been expressed that net local adverse short-term impact to some fishers and fisheries has occurred in areas where a high percentage of fishable area is closed. There is some evidence to support this concern. This has led to reductions in the size of some marine reserves and suggestions that future design should limit their spatial extent. With respect to tourism and other economic benefits, analysis to date has identified some real values and return, but has been limited in scope and focused largely on providing economic return to government. Published work to date is insufficient to fully justify Kenyan reserves economically or design reserves to optimize economic return.

## SMALL COASTAL RESERVES IN CHILE

Chile's spectacular and diverse coastline stretches 6435 km along the Pacific Ocean from the neotropics almost to Antarctica, encompassing both rain for-

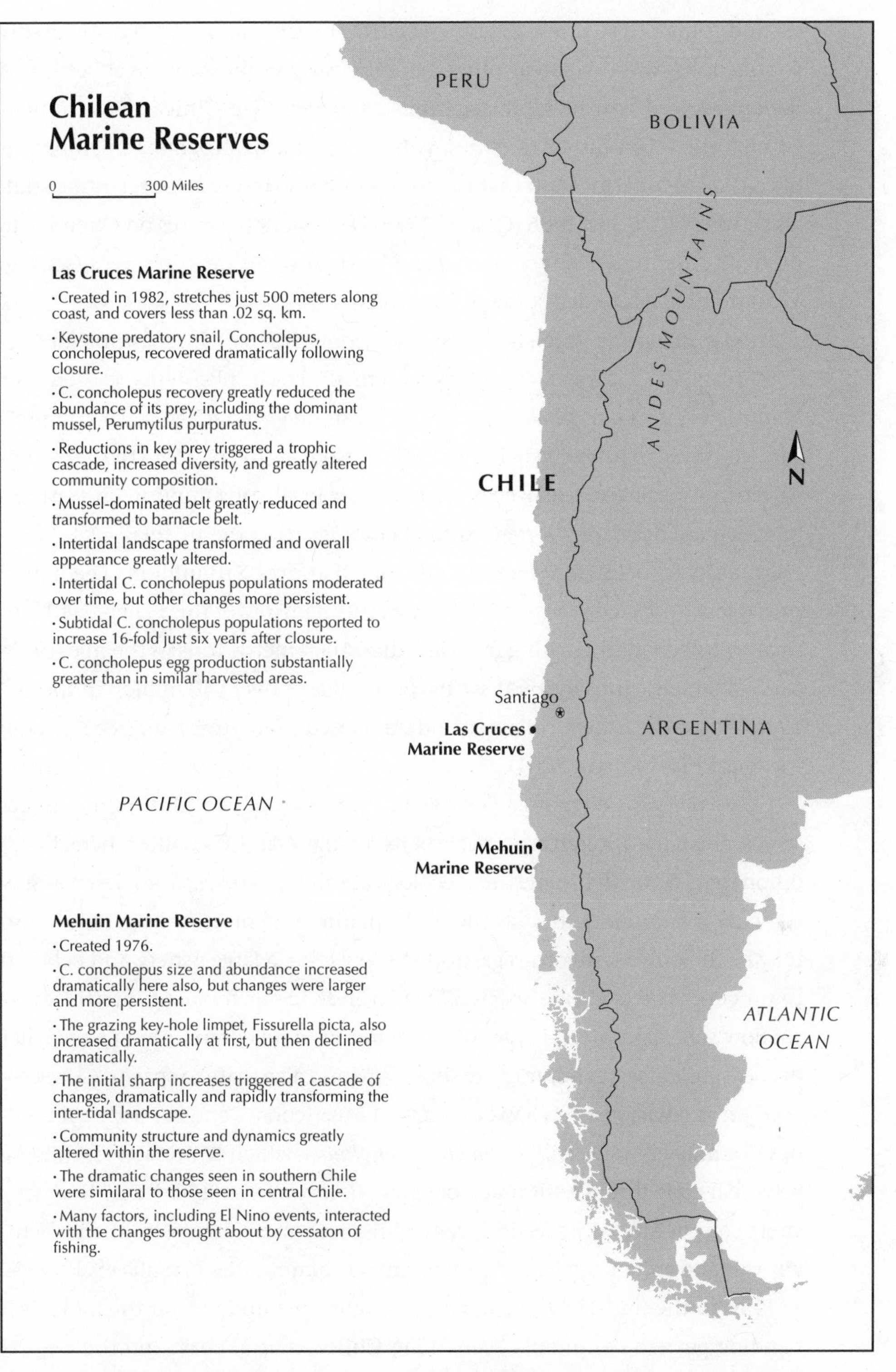


**FIG. 11.4 Chilean Marine Reserves (Las Cruces and Mehuin) Map and Key Points.** Map shows Chile's long coastline and the location of the Las Cruces and Mehuin marine reserves. Key points highlight dramatic changes to carnivorous gastropod, *C. concholepas*, and other species within each reserve, and related cascading or ecosystem effects resulting from their creation reserve. Sources: Castilla (1994, 1986, 1981) and Moreno (2001).

est and some of the driest desert on Earth. The Chilean rocky coast is highly stratified, its intertidal areas often exhibit clearly visible biological belts, and it has received considerable scientific attention. The Chilean government granted three Chilean universities with strong marine programs small parcels of coastal and marine land that function as marine reserves for scientific studies between 1976 and 1986 (Castilla 1986). This section focuses on Chile's rich, productive, and diverse rocky coast and two of these small reserves (see Fig. 11.4) that have revealed some of its secrets.

Humans have exploited food resources from Chile's rocky coastline for over 8000 years, likely a key factor in its settlement. Local subsistence gatherers or *Mariscadores de orilla,* skin divers, and hookah divers attached to air compressors, continue to intensively harvest these resources today. More than sixty invertebrate species and seaweeds are taken for local consumption, at least ten of which are heavily exported. Annual landings from these "small-scale" fisheries can exceed 150,000 metric tons, valued at over $10 million. The highly prized loco (*Concholepas concholepas*) was the mainstay of this fishery, with annual reported landings rising from less than 1000 metric tons in the mid-1940s to 25,000 metric tons by 1980, with a peak value of over $40 million in the late 1980s before the fishery collapsed and was closed a few years later (see Fig. 11.4; Castilla 1994; Moreno 2001).

*C. concholepas* also plays a key role in the stories of two Chilean marine reserves, and a few interesting aspects of its biology warrant mention here. In addition to its cultural, fisheries, and ecological values, *C. concholepas* deserves special conservation attention as the sole remaining member of its genus. It also may be the only carnivorous gastropod supporting a large fishery and believed to function as a keystone species. Smaller individuals frequent intertidal and shallow subtidal waters. Larger ones are found more commonly in somewhat deeper subtidal water but migrate shoreward and aggregate to spawn. *C. concholepas* preys voraciously on a wide variety of invertebrates and has a special fondness for mussels, especially *Perumytilus purpuratus,* which dominates the middle intertidal zone throughout much of Chile. *C. concholepas* "bulldozes" the rocky shore community while feeding, removing many more organisms than it actually eats and enhancing its effect on intertidal communities (Castilla et al. 1994).

Prior to the mid-1980s, ecologists thought they understood the biological banding pattern commonly seen along Chile's central coast and the natural factors responsible for it reasonably well. However, Castilla (1981) noted an unusually high abundance of intertidal *C. concholepas,* a near absence of some of its normally common prey species, and a strikingly different banding pattern along a rare stretch of rocky coast with limited human disturbance in central Chile. The building of a marine lab at Las Cruces, the creation of the Las Cruces marine reserve, and the exclusion of human predators from it beginning in 1982 provided an opportunity to study an area recovering from human disturbance. Observation of this recovery explained the preceding differences, revealed some of the ecological dynamics responsible for them, demonstrated the extent of human perturbation, and dramatically altered the scientific view of what Chile's coastline might look like in the absence of human extraction.

The Las Cruces reserve stretches just 500 meters along the coast and covers just 4152 $m^2$ of intertidal area and an adjacent 44,130 $m^2$ of subtidal area offshore. Scientific studies and monitoring initiated prior to its creation established a baseline for comparison and carefully tracked changes within both the reserve and adjacent control areas. Dramatic changes occurred in the intertidal portion of the reserve following its creation, and its initial recovery was well documented (see Fig. 11.4). *C. concholepas,* initially uncommon in both the reserve and control areas, increased significantly within two years, but only in the reserve area. The initial recovery of *C. concholepas* greatly reduced several of its prey items, including the competitively dominant mussel (*Perumytilus purpuratus*, triggered a cascade, increased diversity, altered community composition, transformed the intertidal landscape, reduced the mussel-dominated belt of the intertidal zone, and dramatically changed its appearance. Over time, intertidal *C. concholepas* abundance did not stay high, but other changes that its temporarily increased predation brought about have been more persistent, including the conversion of the middle intertidal belt from a mussel-dominated community to a barnacle-dominated one. Additional intertidal changes were observed after 1985 but were complicated by a large earthquake and are more difficult to interpret. Subtidal portions of the Las Cruces reserve are logistically more difficult to study, and changes there were less well documented. Nonetheless, researchers reported a sixteen-fold increase in the density of *C. concholepas* within the subtidal portion of the Las Cruces reserve just six years after its creation (Castilla et al. 1994; Castilla and Durán 1985; Durán and Castilla 1989).

Southern Chile's small Mehuin marine reserve and its associated coastal laboratory in southern Chile also conducted extensive long-term research and monitoring of intertidal community structure and dynamics following its establishment in 1976. The protection of the Mehuin reserve's intertidal area from human exploitation produced an increase in *C. concholepas* density and size similar to that documented at Las Cruces, except that it was larger and persisted for at least five years. The grazing key-hole limpet (*Fissurella picta*) showed similar increases within the Mehuin reserve for five years following cessation of fishing but then declined dramatically over the next six years. The sharp increases in both of these consumers also triggered a cascade of changes dramatically and rapidly altering the intertidal landscape, community structure, and dynamics within the reserve. The dramatic community changes seen in southern Chile were similar to those seen in central Chile but not identical. At Mehuin, many factors, including El Niño events, interacted with the changes brought about by the cessation of fishing (Castilla et al. 1994; Godoy and Moreno 1989; Moreno 2001; Moreno et al. 1986).

Much of the current knowledge of Chilean intertidal areas, the dramatic impacts of human extractive activities on them, and their complex community structure and dynamics stems from these two small reserves and may not have been uncovered were it not for their creation. They elucidated and confirmed the keystone role of *C. concholepas,* the important role of human extraction in altering the intertidal landscape, and the cascading effects produced within a reserve when such extractive activities are halted. Furthermore, knowledge derived from these reserves has been translated into recommended and real changes in conservation and management practices for areas external to them. Examples include application to loco, *C. concholepas,* and bull-kelp (*Durvillaea antarctica*) management, and more generally to the development of community-based management plans, including protected areas for Chile's small-scale benthic shellfisheries under its revised fisheries and aquaculture law (Castilla 1994; Castilla and Bustamante 1989; Moreno 2001).

## AUSTRALIA

In many things marine, Australia is both unique and a leader. Among continents, it is the only one composed of a single nation. Among nations, it is the world's largest island nation. Australia's marine waters cover over 16 million km$^2$, more than twice its landmass, giving it one of the world's largest ocean areas. Its tremendous size may be matched or surpassed by its incredible diversity. Australia's vast ocean area houses what may be the world's greatest collection of marine biodiversity, ranging from spectacular coral reefs in its tropical north to lush kelp forests in its temperate south. An array of other special marine environments are scattered off its coasts and throughout its waters, including the world's most expansive collection of sea grass meadows, a large collection of offshore seamounts, and a host of other critical habitats. These marine environments contain more than 4000 fish species and much greater numbers of invertebrates, plants, and microorganisms, many of which are endemic to Australia, placing a special global responsibility on Australia to protect its awesome natural marine heritage (Environment Australia 2003).

No book on MPAs or marine reserves would be complete without some discussion of Australia's existing array and future plans. Earlier sections of this global review chapter focused largely on small numbers of small reserves, typical of the majority of older, better-studied reserves throughout the world. By contrast, Australia possesses the world's largest and best-known MPA, the Great Barrier Reef Marine Park (GBRMP), the world's largest and most numerous marine reserves, and the world's most ambitious plan for developing a broader national system of MPAs and marine reserves. Key lessons to draw from the Australian experience focus on the value of setting marine reserves in a broader context, the importance of a good public process, and the advantages of integrating site-specific development with national-level system planning.

Over the past few decades, Australia has emerged as an international leader in marine conservation, launching and implementing a series of national ocean policy and MPA initiatives relevant to marine reserve development. Enactment of the Great Barrier Reef Marine Park Act of 1975, to protect the world's largest coral barrier reef and one of its natural wonders, was among the first of these. More recent examples include the launching of *Australia's Oceans Policy* (Commonwealth of Australia 1998), *Guidelines for Establishing the National Representative System of Marine Protected Areas* (NRSMPA, ANZECC 1998), and the *Strategic Plan of Action for the NRSMPA* (ANZECC 1999). "The primary goal of the NRSMPA is to establish and manage a comprehensive, adequate and representative system of MPAs to contribute to the long-term ecological viability of marine and estuarine systems, to maintain ecological processes and systems, and to protect Australia's biological diversity at all levels" (ANZECC 1999). Implicit in these policies and plans is a subgoal of developing a national system of no-take marine reserves. They also include an assessment of the status of Australian MPAs and institutional arrangements and commitments to further develop them. By 1999, there were already hundreds of Australian MPAs covering millions of square kilometers. Only a small percentage of this MPA area would qualify as no-take marine reserve, but even this small percentage was substantial.

The GBRMP is the world's largest MPA extending over 2000 km in length and covering 339,750 km$^2$ of ocean area (ANZECC 1999). The GBRMP adopted a multiple-use approach early on and developed strong cooperation with adjacent territorial governments and management plan and zoning plan

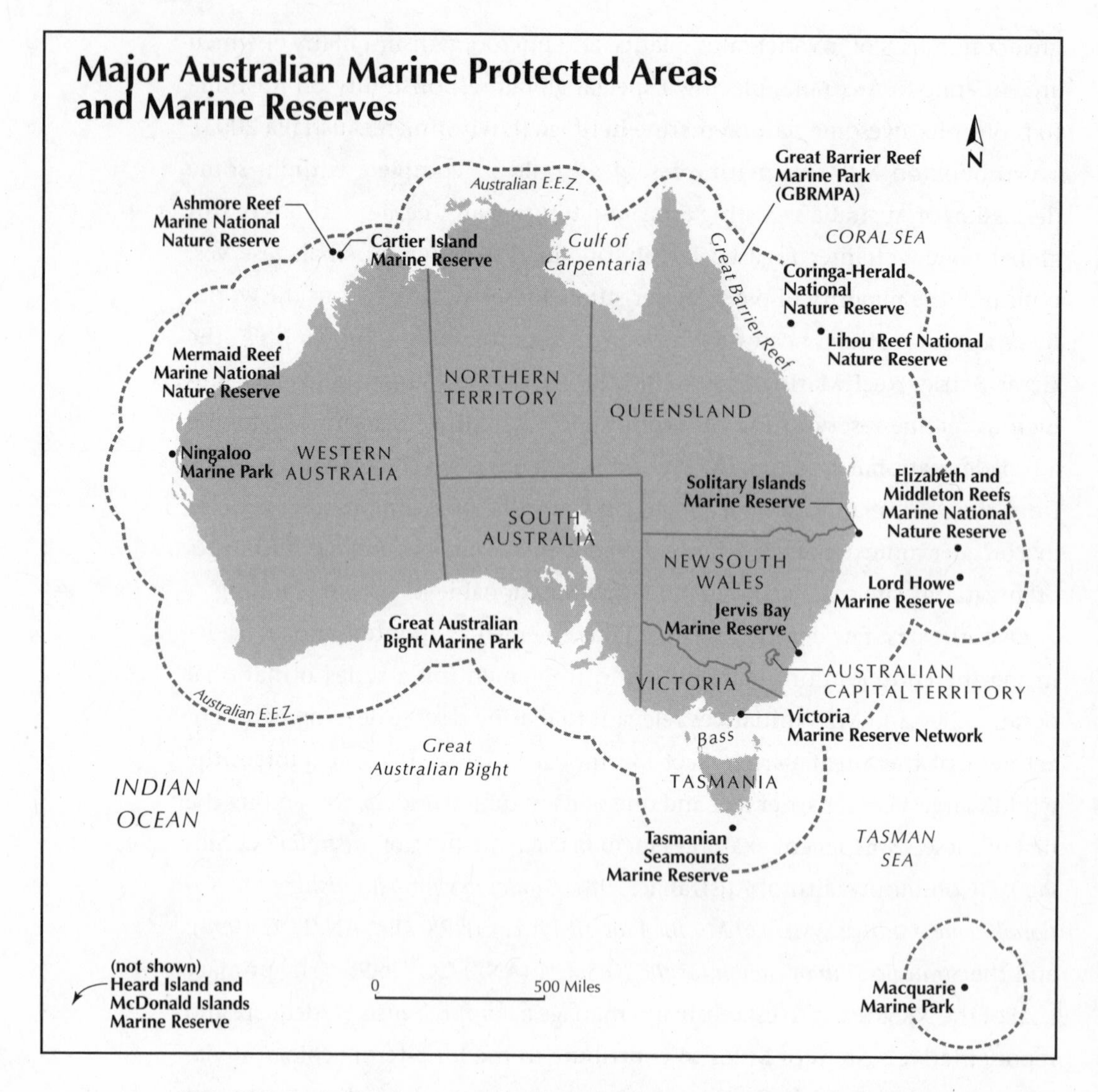


**FIG. 11.5 Overview Map of Major Australian Marine Protected Areas (MPAs) and Marine Reserves.** This map locates major MPA and marine reserve efforts around Australia as identified in its Strategic Plan for MPAs (NRSMPA). Source: ANZECC (1999).

processes to implement it. The ecosystem-oriented, public-driven planning processes developed for managing the GBRMP have often been well-received and viewed as state of the art by many (Kelleher 1999; Salm and Clark 2000). Yet more than a quarter century after its creation, less than 5 percent of GBRMP's area was fully protected from fishing activities, and most of this was in the remote, difficult to access, Far Northern Management Area. Furthermore, only limited research on the areas closed to fishing had been published, and questions had been raised about the adequacy of enforcement (Russ 2002). Nonetheless, those areas previously closed to fishing were also set within a larger marine park zoning and management regime and benefited from buffer areas surrounding them, as well as protection from other threats. Recently approved, profound revisions to the zoning system for the entire GBRMP greatly expand marine reserve coverage throughout the parks, put it at the cutting edge of reserve design, and represent a sea change in its approach (see following).

Although the GBRMP is by far Australia's largest and best-known MPA, significant MPAs, marine reserves, and major initiatives to expand the coverage of both exist throughout Australia's ocean waters (Fig. 11.5). Australia's strategic plan for MPAs (NRSMPA), already mentioned, provides an overview and assessment of these major efforts, including a summary of those focused on both Commonwealth (national) and state and territorial waters (ANZECC 1999). At the state and territorial level, this plan describes major ongoing MPA efforts focused on ocean waters off of New South Wales, the Northern Territory, Queensland, South Australia, Tasmania, Victoria, and Western Australia. We discuss recent marine reserve highlights involving some of these before returning to discuss recent developments regarding the nation's premier GBRMP.

Tasmania sits off the southernmost tip of the Australian mainland with over 5000 km of cool temperate coastline, the most coastline per unit area of any state on the continent. This extensive coastline consists of magnificent rocky reefs, towering sea cliffs, spectacular tide pools, and beautiful beaches, interspersed with scenic bays, sea grass meadows, and estuaries. Further offshore, there is varied bathymetry, including a multitude of large seamounts. Tasmania's marine flora is unsurpassed in diversity and its marine fauna is similarly diverse, including a dizzying array of marine invertebrates and vertebrates ranging from endemic handfish to sharks, seals, penguins, dolphins, and the great whales. Tasmania acknowledges a responsibility to protect its biodiversity for current and future generations (Tasmania Parks and Wildlife 2003).

In 1991, Tasmania designated four relatively small, shallow water, no-take marine reserves (Fig. 11.6) at Governor's Island (0.5 $km^2$), Maria Island (15 $km^2$), Ninepin Point (0.6 $km^2$), and the Tinderbox (0.45 $km^2$). Edgar and Barrett (1999) conducted a multiyear study of changes in the marine biota at these four sites and associated unprotected control sites. They reported that the largest reserve, Maria Island, which covers 7 km of coastline, was most effective for marine conservation and resource protection. The number of fish, invertebrate, and algal species; the density of large fish and lobsters; and the mean size of certain fish and abalone all increased significantly at this reserve relative to control areas. Rock lobster biomass increased by an order of magnitude; large, predatory trumpeter fish (*Latridopsis forsteri*) biomass increased by

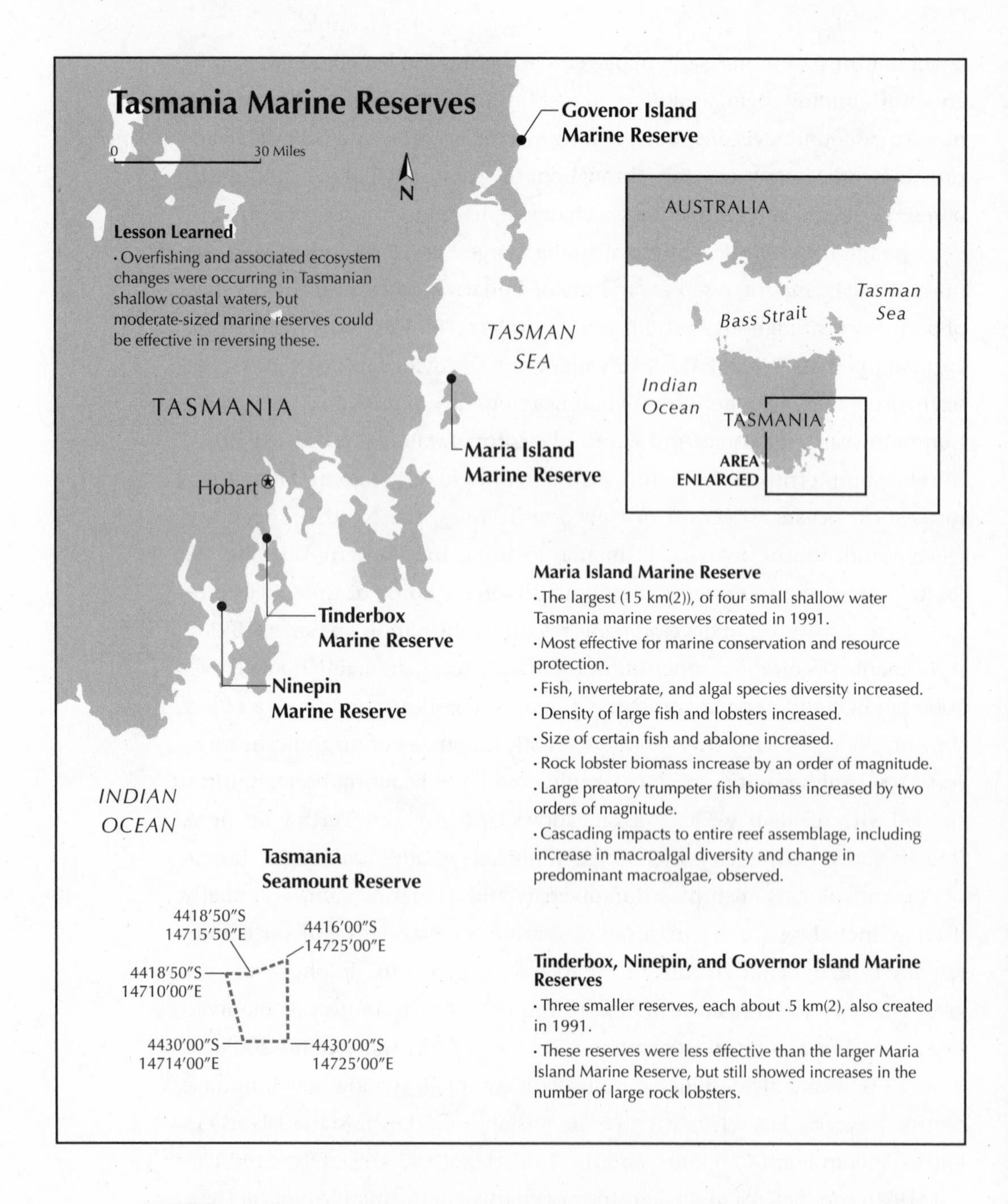


**FIG. 11.6 Tasmanian Marine Reserves Map and Key Points.** This figure shows the location of Tasmania relative to the rest of Australia, an enlarged section of southeast Tasmania, and the location of four coastal reserves and the much larger Tasmania Seamount Reserve offshore. Key points highlight the efficacy of the larger of the four coastal reserves created in 1991 (Maria Island) for species and ecosystem restoration and the more limited efficacy of the three smaller ones for lobster. The newer seamount reserve (1999) covers 370 km$^2$, but is only no-take at depths below 500 meters. Sources: Edgar and Barrett (1999), Koslow et al. (2001), Tasmanian Seamounts Reserve Management Plan (Commonwealth of Australia 2002).

two orders of magnitude; and large abalone became more numerous at this site. Cascading or indirect changes to the entire reef assemblage were also observed at Maria Island, including an increase in macroalgal diversity and a related change in the predominant macroalga from *Cystophora retroflexa* to *Ecklonia radiata*. The results from Maria Island indicated overfishing, and associated ecosystem changes were occurring in shallow coastal waters off Tasmania, but that moderate-sized reserves could be effective in reversing these. The three smaller reserves were less effective, but even these showed increases in large rock lobsters.

In chapter 2, we discussed recent research by Koslow et al. (2001) documenting damage to a group of seamounts south of Tasmania resulting from an orange roughy trawl fishery. Preliminary results from this research convinced the Commonwealth Government to create the Tasmanian Seamounts Marine Reserve in 1999 (Fig. 11.6), encompassing fifteen seamounts, for the purpose of "protecting the unique and vulnerable benthic communities of the seamounts." The Tasmanian Seamounts Marine Reserve Management Plan completed in 2002 to implement it is noteworthy for several reasons. The reserve and its management plan cover and protect about 370 $km^2$ of deep ocean habitat, including fifteen deep water seamounts, from all fishing and extractive activities, but do not provide similar protection to adjacent shallower seamounts or the shallower pelagic environment above the protected sea mounts. The plan essentially defines and treats the "benthic environment" below 500 meters as a separate zone within a larger MPA that extends all the way to the surface. Within the larger MPA, the benthic zone, everything below 500 meters depth is closed to all fishing activities; whereas the pelagic zone above it, less than 500 meters, is treated as a multiple use MPA where fishing is allowed but regulated. The plan does recognize potential impacts to the benthic marine reserve from fishing above and adjacent to it but assumes these will be minimal. It does include plans to monitor such impacts, consider them in future performance assessments, and address them in the future as warranted (Environment Australia 2002).

The Tasmanian government released a state-level MPA strategy (DPIWE 2001) in August 2001 but has yet to add additional state water no-take marine reserves, despite the documented shallow water successes already discussed. Several potential MPAs remain under consideration and may include no-take marine reserves but have yet to overcome political obstacles to their advancement. Meanwhile, several other Australian state level initiatives have rapidly and successfully advanced. Perhaps most notable is the progress made by Tasmania's neighbor to the north, Victoria.

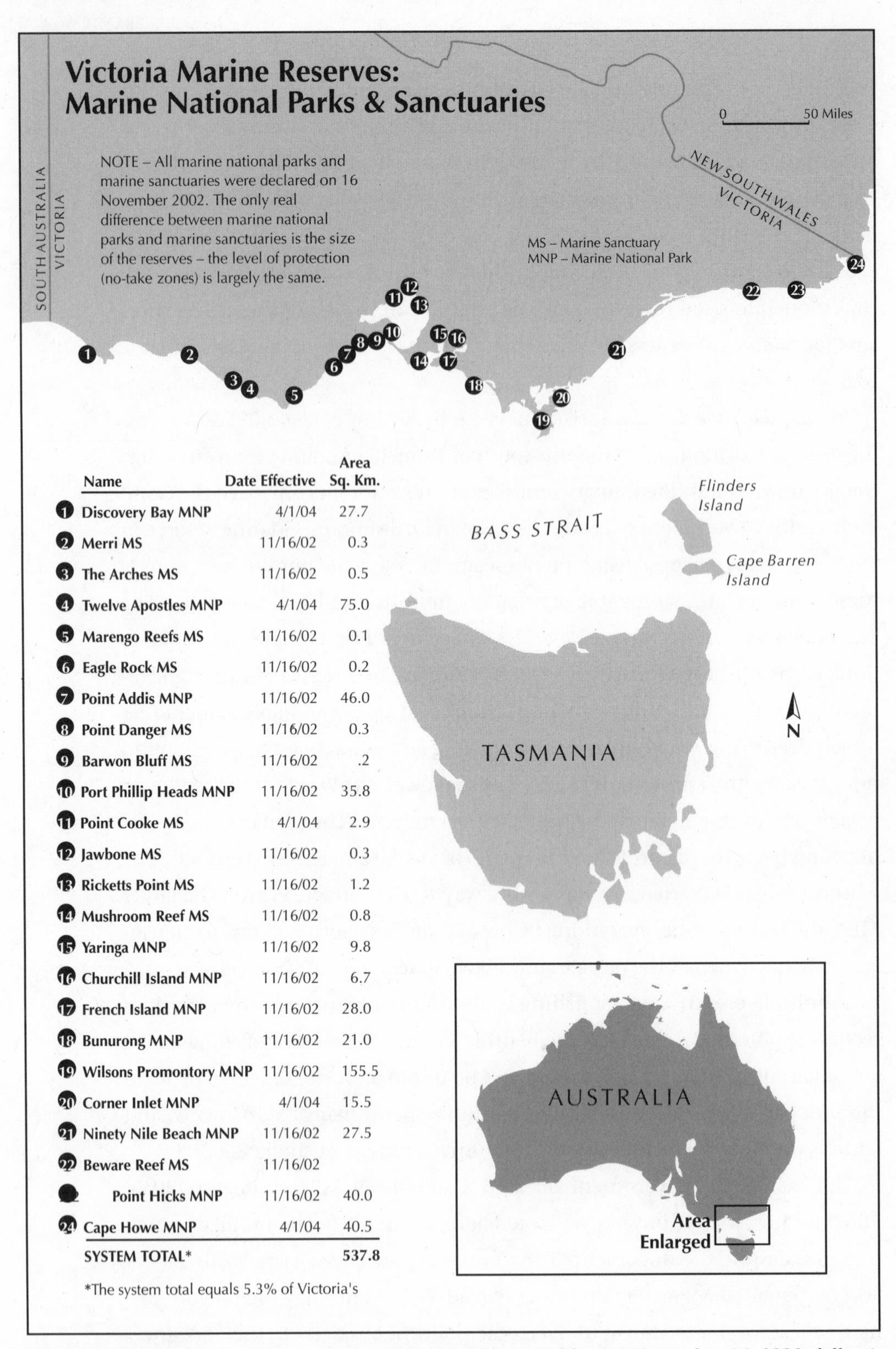


| | Name | Date Effective | Area Sq. Km. |
|---|---|---|---|
| 1 | Discovery Bay MNP | 4/1/04 | 27.7 |
| 2 | Merri MS | 11/16/02 | 0.3 |
| 3 | The Arches MS | 11/16/02 | 0.5 |
| 4 | Twelve Apostles MNP | 4/1/04 | 75.0 |
| 5 | Marengo Reefs MS | 11/16/02 | 0.1 |
| 6 | Eagle Rock MS | 11/16/02 | 0.2 |
| 7 | Point Addis MNP | 11/16/02 | 46.0 |
| 8 | Point Danger MS | 11/16/02 | 0.3 |
| 9 | Barwon Bluff MS | 11/16/02 | .2 |
| 10 | Port Phillip Heads MNP | 11/16/02 | 35.8 |
| 11 | Point Cooke MS | 4/1/04 | 2.9 |
| 12 | Jawbone MS | 11/16/02 | 0.3 |
| 13 | Ricketts Point MS | 11/16/02 | 1.2 |
| 14 | Mushroom Reef MS | 11/16/02 | 0.8 |
| 15 | Yaringa MNP | 11/16/02 | 9.8 |
| 16 | Churchill Island MNP | 11/16/02 | 6.7 |
| 17 | French Island MNP | 11/16/02 | 28.0 |
| 18 | Bunurong MNP | 11/16/02 | 21.0 |
| 19 | Wilsons Promontory MNP | 11/16/02 | 155.5 |
| 20 | Corner Inlet MNP | 4/1/04 | 15.5 |
| 21 | Ninety Nile Beach MNP | 11/16/02 | 27.5 |
| 22 | Beware Reef MS | 11/16/02 | |
| | Point Hicks MNP | 11/16/02 | 40.0 |
| 24 | Cape Howe MNP | 4/1/04 | 40.5 |
| | SYSTEM TOTAL* | | 537.8 |

*The system total equals 5.3% of Victoria's

**FIG. 11.7 Victoria Marine Reserve Network Map and Data Table.** On November 16, 2002, following more than a decade of consultation, The Government of Victoria created this extensive network of no-take marine reserves, covering more than 500 km² and 5 percent of its territorial waters and including a total of twenty-four marine national parks and sanctuaries. The twenty-four sites are mapped and listed here, along with their size and effective date. Source: Victoria's System of Marine National Parks and Marine Sanctuaries Management Strategy 2003–2010 (Parks Victoria, 2003)

The State of Victoria lies on the southern tip of the Australian mainland across from the island(s) of Tasmania. Victoria possesses a similarly spectacular diverse coastline stretching along 2000 miles of ocean and hosting an assortment of habitats ranging from sandy beaches, mud flats, and sea grass meadows to kelp and mangrove forests and rocky coasts and reefs. Victoria's diverse and distinctive marine environment supports a myriad of marine life, much of it endemic to southern Australia, and is a key piece of the state's natural heritage. Victoria's Parliament recognized this on June 11, 2002, passing legislation to create an extensive network of marine reserves to protect representative examples of its marine ecosystems. On November 16, 2002, the Government of Victoria finalized action to protect these marine assets, proclaiming the creation of thirteen new marine national parks and eleven new marine sanctuaries, all of which fit the definition of *no-take marine reserve* used in this book (see Fig. 11.7). Collectively, they cover 540 $km^2$ or about 5.3 percent of Victoria's marine waters. The largest park is roughly 150 $km^2$ (Parks Victoria 2003).

Although final enactment of the legislation and the proclamation formally establishing Victoria's system of no-take marine national parks and sanctuaries occurred rapidly once a final agreement was hammered out, the creation of this reserve network certainly did not occur overnight. Rather, like most successful marine reserve efforts, the concept of a system of such reserves for Victoria was carefully considered and negotiated over a long period of time with extensive scientific, public, stakeholder, and political input. The process took more than a decade, produced six public reports, provided six periods of public comment, received approximately 5000 public submissions (comments), survived three state governments and the disbanding of the agency originally assigned to it, incorporated a shift in attitudes regarding the relative value of no-take versus multiple-use MPAs, held hundreds of meetings, and, not surprisingly, withstood or perhaps more accurately benefited from a robust level of community debate. Despite this, the proposal still initially failed in Parliament when consensus on it could not be reached between the ruling (majority) party and the opposition (minority) party, and succeeded the second time only after concessions were made (Allen, 2003; MPA News 2003).

The Victoria effort began in 1991 when the Land and Conservation Council (LCC), an independent statutory authority responsible for land-use planning, was assigned the task of making recommendations for a representative system of marine parks. In 1996, they released a draft report for review proposing twenty-one multiple-use MPAs, with only limited no-take zones in each. State government disbanded the LCC while it was in the process of preparing final recommendations, created another independent body, the Environmental Conservation Council (ECC), and reassigned the task to it. In 1998, the ECC, composed of three members from the fields of academia, agriculture, and finance, each having extensive experience in natural resource management, released another interim report focusing on multiple use MPAs for public review. However, by then, both the ECC and the public were questioning the ability of multiple-use MPAs on their own to protect Victoria's biodiversity. Based on science, the ECC concluded that no-take areas provided better protection and were warranted. In 1999, they released yet another report for public comment proposing a network of no-take parks and sanctuaries. Following additional public comment and consultation, the ECC provided a final report to government calling for thirteen marine national parks and eleven marine sanctuaries. After government further modified the proposal, altering some boundaries and adding a compensation package for fishers, a bill was introduced in 2001 but was blocked by the opposition. The bill was again revised, boundaries adjusted, and fisher compensation increased, before being reintroduced and finally enacted in 2002 (MPA News 2003).

Researchers associated with the University of Washington School of Marine Affairs conducted interviews with six individuals who were intimately involved in the creation of the Victoria reserve network following its successful conclusion. Those interviewed spanned the government, nongovernment, conservation, and recreational and commercial fishing sectors. Based on those interviews, the following lessons were distilled: (1) opposition is part of consultative processes on no-take areas; (2) advisory bodies should strive for independence, fairness, and credibility; (3) educating the public on the need to protect biodiversity is critical; (4) negotiation is sometimes a better strategy for opponents than total opposition; and (5) an "all-at-once" strategy for creating a system of no-take reserves has some advantages (MPA News 2003). Experience elsewhere clearly provides strong support for at least the first three of these. Additionally, active engagement in marine reserve processes by nongovernment supporters is also critical to their success.

Elsewhere around Australia, both the Commonwealth and individual state governments have established marine reserves and/or are advancing marine reserve and MPA initiatives. For example, the State of New South Wales (NSW) and the Commonwealth recently finalized management plans for several large multiple-use MPAs that include significant no-take zones within them and have several similar plans in draft form. NSW lies on the east coast of Australia sandwiched between Victoria to the south and Queensland and the GBRMP to the north. In 1997, the NSW Marine Parks Authority was created to "manage

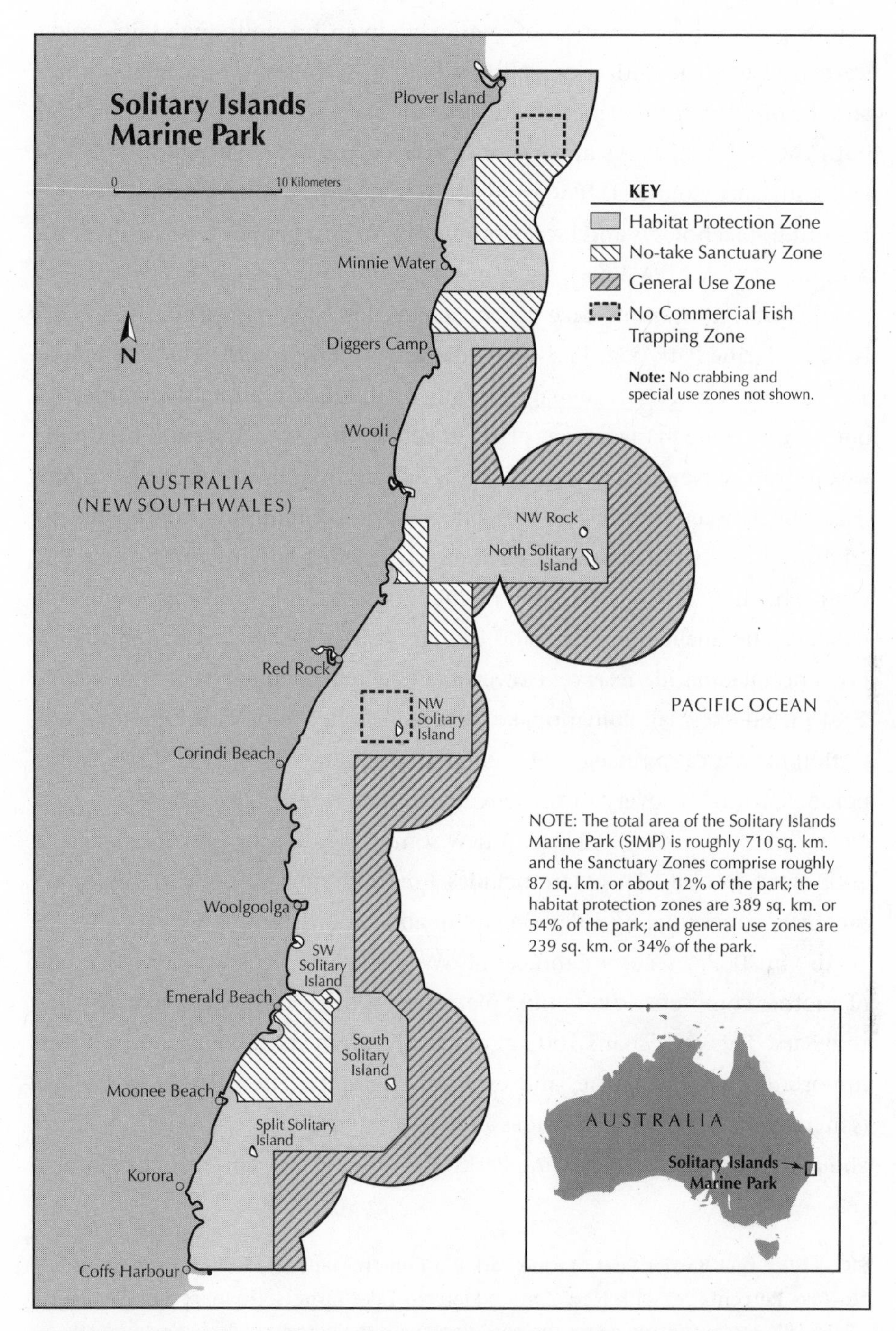


**FIG. 11.8 Solitary Islands Marine Park and Zoning Plan Map.** The new Solitary Islands Marine Park Zoning Plan depicted here was approved in 2001 and implemented in 2002 creating New South Wales first significant no-take system. The park, designated in 1998, covers a total of 710 km$^2$. Under the new plan, about 86 km$^2$ or 12 percent is now zoned no-take as "sanctuary zones." A second no-take system was also approved for the Jervis Bay Marine Park in 2002, and efforts are underway to implement similar systems for Cape Byron and Lord Howe Marine Parks. Sources: NSW MPA (2003) and Environment Australia (2001).

marine parks for conservation of marine biodiversity and to maintain ecological processes." The Authority reports to both the Minister for the Environment and the Minister for Fisheries and Agriculture and includes representation from both the National Parks and Wildlife Service and NSW Fisheries. A Marine Parks Advisory Council (MPAC) and additional local advisory committees, representing stakeholders and local community interests, support the work of the Authority (NSW MPA 2003).

The first significant no-take system achieved in NSW lies within the Solitary Islands Marine Park (Fig. 11.8) and adjacent Solitary Islands Marine Reserve. The Park and Reserve are managed jointly by the NSW Marine Parks Authority, but have separate management plans. Together, they span State and Commonwealth waters. New South Wales and the Solitary Islands region lie in a transition zone between tropical and temperate marine communities, contain diverse and unique assemblages of tropical and temperate species and diverse and unique habitats, and are critical to maintaining rare and vulnerable species and habitats. The smaller (160 km$^2$) and further offshore Reserve, created in 1993, is not a no-take marine reserve. However, a new management plan approved in 2001 includes a small, fully no-take "sanctuary zone" to provide high-level protection for critical pinnacle reef habitat around Pimpernel Rock of special importance to the recovery of the threatened grey nurse shark. The larger park, created in 1998, covers 710 km$^2$. A new zoning plan for the park took effect in 2002 (see Fig. 11.8). This plan includes a more significant network of no-take sanctuary zones totaling about 86 km$^2$ or about 12 percent of the park area.

Also in 2002, a second significant NSW no-take network was created as part of another comprehensive zoning plan approved for the coastal Jervis Bay Marine Park. This park spans 100 km of coastline, covers 220 km$^2$, and protects important coastal, marine, and estuarine habitats. The park's new no-take (sanctuary zone) network includes a representative 20 percent of its habitats or about 43 km$^2$. The NSW Marine Parks Authority is also currently developing

**FIG. 11.9 Great Barrier Reef Marine Park Bio-Region Maps and Comparison of Prior No-Take Percentages with New Zoning Plan No-Take Targets.** (A) maps and lists the 32 Reef Bio-regions identified for the park along with the percent of each zoned no-take prior to 2004 and the percent that received no-take protection under the new Zoning Plan of 2004. (B) provides the same information for the 41 nonreef bioregions identified for the park. The new zoning plan received final approval in 2004 and protects a representative 30 percent of the entire park in no-take "Marine National Park Zones." The total area protected exceeds 100,000 km$^2$ dwarfing any prior no-take network. Source: GBRMPA RAP (2003).

**Reef Bioregions**

1. **Torres Straits Influence Mid-Shelf Reefs** 0% (21.6%)
2. **Deltaic Reefs** 41.1% (42.7%)
3. **Far Northern Open Lagoon Reefs** 28.7% (30.2%)
4. **Far Northern Outer Mid-Shelf Reefs** 7.4% (26.1%)
5. **Coastal Far Northern Reefs** 28.8% (35.4%)
6. **Far Northern Protected Mid-Shelf Reefs and Shoals** 42.2% (47.6%)
7. **Outer Barrier Reefs** 38.1% (42.2%)
8. **Northern Open Lagoon Reefs** 6.7% (51.5%)
9. **Coastal Northern Reefs** 43.7% (46.7%)
10. **Sheltered Mid-Shelf Reefs** 25.4% (30.0%)
11. **Outer Shelf Reefs** 33.2% (56.7%)
12. **Coastal Central Reefs** 1.7% (21.5%)
13. **Exposed Mid-Shelf Reefs** 7.0% (20.2%)
14. **High Continental Islands** 10.6% (24.0%)
15. **Central Open Lagoon Reefs** 8.9% (21.9%)
16. **Strong Tidal Mid-Shelf Reefs (West)** 17.1% (27.8%)
17. **Strong Tidal Outer Shelf Reefs** 10.4% (19.7%)
18. **Hard Line Reefs** 13.3% (21.1%)
19. **High Continental Islands** 10.6% (24.0%)
20. **Strong Tidal Inner Shelf Reefs** 1.3% (22.7%)
21. **Strong Tidal Inner Mid-Shelf** Reefs 1.3% (24.6%)
22. **High Tidal Fringing Reefs** 1.1% (24.5%)
23. **High Tidal Fringing Reefs** 1.1% (24.5%)
24. **Incipient Reefs** 9.1% (33.9%)
25. **Tidal Mud Flat Reefs** 0.3% (23.3%)
26. **Coral Sea Swains Northern Reefs** 40.9% (70.5%)
27. **Strong Tidal Mid-Shelf Reefs (East)** 15.1% (20.9%)
28. **Swains Mid-Reefs** 26.2% (36.6%)
29. **Swains Outer Reefs** 3.1% (20.3%)
30. **Coastal Southern Fringing Reefs** 3.5% (24.5%)
31. **Capricon Bunker Mid-Shelf Reefs** 4.6 (35.5%)
32. **Capricorn Bunker Outer Reefs** 13.5% (20.2%)



# Great Barrier Reef Marine Park Bioregions (A)

NOTE: For each bioregion, the first percentage represents the portion of it included in no-take prior to the Zoning Plan of 2003. The second percentage (in parens) represents the new portion that will recieve no-take protection under this zoning plan, which take effect in July 2004.

## Non-Reef Bioregions

1. **Outer Far Northern Inter-Reef** 62.2% (74.1%)
2. **Outer Shelf Lagoon** 26.3% (47.0%)
3. **Mid-Shelf Inter-Reef** 27.0% (36.7%)
4. **Eastern Plateau** 14.4% (56.7%)
5. **Mid-Shelf Inter-Reef Seagrass** 17.2% (27.3%)
6. **Hallmeda Banks–Some Coral** 12.7% (44.0%)
7. **Inshore Muddy Lagoon** 24.8% (29.9%)
8. **Far Northern Offshelf** 0.0% (21.7%)
9. **Hallmeda Banks** 4.9% (25.4%)
10. **Mid-Shelf Sandy Inter-Reef** 4.8% (32.9%)
11. **Princess Charlotte Bay Outer Shelf** 0.0% (28.2%)
12. **Steep Slope** 1.1% (32.6%)
13. **Princess Charlotte Bay** 2.7% (28.2%)
14. **Outer Shelf Algae and Seagrass** 18.0% (39.7%)
15. **Mid-Shelf Seagrass** 3.6% (57.1%)
16. **Coastal Strip** 30.7% (35.2%)
17. **Offshore Queensland Trough** 0.0% (54.6%)
18. **Outer Shelf Seagrass** 3.0% (30.0%)
19. **High Nutrients Coastal Strip** 0.5% (19.8%)
20. **Outer Central Inter-Reef** 0.5% (35.8%)
21. **Inner-Shelf Seagrass** 0.2% (23.0%)
22. **Queensland Trough** 0.0% (22.2%)
23. **Intermediate Broad Slope** 0.0% (28.0%)
24. **Central Offshelf** 0.0% (64.8%)
25. **Outer Shelf Inter-Reef Central** 3.2% (27.0%)
26. **Inner Mid-Shelf Lagoon** 0.2% (20.3%)
27. **Terraces** 0.1% (22.1%)
28. **Western Pelagic Platform** 0.0% (48.5%)
29. **Outer Shelf Inter-Reef Southern** 4.5% (24.8%)
30. **Central Inner Reef** 3.3% (32.2%)
31. **Mid-Shelf Lagoon** 0.1% (25.3%)
32. **Inner Shelf Lagoon Continental Islands** 0.1% (31.8%)
33. **Eastern Pelagic Platform** 0.0% (99.9%)
34. **Swains Inter-Reef** 7.6% (27.6%)
35. **Inshore Terrigenous Sands** 0.5% (28.6%)
36. **Terraces** 0.1% (22.1%)
37. **Capricorn Trough** 0.0% (40.7%)
38. **Capricorn Bunker Banks** 0.0% (27.3%)
39. **Capricorn Bunker Inter-Reef** 1.9% (40.3%)
40. **Capricorn Bunker Lagoon** 0.0% (20.2%)
41. **Southern Embayment** 0.0% (43.5%)

AUSTRALIA (QUEENSLAND)

CORAL SEA

Great Barrier Reef

N

## Great Barrier Reef Marine Park Non-Reef Bioregions (B)

0 100 Miles

NOTE: For each bioregion, the first percentage represents the portion of it included in no-take prior to the Zoning Plan of 2003. The second percentage (in parens) represents the new portion that will recieve no-take protection under this zoning plan, which takes effect in July 2004.

zoning plans that will include no-take networks for the 480 km$^2$ Lord Howe Island Marine Park created in 1999 and the 227 km$^2$ Cape Byron Marine Park created in 2002. The range of options currently under consideration for the Lord Howe Island Marine Park range would designate between 17 and 51 percent or 82 to 245 km$^2$ of its area as no-take. Since the creation of the NSW Marine Parks Authority and its associated advisory bodies, NSW has put together an impressive and growing system of no-take reserves (Environment Australia 2001; NSW MPA 2003).

Given the advancing tide of marine reserves elsewhere along Australia's east coast, a casual observer in early 2003 might have wondered what was happening with its premier GBRMPA and no-take marine reserves. GBRMPA provided a resounding answer to that question in June 2003, announcing a new draft comprehensive zoning plan that would fully protect a representative 30 percent of the GBRMP in no-take reserves termed Marine National Park Zones (MNPZs; Figs. 11.9a and b; GBRMPA RAP 2003). The new draft zoning plan covered the entire GBRMP, a departure from prior zoning plans that were done piecemeal, section-by-section; includes twenty-eight recently added coastal sections; and views the park as seventy interconnected bioregions. The total area within the GBRMP proposed for no-take status was 101,925 km$^2$, dwarfing any prior no-take network and likely the sum total of all other no-take areas in the world! Many scientists have supported no-take status for 20 to 30 percent of representative habitats as a reasonable estimate of what may be necessary to protect their ecological integrity, and similar figures have been applied elsewhere in creating no-take marine reserve networks for smaller areas, but never at anything approaching this scale.

The GBRMP's Representative Areas Program (RAP) is responsible for developing and implementing the zoning plan, one of several recent actions by the Australian government to increase protection for this national landmark. It comes on the heels of a recently released draft Water Quality Action Plan designed to address land-based sources of pollution. The stated purpose of the RAP and proposed zoning plan is to "Increase the protection of biodiversity within the Marine Park through increasing the extent of Marine National Park Zones (MNPZs), also called green zones or no-take zones, to:

- maintain biological diversity of the ecosystem, habitat, species, population, and genes;
- allow species to evolve and function undisturbed;
- provide an ecological safety margin against human-induced disasters;

- provide a solid ecological base from which threatened species or habitats can recover or repair themselves; and
- maintain ecological processes and systems.

Two independent steering committees were created to advise the RAP on:

- biological and physical aspects of the Great Barrier Reef; and
- social, economic, cultural, and management feasibility of human use of the Park.

These steering committees developed zoning principles largely consistent with the criteria put forth earlier in this book (GBRMPA RAP 2003).

Why did this plan move forward now? Scientific, socioeconomic, public, and political support are all contributing to its advancement. Much groundwork was done prior to the release of the draft zoning plan, including an extensive biodiversity and habitat mapping initiative, a series of State of the Great Barrier Reef environmental status reports, and extensive consultation. The weight of scientific evidence indicated that the prior 4.5 percent of the GBRMP closed to extractive activities was not enough. Strong science supported the proposed management plan. A recognized decline in reef resources, increased human pressures, and the efficacy of existing closed areas prompted the action to increase protection. Analyses of marine-based industries dependent on a healthy GBR ecosystem demonstrate its annual value in the billions of dollars and that the great majority of this is generated by nonextractive activities. The vast majority of more than 10,000 public submissions (comments) supported stronger protection, and the high level of response indicated strong public interest. Members of both the ruling and opposition parties supported the draft zoning plan. The zoning plan was finalized and approved by Parliament March 25, 2004 and will come into force on July 1, 2004 (GBRMPA RAP, 2003; Kemp, 2004). The final approved plan mirrors the draft plan targets (Fig. 11.9a&b) and increases the percentage of the GBRMPA protected in "Green Zones" or marine reserves (areas in which all fishing and other extractive activities are prohibited) from less than 5 percent to over 30 percent of the park (Fig. 11.10a&b). The total "Green Zone" or marine reserve area approved covers more than 100,000 sq. kms!

Why was this ambitious proposal successful? The factors discussed above including diverse support, groundwork, sound science, mapping, status reports, and extensive consultation all contributed to its success. In addition, the GBRMPA's history and experience in dealing with complicated zoning and other issues, conducting extensive community outreach, and developing good public process unquestionably helped. Finally, top-down national-level pol-

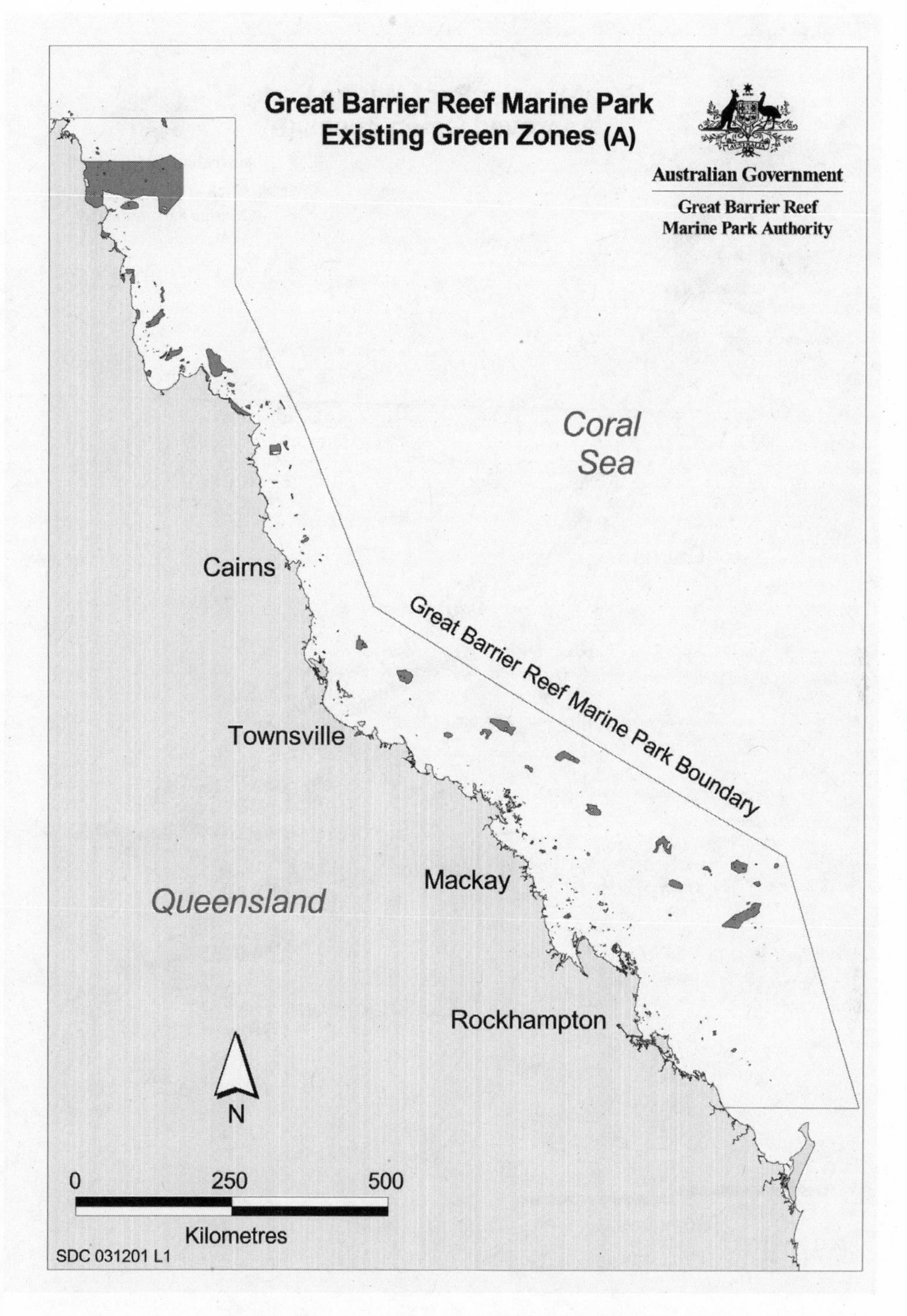


**Figure 11.10 Great Barrier Reef Marine Park (GBRMP) Green Zones.** (A) shows no-take Marine Reserves as of 2003, prior to adoption of new zoning plan in 2004. These pre-existing zones covered <5% of the GBRMP, mostly in the Far Northern Section. (B) shows no-take Marine Reserve zones approved and effective July 1, 2004. These new no-take zone constitute >30% of the GBRMP and are more representative than early ones. Source: Kemp, 2004; GBRMP RAP, 2003).

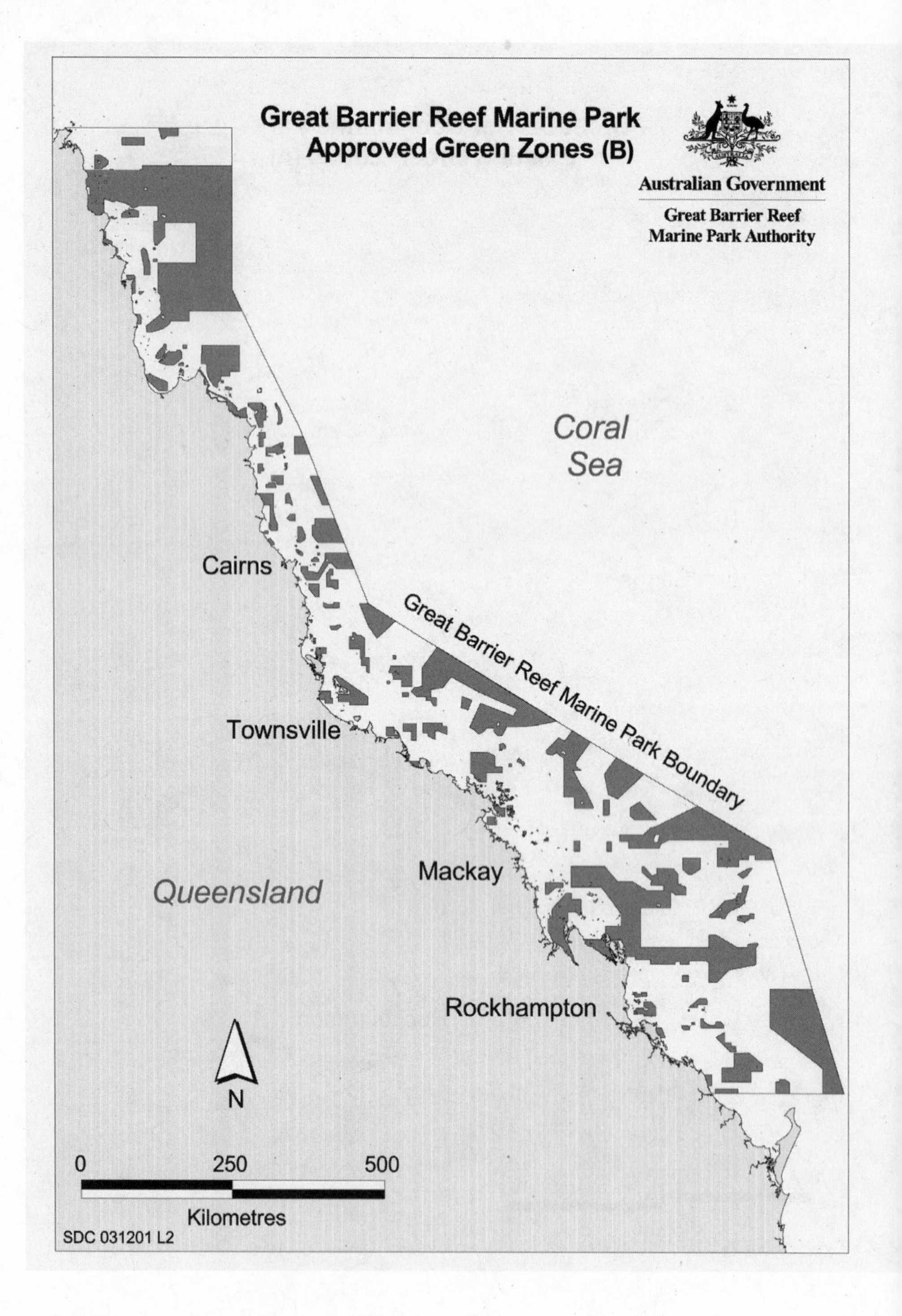

Great Barrier Reef Marine Park
Approved Green Zones (B)
Australian Government
Great Barrier Reef
Marine Park Authority
Coral Sea
Great Barrier Reef Marine Park Boundary
Cairns
Townsville
Mackay
Rockhampton
Queensland
N
0
250
500
Kilometres
SDC 031201 L2

icy support and consistency with this plan developed over a number of years through extensive public consultation and outreach and a national commitment to safeguard Australia's ocean resources (e.g., Australia's National Ocean Policy and Strategic Plan for Establishing the National Representative System of Marine Protected Areas; see above) were essential to finalizing it. The resulting strong public recognition of the reef's value for more than just extractive use, the threats facing it, and the need to protect it were critical. What are the prospects for the plan successfully protecting the Great Barrier Reef? Making predictions about the outcome of marine conservation is risk-prone and threats to reefs continue to mount. The new zoning plan with its expanded no-take areas provides part of the solution, but cannot do it alone. Fortunately, the new reserves are set in the context of and buffered by the larger GBRMP and newly enhanced water quality protection.

## CONCLUSION

Based on the global experience to date, the most compelling evidence on the efficacy of marine reserves concerns their role in ecosystem protection and biodiversity conservation. This, coupled with the fact that many other reserve benefits derive from ecosystem protection, suggests it as the central organizing principle around which marine reserves should be built. This case is strengthened when one considers that ecosystem protection, properly presented, probably has the greatest appeal to the broadest public sector.

Supportive arguments for reserves built on improved fishery yields, expanded knowledge and understanding, and enhanced nonconsumptive opportunities are also valid and should be utilized but should be considered secondary. These arguments may have special appeal to more limited sectoral interests. Even when such groups are not leading efforts to establish reserves, cultivating their support or even tempering their opposition is a worthy endeavor. In particular, support from fishers can be critically important. The theoretical and common sense arguments for the role of marine reserves in maintaining and increasing fisheries yields have always been very strong, and empirical evidence is now increasingly mounting to support them. Such evidence will continue to mount the longer existing reserves remain in place, the more design improvements are implemented, the greater the size and number of reserves, and the greater the percentage of the ocean they occupy.

The examples highlighted in this global review share much in common with the more detailed case studies presented earlier. Across these jurisdictions, a growing trend is evolving toward successful development of larger individual marine reserves and reserve networks utilizing improved design and public process principles tailored to local situations. These larger reserves and reserve networks provide opportunities to greatly expand and document reserve benefits related to both ecosystem protection and fisheries conservation.

## THE FUTURE OF MARINE RESERVES

A sea change is occurring with respect to marine reserve development around the world. The examples highlighted in the global review share much in common with the more detailed case studies presented earlier. Across these jurisdictions, a growing trend is evolving toward successful development of larger individual marine reserves and reserve networks utilizing improved design and public process principles tailored to local situations. These larger reserves and reserve networks provide opportunities to greatly expand and document reserve benefits related to both ecosystem protection and fisheries conservation.

In the United States, development of marine reserves off the Florida Keys, including the larger Tortugas Ecological Reserve; the more extensive network of reserves off of California's Channel Islands; and of several remote island marine reserves in the Caribbean and Central Pacific are especially noteworthy and indicative of this trend. Ongoing efforts to develop a strong network of marine reserves surrounding the remote, undeveloped Northwestern Hawaiian Islands and set within a marine protected area covering a very large spatial scale afford a tremendous opportunity to advance this trend to a new level. Among other more developed countries, the continued expansion of New Zealand's national network of marine reserves and the more recent and extensive advances in development of marine reserve networks in Australia likely lie at the leading edge of marine reserve progress. Similarly, developing national marine reserve networks in the Bahamas and in Belize are representative of the forefront of marine reserve progress among less developed countries.

Maximizing the conservation and other benefits afforded by potentially larger marine reserves and reserve networks will require applying the lessons learned to date around the world and adapting them as appropriate to these new opportunities. The following lessons summarized here are among those worthy of application elsewhere:

- Even as human alteration of marine ecosystems accelerates outside their boundaries, marine reserves of all sizes are proving effective in stemming impacts, reversing declines, and protecting resources from degradation within their boundaries.

- Larger reserves and reserve networks provide a greater potential for a broad range of ecosystem level benefits than smaller ones, especially extending fishery and other benefits beyond their borders, but even smaller individual ones can contribute significantly with respect to some benefits.
- Longer time frames, at least 20–30 years, are necessary to maximize and see the full range of marine reserves benefits, though some reserve benefits are often seen fairly quickly, within a few years.
- Public support and community involvement, involving both fishers and other interests, are normally essential to reserve success, especially in populated areas.
- Lively and vigorous discussion regarding marine reserve issues is often beneficial. Attempts to stifle debate often backfire, though tools for keeping it civil, respectful, and constructive are warranted. Good, open public process is highly desirable and effective.
- Marine reserve benefits involve much more than just fisheries and must be considered with more in mind.
- Strong research and monitoring programs, including natural and social sciences, can be critical to marine reserve success, evaluation, and expansion.
- Protecting marine reserves from other forms of human alteration than fishing can also be critical to their success.
- Recent marine reserve advances in Australia and New Zealand suggest that sound top-down or national level approaches, policies, and support combined with more localized, on the ground, implementation offer great prospects for success.

Marine reserves will likely remain controversial and contentious in many places and among some stakeholders, despite, and in some cases because of, the considerable progress made to date in many areas with the participation of many stakeholders. In the United States, there has been some backlash within certain user communities to the successful establishment of marine reserves in the Florida Keys and Channel Islands. Yet, in the long run, the resulting public debate on marine reserves will likely be a net benefit, and recent progress on marine reserve science, design, and use will continue and likely accelerate further. Discussion of marine reserve issues among many constituencies, across multiple public sectors, and at a variety of levels is highly desirable, much needed, and likely essential to their continued success as a key marine policy tool.

Support from fishers can be extremely helpful in this respect. The theoretical and common sense arguments for the role of marine reserves in maintaining and increasing fisheries yields have always been very strong. Many knowledgeable and progressive fishers know that existing management tools are not working and are frequently among those most aware of declining resources and ecosystems. Empirical evidence is now accumulating to support marine reserves potential for benefiting fisheries. Such evidence will continue to mount the longer existing reserves remain in place, the more design improvements are implemented, and the greater the percentage of the ocean they occupy.

And, as human alteration of the oceans and our recognition of it continue to expand, the need and support for, and our use of marine reserves will also continue to grow. Expanded use and study of marine reserves will continue to improve our understanding of their efficacy and our ability to design and implement even more effective marine reserves and marine reserve networks. Already, we know that marine reserves and marine reserve networks can provide a broad range of benefits falling into four broad categories:

- Ecosystem protection,
- Improved fisheries,
- Expanded knowledge and understanding, and
- Enhanced non-consumptive use.

Until recently, most of the world's experience with marine reserves derived from relatively few, fairly small, isolated marine reserves. In both developed and developing countries, most marine reserves, including nearly all of the older, better known, and well studied ones, were on the order of a few square miles, square kilometers, or less. Most of these were single reserves, designed, managed, and studied individually and not part of a larger group, system, or network. Despite these limitations, these existing reserves have proven remarkably effective in delivering certain benefits, demonstrating their potential to achieve others, and documenting both. Now, we have overwhelming empirical evidence for the efficacy of marine reserves to deliver many of these benefits and strong theoretical evidence and common sense arguments support the others.

That is not to say marine reserves are a silver bullet. Marine reserves are necessary, but not sufficient for ocean conservation. As Leopold suggested, we can't afford to continue losing the pieces, even those whose value we don't yet recognize. A more precautionary, ecosystem-based approach is needed to prevent such loss. Marine reserves are an essential piece of such an approach, but not the whole answer. In the long-term, marine reserves will need to be implemented together with improved water quality protection, fisheries management, climate change mitigation, larger scale and more comprehensive marine zoning, and a full array of effective marine protected areas to maximize their effectiveness and meet conservation goals. However, the great should not be the enemy of the good. There is plenty of evidence that even small, isolated marine reserves can provide real benefits. Further, we should avoid the false dichotomy between marine reserves and other effective tools; both are necessary.

Globally and in individual countries, marine reserves have come a long way in recent years. Yet, we have still barely scratched the surface of their potential to assist in meeting a broad range of societal goals and providing a full suite of public benefits. Human activities continue to clearly, rapidly, and profoundly alter the world's oceans, its incredibly vibrant marine ecosystems, and their diverse myriad of life in tangible, fundamental, and potentially irreversible ways. Such changes are not new, but they are rapidly accelerating and will likely continue to do so for the foreseeable future. New ways of thinking and acting, changes to the status quo, and rapid action are urgently needed to address these impacts, reverse current trends, and prevent additional changes from becoming irreversible. They are needed to restore some of what's been lost, hold on to some of what's left, protect our options, understand our choices, improve our management decisions, and preserve some of the ocean's wilderness and the awe it inspires within us for the continued enjoyment of current and future generations.

Human alteration of marine ecosystems and their living inhabitants continues to accelerate and expand, but increased public awareness of such anthropogenic change and related changes in societal values and ethics offer some hope for the oceans future. These two factors combined with the continued failure of other existing management tools to successfully address the former and adapt to or reflect the latter, fuel our belief that the use of marine reserves will continue to advance. Marine reserves and the debate about their use are at least as much about societal goals, values, and ethics related to marine resource use as their science and design, though debate over the latter often masks more fundamental disagreement over the former. Nonetheless, such discord will likely ameliorate somewhat as the needs of ecosystem protection and more traditional fisheries management increasingly converge. A new ocean ethic focused more on what we must leave in ocean ecosystems to preserve their health and productivity, rather than simply maximizing what we extract, may yet emerge.

## REFERENCES

Asian Development Bank (ADB). 1993. *Fisheries Sector Profile of the Philippines*. Manila, Philippines.

Alcala, A. C. 1988. Effects of marine reserves on coral fish abundances and yields of Philippine coral reefs. *Ambio* 17(3):194–199.

Alcala, A. C., and G. R. Russ. 1990. A direct test of the effects of protective management on abundance and yield of tropical marine resources. *J. Cons. Ciem.* 47(1):40–47.

Allen, Tim. 2003. Personal Communication. Victorian Coordinator, Marine and Coastal Community Network. East Melbourne, Victoria, 3002 Australia.

Australia and New Zealand Environment and Conservation Council (ANZECC) Task Force on Marine Protected Areas, 1998. *Guidelines for Establishing the National Representative System of Marine Protected Areas*, Environment Australia. Canberra.

Australia and New Zealand Environment and Conservation Council (ANZECC) Task Force on Marine Protected Areas, 1999. *Strategic Plan of Action for the National Representative System of Marine Protected Areas*, Environment Australia. Canberra.

Babcock, R. C., S. Kelly, N. T. Shears, J. W. Walker, and T. J. Willis. 1999. Changes in community structure in temperate marine reserves. *Marine Ecology Progress Series* 189:125–134.

Ballantine, W. J. 1991. *Marine Reserves for New Zealand.* Auckland: University of New Zealand.

———. 1997. "No-take" marine reserve networks support fisheries. In Hancock, D. A., D. C. Smith, A. Grant, and J. P. Beumer, eds. *Developing and Sustaining World Fisheries Resources: The State of Science and Management,* 702–706. Proceedings from the Second World Fisheries Congress, Brisbane, Australia. Collingwood, Australia: CSIRO Publishing.

———. 1999. Marine reserves in New Zealand: The development of the concept and the principles. In *Proceedings of an International Workshop on Marine Conservation for the New Millenium,* 3–38. Cheju Island, Korea: Korean Ocean Research and Development Institute.

Barber, C. V., and V. R. Pratt. 1997. *Sullied Seas: Strategies for Combating Cyanide Fishing in Southeast Asia and Beyond.* Washington, DC: World Resources Institute.

———. 1998. Poison and profits. *Environment* (Washington DC) 40(8):4–9.

Brown, P. J., and R. B. Taylor. 1999. Effects of trampling by humans on animals inhabiting coralline algal turf in the rocky intertidal. *Journal of Experimental Marine Biology and Ecology* 235(1):45–53.

Castilla, J. C. 1981. Perspectives for investigations of the structure and dynamics of rocky intertidal communities from Central Chile, II: High trophic level predators. *Medio Ambiente. Valdivia* 5(1–2):190–215.

———. 1986. Sigue Existiendo la necesidad de establecer parques y reservas maritimas en Chile? *AMB. y DES.* 11(2):53–63.

———. 1994. The Chilean small-scale benthic shellfisheries and the institutionalization of new management practices. *Ecology International Bulletin* 21:47–63.

Castilla, J. C., and R. H. Bustamante. 1989. Human exclusion from rocky intertidal of Las Cruces, central Chile: Effects on *Durvillaea antarctica* (Phaeophyta, Durvilleales). *Marine Ecology Progress Series* 50(3):203–214.

Castilla, J. C., and L. R. Durán. 1985. Human exclusion from the rocky intertidal zone of central Chile: The effects on *Concholepas concholepas* (Gastropoda). *Oikos* 45(3):391–399.

Castilla, J. C., G. M. Branch, and A. Barkai. 1994. *Exploitation of Two Critical Predators: The Gastropod,* Concholepus concholepus, *and the Rock Lobster,* Jasus lalandii. Heidelberg, Germany: Springer-Verlag.

Cole, R. G., E. Villouta, and R. J. Davidson. 2000. Direct evidence of limited dispersal of the reef fish *Parapercis colias* (Pinguipedidae) within a marine reserve and adjacent fished areas. *Aquatic Conservation: Marine and Freshwater Ecosystems* 10(6):421–436.

Commonwealth of Australia. 2002. *Tasmanian Seamounts Marine Reserve Management Plan*. Environment Australia, Canberra.

Creese, R. G., and A. Jeffs. 1993. Biological research in New Zealand marine reserves. In Battershill, C. N. et al. *Proceedings of the Second International Temperate Reef Symposium, 7–10 January 1992, Auckland, New Zealand,* 15–22. Wellington, New Zealand: NIWA Marine.

Denny, C. M., T. J. Willis, R. C. Babcock. 2003. Effects of Poor Knights Marine Reserve on demersal fish populations. DOC Science Internal Series 142. Dept. of Conservation, Wellington.

Department of Primary Industries, Water and Environment (DPIWE), 2001. *Tasmanian Marine Protected Areas Strategy*. Available at http://dpiwe.tas.gov.au/inter.nsf /WebPages/BHAN-54983Z?open. Last accessed 6/23/2003.

Durán, L. R., and J. C. Castilla. 1989. Variation and persistence of the middle rocky intertidal community of central Chile, with and without human harvesting. *Marine Biology* 103(4):555–562.

Edgar, G. J. and N. S. Barrett. 1999. Effects of the declaration of marine reserves on Tasmanian reef fishes, invertebrates and plants. *Journal of Experimental Marine Biology and Ecology* 242: 107–144.

Francour, P., J. G. Harmelin, D. Pollard, and S. Sartoretto. 2001. A review of marine protected areas in the northwestern Mediterranean region: Siting, usage, zonation and management. *Aquatic Conservation: Marine and Freshwater Ecosystems* 11:155–188.

Godoy, C., and C. Moreno. 1989. Indirect effects of human exclusion from the rocky intertidal in southern Chile: A case of cross-linkage between herbivores. *Oikos* 54(1): 101–106.

Gomez, E. D., A. C. Alcala, and A. C. San Diego. 1981. Status of Philippine coral reefs—1981. In *Proceedings of the Fourth International Coral Reef Symposium, Manila, Philippines,* 275–282. Manila, Philippines.

Gomez, E. D., P. M. Alino, H. T. Yap, and W. R. Y. Licuanan. 1994. A review of the status of Philippine reefs. *Marine Pollution Bulletin* 29:62–68.

Great Barrier Reef Marine Protected Authority Representative Areas Program (GBRMPA RAP), 2003. Website. Available through http://www.reefed.edu.au/rap. Last accessed 6/20/03.

Jennings, S., and J. M. Lock. 1996. Population and ecosystem effects of fishing. In Polunin, N. V. C., and C. M. Roberts, eds. *Reef Fisheries,* 193–218. London: Chapman and Hall.

Kelleher, Graeme. *Guidelines for Marine Protected Areas*. 1999. Gland, Switzerland & Cambridge, UK, IUCN.

Kelly, S., D. Scott, A. B. MacDiarmid, and R. C. Babcock. 2000. Spiny lobster, *Jasus edwardsii,* recovery in New Zealand marine reserves. *Biological Conservation* 92(3): 359–369.

Kemp, D. 2004. *Historic protection for the Great Barrier Reef.* Media Release for Australian Minister for the Environment and Heritage David Kemp. March 24, 2004. Available at: http://www.deh.gov.au/minister/env/2004/mr25mar04.html. Last accessed 5/5/04.

Koslow, J. A., K. Gowlett-Holmes, J. K. Lowry, T. O'Hara, G. C. B. Poore, and A. Williams. 2001. Seamount benthic macrofauna off southern Tasmania: Community structure and impacts of trawling. *Marine Ecology Progress Series* [*Mar. Ecol. Prog. Ser.*]. Vol. 213, 111–125.

MacDiarmid, A. B. 1989. Size at onset of maturity and size-dependent reproductive output of female and male spiny lobsters *Jasus edwardsii* (Hutton) (Decapoda, Palinuridae) in northern New Zealand. *Journal of Experimental Marine Biology and Ecology* 127(3):229–243.

MacDiarmid, A. B., and P. A. Breen. 1992. Spiny lobster population changes in a marine reserve. In Battershill, C. N., et al. *Proceedings of the 2nd International Temperate Reef Symposium,* 47–56. Wellington, New Zealand: NIWA Marine.

McClanahan, T. R. 1990. Are conservationists fish bigots? *Bioscience* 40(1):2.

———. 1995. Fish predators and scavengers of the sea urchin *Echinometra mathaei* in Kenyan coral-reef marine parks. *Environmental Biology of Fishes* 43:187–193.

———. 1999. Is there a future for coral reef parks in poor tropical countries? *Coral Reefs* 18:321–325.

———. 2000. Recovery of a coral reef keystone predator, *Balistapus undulatus,* in East African marine parks. *Biological Conservation* 94:191–198.

McClanahan, T. R., and R. Arthur. 2001. The effect of marine reserves and habitat on populations of East Aftrican coral reef fishes. *Ecological Applications* 11:559–569.

McClanahan, T. R., and B. Kaunda-Arara. 1996. Fishery recovery in a coral-reef marine park and its effect on the adjacent fishery. *Conservation Biology* 10:1187–1199.

McClanahan, T. R., and S. Mangi. 2000. Spillover of exploitable fishes from a marine park and its effect on the adjacent fishery. *Ecological Applications* 10(6):1792–1805.

———. 2001. The effect of a closed area and beach seine exclusion on coral reef fish catches. *Fisheries Management and Ecology* 8(2):107–121.

McClanahan, T. R., and N. A. Muthiga. 1988. Changes in Kenyan coral reef community structure and function due to exploitation. *Hydrobiologia* 166:269–276.

McClanahan, T. R., and D. Obura. 1995. Status of Kenyan coral reefs. *Coastal Management* 23:57–76.

McClanahan, T. R., and S. H. Shafir. 1990. Causes and consequences of sea urchin abundance and diversity in Kenyan coral reef lagoons. *Oecologia* 83:362–370.

Moreno, C. A. 2001. Community patterns generated by human harvesting on Chilean shores: A review. *Aquatic Conservation: Marine and Freshwater Ecosystems* 11(1): 19–30.

Moreno, C. A., K. M. Lunecke, and M. I. Lepez. 1986. The response of an intertidal *Concholepas concholepas* (gastropods) population to protection from man in southern Chile and the effects on benthic sessile assemblages. *Oikos* 46(3):359–364.

MPA News. 2003. *Balancing Ecology and Economics, Part II: Lessons Learned from Planning an MPA Network in Victoria, Australia.* Vol.4(7): p.1–4. February 2003. Marine Affairs Research and Education (MARE) and School of Marine Affairs, Univ. of Washington. Seattle, Washington.

New South Wales (NSW) Marine Parks Authority. 2003. Marine Parks Authority NSW Website. Available through http://www.mpa.nsw.gov.au/html. Last accessed 6/24/03.

New Zealand Department of Conservation (NZDOC). 2004. New Zealand Department of Conservation Web site. Available at: http://www.doc.govt.nz. Accessed 3/31/04.

New Zealand Marine Sciences Society (NZMSS). 2002. New Zealand Marine Sciences Society Web site. Available at: http://nzmss.rsnz.org. Accessed 12/31/02.

Parks Victoria. 2003. *Victoria's System of Marine National Parks and Marine Sanctuaries Management Plan* 2003–2010. Parks Victoria. Melbourne, Australia.

Ramos-Espla, A. A., and S. E. McNeill. 1994. The status of marine conservation in Spain. *Ocean and Coastal Management* 24(2):125–138.

Roberts, C. M., W. J. Ballantine, C. D. Buxton, P. Dayton, L. B. Crowder, W. Milon, M. K. Orbach, D. Pauly, and J. Trexler. 1995. *Review of the Use of Marine Fishery Reserves in the U.S. Southeastern Atlantic.* National Marine Fisheries Service, Southeast Fisheries Center; NOAA Technical Memorandum, v. NMFS-SEFSC-376.

Rubec, P. J. 1988. The need for conservation and management of Philippine coral reefs. *Env. Biol. Fish.* 23:141–154.

Russ, G. R. 1991. Coral reef fisheries: Effects and yields. In Sale, P. F. *The Ecology of Fishes on Coral Reefs,* 601–635. San Diego: Academic Press.

———. 2002. Chapter #19, Marine reserves as reef fisheries management tools: Yet another review. *Coral Reef Fishes.*

Russ, G. R., and A. C. Alcala. 1989. Effects of intense fishing pressure on an assemblage of coral reef fishes. *Marine Ecology Progress Series* 56(1–2):13–27.

———. 1994. Sumilon Island reserve: 20 years of hopes and frustration. *Naga* 17(3):8–12.

———. 1996a. Do marine reserves export adult fish biomass? Evidence from Apo Island, Central Philippines. *Marine Ecology Progress Series* 132(1–3):1–9.

———. 1996b. Marine reserves: Rates and patterns of recovery and decline of large predatory fish. *Ecological Applications* 6(3):947–961.

———. 1998. Natural fishing experiments in marine reserves 1983–1993: Roles of life history and fishing intensity in family responses. *Coral Reefs* 17(4):399–416.

———. 1999. Management histories of Sumilon and Apo Marine Reserves, Philippines, and their influence on national marine resource policy. *Coral Reefs* 18(4):307–319.

Russ, G. R., A. C. Alcala, and A. S. Cabanban. 1992. Marine reserves and fisheries management on coral reefs with preliminary modelling of the effects on yield per recruit. In *Proceedings of the Seventh International Coral Reef Symposium, Guam, 1992,* 978–985.

Sale, P. F. 1980. The ecology of fishes on coral reefs. *Oceanography and Marine Biology, an Annual Review* 18:367–421.

Salm, R. V. and John R. Clark. 2000. *Marine and Coastal Protected Areas: A Guide for Planners and Managers.* World Conservation Union (IUCN) Gland, Switzerland.

Tasmania Parks and Wildlife. 2003. Information available through Tasmania Parks and Wildlife Home Page at http://www.dpiwe.tas.gov.au/inter.nsf/ThemeNodes/SSKA-4X33SG. Last accessed 6/23/03.

United Nations Environmental Program/International Union for the Conservation of Nature (UNEP/IUCN). 1988. Kenya. In *Indian Ocean, Red Sea, and Gulf,* 153–162. *Coral Reefs of the World,* vol. 2. Gland, Switzerland: IUCN; Cambridge: UK/UNEP, Nairobi, Kenya. 389 pp.

White, A. T. 1986. Philippine marine park pilot site: Benefits and management conflicts. *Environmental Conservation* 13:355–359.

———. 1989. Two community-based marine reserves: Lessons for coastal management. In *Coastal Area Management in Southeast Asia: Policies, Management Strategies and Case Studies*, 85–96. ICLARM Conference Proceedings, no. 19. Manila: ICLARM.

White, A. T., H. P. Vogt, and T. Arin. 2000. Philippine coral reefs under threat: The economic losses caused by reef destruction. *Marine Pollution Bulletin* 40(7):598–605.

Wilkinson, C. R. 1998. *Status of Coral Reefs of the World: 1998*. Townsville, Australia: Global Coral Reef Monitoring Network and Austalian Institute of Marine Science.

Wilkinson, C. R. 2000. *Status of Coral Reefs of the World: 2000*. Townsville, Australia: Global Coral Reef Monitoring Network and Austalian Institute of Marine Science.

Wilkinson, C. R. 2002. *Status of Coral Reefs of the World: 2002*. Townsville, Australia: Global Coral Reef Monitoring Network and Austalian Institute of Marine Science.

# ABOUT THE AUTHORS

*Jack Sobel*'s work spans the often-formidable gap between ocean science and policy and helps fill it by developing sound science-based policy and providing the science needed to support it. His current work as Director of Strategic Conservation Science and Policy for The Ocean Conservancy (TOC) integrates strong science with effective policy to protect marine ecosystems. He previously directed TOC's ecosystem, marine protected area, and habitat programs and served as its Senior Ecosystem Scientist. His "hands-on" marine reserve experience includes work in Florida, California, and the Caribbean. He chaired the South Atlantic Fisheries Management Council's Marine Reserves Advisory Panel and serves on the World Conservation Union's Commission on National Parks and Protected Areas. Jack covered a range of marine issues for the U.S. Senate National Ocean Policy Study, directed a conch research project in Belize, conducted field-based coral reef research, and is an accomplished scientific diver. Jack is an ocean and outdoor enthusiast and an avid recreational fisher, boater, and diver.

*Craig Dahlgren, Ph.D.*, lives in the Bahamas where he is the Science Director for the Perry Institute for Marine Science's Caribbean Marine Research Center on Lee Stocking Island. He has been conducting research in both temperate and tropical marine systems for 15 years, with a focus on community ecology and the population dynamics of important fishery species. Much of his current research involves evaluations of the design and effects of no-take marine reserves and reserve networks. He has also served as an advisor to governmental agencies of several Caribbean nations on ecological issues related to the creation of marine protected areas. He is also an avid recreational fisherman and conservationist.

# ABOUT THE CONTRIBUTORS

*Joshua Sladek Nowlis, Ph.D.*, strives to find solutions to marine problems through scientific study and consensus building. He is best known for his innovative research on marine reserves and other ocean management tools, and for effectively bridging the communication barriers that often separate science and policy. His contributions to this book predated his current job with the National Marine Fisheries Service.

*Alan M. Friedlander, Ph.D.*, is actively engaged in marine research and management with over 20 years of extensive experience in the tropical Atlantic and Indo-Pacific oceans. Currently, he works as a Pacific Coral Reef Science Coordinator for NOAA and as a Fisheries Ecologist with the Oceanic Institute in Hawaii. His research interests focus on coral reef ecosystem conservation including species-habitat interactions, the ecological impact of fishing, use of traditional fishing controls for sustainable resource use, and the use of mapping and monitoring in designing and assessing marine reserves.

*Michael B. Mascia, Ph.D.*, is a social scientist with the U.S. Environmental Protection Agency and a World Wildlife Fund Science Fellow. He is both a researcher and policy practitioner, with numerous scientific articles on the social dimensions of biodiversity conservation and environmental policy experience in the United States and abroad. His graduate research in environmental politics and policy focused on marine protected area governance in the Wider Caribbean.

*Janet Gibson* has contributed extensively to the planning and management of marine protected areas in Belize since 1985 through her work with the

Coastal Zone Management Institute and the Wildlife Conservation Society. She was critical to the development of the Hol Chan Marine Reserve, Belize's first and subsequent efforts. Her efforts are presently focused on the Glover's Reef Marine Reserve, which she hopes will serve as an example of an MPA that sustains fisheries resources and conserves outstanding marine biodiversity.

# Index

*Italicized page numbers refer to boxes, figures, and tables.*